SPACE SCIENCE, EXPLORATION AND POLICIES

THE SPACE EDUCATION PHENOMENON AT NASA, BRAZIL AND BEYOND

SPACE SCIENCE, EXPLORATION AND POLICIES

Additional books in this series can be found on Nova's website
under the Series tab.

Additional e-books in this series can be found on Nova's website
under the e-book tab.

SPACE SCIENCE, EXPLORATION AND POLICIES

THE SPACE EDUCATION PHENOMENON AT NASA, BRAZIL AND BEYOND

NORMA TERESINHA OLIVEIRA REIS
EDITOR

New York

For permission to use material from this book please contact us:
Telephone 631-231-7269; Fax 631-231-8175
Web Site: http://www.novapublishers.com

Library of Congress Cataloging-in-Publication Data

The space education phenomenon at NASA, Brazil and beyond / editor, Norma Teresinha Oliveira Reis.
p. cm.
Includes bibliographical references and index.
ISBN: 978-1-62618-188-5 (soft cover)
1. Space sciences--Study and teaching--United States. 2. Space sciences--Study and teaching--Brazil. I. Reis, Norma Teresinha Oliveira.
QB500.266.U6S615 2013
500.5071'073--dc23
2013005770

Published by Nova Science Publishers, Inc. † New York

In grateful memory of my mother

Antonia Reis
(1920-2000)

For my father Antonio Oliveira Reis

For those who believe that children's dreams can come true.
For those who help people's dreams become reality.
And for everyone who is crazy enough to follow one's own star!

CONTENTS

FOREWORD

One may ask *"What is so special about Space Education?"* Certainly, no particular topic, of course, should have a monopoly on the interests of educators. Nevertheless, space science and technology is unique in its potential as a unifying framework for education. In terms of spatial domains, it is radically inclusive (Earth system science, space physics and planetary science, astrophysics and cosmology). Further, it's exceptionally inclusive of disciplines (physical and biological sciences, mathematics, computer science…).

Temporal inclusiveness can be seen as well through new scientific and historical research that cuts across multiple temporal scales. For example, the new "Big History" framework is very useful in helping students, and all of us, recognize the fundamental historical context of not just all humans and civilizations, but also of natural environments and systems at all scales. In particular, Eric Chaisson (2007) has introduced seven key thresholds, in both time and system complexity, for both natural and human history, inclusively ["creation of" universe, stars, planets, solar system/Earth, life, agriculture, and the modern revolution]. As argued by David Christian and other scholars, Big History provides a useful integrative framework for discussing both human and natural perspectives on evolution and environments, multi-scale systems, emergence, and the interplay of the local and the global, at multiple scales (Christian, 2011). Of course, human perspectives on these issues draw in essentially all of the humanities: anthropology, sociology, psychology, philosophy, religion... With this in mind, I can imagine that a unifying theme for all education (from history and literature to physics and chemistry) could be that of "understanding environments."

> From "environments" to "Big History," Space Education has a truly unique and expansive "big tent."

Thus, although there remains no substitute for excellent teachers, teacher education, and support for teachers and educational infrastructure, there is a unique capacity of space science as a robust, unifying framework for education. However, now that I have pointed out this conceptual "big tent," let me now illustrate through the use of just one example from the exciting suite of educational tools and methodology that are featured in this volume within the rubric of "Space Education."

The Space Weather Action Center (SWAC), initially developed at NASA's Goddard Space Flight Center, is described in detail within Part I, Section 4, by this volume's editor,

Norma Teresinha Oliveira Reis. Using a relatively low-cost computer and background screen, students using the SWAC system can create "Space Weather" reports that are very much analogous to standard terrestrial weather reports. Such reports involve students in an active learning environment in which they learn about our Sun, its particle and fields environment, in addition to its photon output, and then learn about how these inputs to Earth's space environment can impact both technological systems (especially spacecraft and communication systems) and astronauts. Thus, while the students get excited and engaged with creating their own "weather" report, they simultaneously learn very actively the educational content of interest, along with applying some measure of analysis and critical thinking. Further, these reports can be created as blogs, podcasts, or vodcasts, or other venue of the new media. However, you might ask, what does "Space Weather" have to do with my next lesson plan? Here, as with many examples of state-of-the-art tools and methodologies discussed in this volume, one can simply replace "Space Weather" with other content – agriculture, forestry, urban planning, or some other topic , either more focused or broader – perhaps "Physical Environment Action Center & Extensions" (PEACE) – it's all your choice as educator.

> Harness the power of new educational tools and technologies to meet the educational needs of your students; e.g., consider the use of Action Centers for active learning.

In addition to reports featuring some topic in the physical or biological sciences, such as environmental sciences, the Action Center concept can be used to enable active learning for topics in the humanities as well; for example, as a complement to covering some scholarly debate or discussion or documentary film activities. As noted above, the last major threshold currently highlighted in Big History studies is the "modern revolution." However, such modernity incorporates political and economic frameworks that are well documented as being unsustainable. This reality highlights the critical need to move beyond modernity to longer-term sustainable modes for humans, human civilizations, and their environments. Recent scholarship has sometimes termed this next major phase (Big History threshold?) "Ecological Civilization" (e.g., see Kallis and Norgaard, 2010, Norgaard, 2010, and Gare, 2011). I can readily imagine a very broad range of outstanding Action Center projects that focus on various aspects (science/technology/sociology/ psychology/economics…) of this important future transition to ecological civilization. After all, *our students truly are our future*.

From very specific educational tools and methodologies to topics and applications that have transformative potential for education, and even our very civilization, this volume presents some important inspiration and starting points.

Timothy E. Eastman (PhD, Geophysics)
Wyle Information Systems at NASA Goddard
Space Flight Center, Greenbelt, MD, USA

References

Eric J. Chaisson, *Epic of Evolution: Seven Ages of the Cosmos*, New York City: Columbia University Press, 2007.

David Christian, *Maps of Time: An Introduction to Big History*, Berkeley: University of California Press, 2nd edition, 2011.

Arran Gare, From Kant to Schelling to Process Metaphysics: On the way to Ecological Civilization, *Cosmos and History: The Journal of Natural and Social Philosophy* [www.cosmosandhistory.org], vol. 7, no. 2, 2011.

Giorgos Kallis and Richard B. Norgaard, Coevolutionary ecological economics, *Ecological Economics*, vol. 69, issues 4, 690-699, 15 Feb. 2010.

Richard Norgaard, "A Coevolutionary Interpretation of Ecological Civilization," *Fourth International Symposium on Ecological Civilization*, Claremont, California, May 3-4, 2010.

ACKNOWLEDGMENTS

First of all, my special and foremost thanks to God, to Whom I owe my life, my parents, health, unique educational and professional opportunities, and the inspiration and strength to have carried forward the idea of writing this book. I am heartly thankful to my mother Antonia Reis (1920-2000), for her unconditional and infinite love, good example, for helping me read and write my very first words and stories, and for believing in my 'space' childhood dreams. My hearlty thanks to my father Antonio Oliveira Reis, for all love, encouragement, advices, companionship, for teaching me to always follow my star, and for all his support, recommendations, and dialogues that helped me a lot prepare this book.

There are several other special people to whom I would like to express my deep gratitude. Some of them helped me and inspired me since I was a school student, others while I was a university student and a schoolteacher in my home state Parana (South Brazil), some are from my workplace in the capital Brasilia, and others I came to know on a beautiful pathway to NASA Goddard Space Flight Center, as a summer graduate ISU intern, in 2008. These persons are very special to me in several ways. If I were to describe how much each of them helped me on my pathway, I would seriously risk unintentionally omitting important information, as they have helped me a lot, in several moments of my life, and in unique ways. For this reason, let me just express that the persons whose names are below have a truly special meaning to me in a singular way, that I deeply admire them, and learned a lot from them. For each of them, thank you very much indeed for trusting me, and for all concrete and continued support to help me make my childhood 'space dream' come true, and grow up as a vision. The names are in chronological order in which these exceptional people came into my life.

Olga Anna W. Gioppo
(Soroptimist International of the Americas – Brazil Region)
Nilson Marcos Dias Garcia
(Universidade Tecnológica Federal do Parana, Brazil)
Meireluce Fernandes da Silva
(Agência Espacial Brasileira – AEB)
Marco Antonio Bologna
(TAM Linhas Aéreas S.A., Brazil)

Ivo Hering
(Cia. Hering do Brasil)
Ivette Maria Soares Rodrigues
(Agência Espacial Brasileira – AEB)
Himilcon de Castro Carvalho
(Agência Espacial Brasileira – AEB)
Sérgio Gaudenzi
(Agência Espacial Brasileira – AEB)
Nadia Repussard
(International Space University – ISU)
Norma Maria do Valle
(Soroptimist International of the Americas – Brazil Region)
Paulo de Tarso Milani
(Conselho Nacional de Desenvolvimento Científico e Tecnologico, Brazil)
Lucia Helena Lodi
(Brazilian Education Ministry)
Marcos Cesar Pontes
(First Brazilian Astronaut)
Maria Elisa Paciornik
(Ferraz Paciornik Consultoria e Planejamento Ltda.)
Jean-Pierre Courtiat
(Embaixada da França no Brasil)
Timothy E. Eastman
(NASA Goddard Space Flight Center)
Walter Peeters
(International Space University - ISU)
Hugh Hill
(International Space University - ISU)
Louis A. Mayo
(NASA Goddard Space Flight Center)
James R. Thieman
(NASA Goddard Space Flight Center)
Margareth Maher
(NASA Goddard Space Flight Center)
Steve Ambrose
(NASA Headquarters)
Maria Izabel H. Mendes
(Soroptimist International of the Americas – Brazil Region)
Nazareth Tunholi
(Brazilian producer and journalist)
Ciclamio Barreto
(Universidade Federal do Rio Grande do Norte – UFRN)
Gilvan Luiz Borba
(Universidade Federal do Rio Grande do Norte – UFRN)

This book derives from my research carried out as an ISU intern at NASA/GSFC. Thus, I would like to extend my gratitude here to my ISU studies' sponsors: the International Space University – ISU; TAM Airlines S.A. (http://www.tam.com.br/b2c/jsp/defaultBrasil.jsp); In Marsat (http:// www.inMarsat.com/); The Embassy of France in Brazil (http://www. ambafrance.org.br/); Hering S.A. (http://www.hering.com.br/); Soroptimist International of the Americas – Brazil Region (http://www.soroptimistbrasil. org/si.htm); The Legislative Chamber of Parana State (http://www.alep.pr. gov.br), Brazil; and The Region Alsace (https://ssl.region-alsace.fr/boussole/), France. The support of these institutions has been fundamental to make my 'space education dream' a reality.

Further, I am heartly thankful to Nova Science Publishers, Inc. for the opportunity of sharing my research in this book, and in special to Dr. Carra Feagaiga, for all her substantial support and guidance with regard to several aspects related to the publication process. Finally, I would like to express my special gratitude to all authors who joined this enterprise by contributing invited chapters in Part II. Thank you very much indeed for believing in this effort, and for your time to share your experience in these pages.

I hope that this book will inspire you, and contribute to have more educators, researchers, and project managers dreaming without limitations towards bringing innovative, engaging space education solutions inside the classroom. Mission success belongs to those who trust that they can reach their own star, by following the messages that come from the heart. An educator's heart is always in tune with creating scenarios in which his or her student could overcome oneself, feel challenged and capable, and identify possibilities of reaching higher skies. May God allow that this book becomes a resource to assist educators in the mission of identifying and developing 'out-of-this-world' pedagogical practices that will inspire more and more 'space-kids' to nurture and accomplish possible dreams! And may the power of accomplishment through space education be with you!

Norma Teresinha Oliveira Reis
Editor

INTRODUCTION

It is difficult to say what is impossible,
For the dream of yesterday is the hope of today
And the reality of tomorrow.
- *Robert H. Goddard (1882-1945), father of modern rocketry*

The quote above belongs to the father of modern rocketry, Dr. Robert H. Goddard (1882-1945), who had the largest NASA center named after him in 1959. Goddard's dream, passion, hard work, and geniality contributed to enable us today send machines such as satellites, telescopes, and spacecrafts into space, widening horizons that allow us learn more about the Universe and our home planet, and to turn into reality humanity's space exploration desire. What about you? Do you have a dream? Almost all children one day dreamed of becoming an astronaut and discovering new worlds. Being an educator somewhat means mastering the art of bringing into existence our students' deepest dreams, like it is the soil role to provide the ingredients to have beautiful trees flourishing from small seeds. The act of dreaming is challenging because it is an invitation to explore unknown worlds, and identify new possibilities. It is our role as educators to help our students find out their dreams, which will inspire them to learn and flourish. As in every exploration we may be afraid of what we will encounter on the way. We may find unexpected situations, and we may need further learning. The 'space education phenomenon' is the object of this book, which is all about cultivating real dreams inside the classroom through innovative, passionate educational practices.

Space science and exploration capture interest and curiosity of individuals at all ages, and educators can use this motivation to encourage students to become real explorers and innovators through space education. But what is space education anyway? It stands for a wide range of pedagogical practices that take advantage of the amazement that space matters exert on students' imagination, thus capturing early interest in Science, Technology, Engin-eering, and Mathematics (STEM) subjects; elucidating scientific knowledge in a non-abstract approach; spreading the benefits of space programs; and ultimately encouraging STEM careers. In this publication you will find information regarding space education projects on subjects such as radio astronomy, space weather, robotics, Mars, astronautics, along with information related to technologies such as videoconference, virtual classes, mobile devices, and podcasts. Educators can use this book as a reference to identify innovative possibilities to enrich classes at precollege or college level. Researchers in STEM education and/or space education will find here interesting pedagogical analyses of this relatively recent field of investigation (first space education initiatives arose in the world in the 1980's). Further, this book is an important resource for space education project managers as it provides valuable information concerning structure and functioning of the complex NASA Education enterprise. Historians, astronomers, and the general public may also enjoy this publication.

Content here is divided in two parts. In the first one, this book features a thesis entitled "Pedagogical Analyses of Current Space Science Education Practices at NASA and the Brazilian Space Agency." The object was a sample of successful NASA and Brazilian Space Agency space (science) education practices. I (*Norma*) wanted to investigate NASA Education because it is the largest space education enterprise in the world, comprising initiatives for all educational levels, and countless types of products and services. The agency invests massively in capturing students' interest in STEM and in its own missions. NASA has been benefiting U.S. students and also international communities, as there are important NASA Education projects carried out in cooperation with other countries. Last but not at all least, I wanted to find out how space (science) education is capable of effectively contributing to STEM education betterment. This was also the primary motivation for the thesis, which has been divided in five main segments: (a) space and education concepts; (b) space education at NASA, its structure, functioning, and resources; (c) examples of state-of-the-art practices in space science education at NASA and the Brazilian space agency (d) a few pages about my personal history in space, and my passion for NASA; (e) and in Appendixes, an article entitled "Eclipses through the Centuries," which provides readers with several curious eclipses stories showing how humanity reacted to these phenomena through the ages.

In order to carry out investigations that derived on this thesis, I had the opportunity to visit NASA Goddard Space Flight Center in summer 2008, as an International Space University (ISU)'s intern. During that summer, I carried out under supervision of the Sun-Earth Connection team, a comprehensive historical and scientific research entitled "Eclipses through the Centuries," as part of NASA Sun-Earth Day project's activities related to the total solar eclipse of August 1, 2008 in China. For that, I interviewed renowned NASA specialists and worked for several weeks in the Library of Congress (LOC), in order to select and offer to readers a noteworthy collection of stories. As part of my graduation requirements, I carried out qualitative research on a sample of space science education pedagogical practices at NASA and the Brazilian Space Agency, and interviewed several NASA and Brazilian Space Agency education experts in order to analyze the space (science) education phenomenon; to study NASA Education structure and functioning; to access state-of-the-art projects/programs; to discover what makes these initiatives successful; and to identify challenges and trends. Some projects investigated: Space Weather Action Center (SWAC), Radio JOVE (radio astronomy), NASA education initiatives targeted to impaired students, e.g. the Educational Kiosk, and the Brazilian "AEB School Program."

In the second part of this book, you will find treasured resources: special chapters regarding selected space education projects/programs by outstanding space education specialists, from NASA, Brazil, and beyond. It comprises practical examples of space education projects in several themes. For example, there is a radio astronomy education project in which readers will learn how students and other interested individuals can build an inexpensive radio telescope to receive and analyze radio waves from the Sun, Jupiter, the galaxy, the Earth, etc. Another chapter presents guidelines for using mobile devices to support science and technology learning, data sharing, and collaboration for experts and learners. There is a chapter that analyzes three types of opportunities for STEM professionals: two research and development experiences for science and engineering undergraduate and graduate students, and one for STEM professionals at informal learning centers. It also provides insights into their effective design and management. Further, there is a chapter entitled "Space Technology in Education – Indian Perspective" that depicts the initiatives by the Indian space program to promote the space education with the effective usage of space technology and the successful outcomes of this effort. There is another special chapter that features an exclusive educational analyzes by the first Brazilian astronaut (currently also educator). Finally, there is a chapter about an important Brazilian Olympics in Astronomy and Astronautics, which unites students and teachers in Brazil to study and explore space-related topics. Each author presents detailed information regarding the enterprise, and explains how you can get your students involved.

The decision to write a book about the 'space education phenomenon' was due to some combined factors. There is not so much literature regarding space education, and although there are several enterprises in space (science) education in the entire world, there are still insufficient pedagogical analyses of these practices. Further, national governments and international organizations have been working hard towards identifying strategies to promote STEM education betterment, and effective space education practices integrated to the curriculum represent an interesting solution in this scenario. Featuring content about innovative pedagogical approaches for STEM education, this book represents a valuable resource for identifying possibilities to enhancement of pedagogical practices in STEM. Another relevant differential is that this publication provides practical examples of space education programs/projects in which educators and students can participate. And the book represents an effort by authors working in several different realities and facing specific challenges, with expertise in implementing space education programs. Finally, this wide range of space (science) education solutions helps teachers, researchers, and managers think about solutions for their specific context.

We hope this book will offer you valuable insights and practical ideas on how to apply space (science) education practices to turn STEM education more accessible, challenging, meaningful, and engaging. And we do hope that you could inspire your students to dream deeper, unlock their potential, and fly higher. The sky is not the limit and, as Dr. Robert H. Goddard has stated, it is difficult to say what is an impossible dream!

Norma Teresinha Oliveira Reis,
Editor

The original thesis material that I adapted for the purpose of this publication has been written as part of my ISU internship at NASA Goddard Space Flight Center, under the distinguished mentorship of Dr. Timothy E. Eastman, Dr. Louis A. Mayo, and Dr. James R. Thieman, and under the distinguished academic advisory of Dr. Walter Peeters (ISU).

PART I

PEDAGOGICAL ANALYSES OF CURRENT SPACE SCIENCE EDUCATIONAL PRACTICES AT NASA AND THE BRAZILIAN SPACE AGENCY[1]

Norma Teresinha Oliveira Reis[*]

ABSTRACT

A systematic, yet qualitative research in space science education is presented here. Scientific, technological development, and innovation are exceptionally important for both industrialized and developing countries, for the achievement and maintenance of sovereignty and sustainable development in the knowledge society. Preparation of competent professionals in Science, Technology, Engineering, and Mathematics (STEM) constitutes a high priority in this framework; thus education in STEM plays a vital role in governmental agendas of a continuously increased number of nations. The effective success of the teaching-learning process of STEM challenges educators and policymakers worldwide. The teacher-student-knowledge interaction is complex, multifaceted, dynamic, and permeated by difficulties intrinsic and extrinsic to the scholarly educational process. The integrated understanding of sciences as part of everyone's life has not achieved expected results in many countries with different levels of socioeconomic development. This is partially demonstrated by international standardized evaluations, such as the Programme for International Student Assessment (PISA). Space education for precollege students has been carried out for more than three decades at NASA, and consists of a portfolio of practices that use the singular fascination of space as a vehicle that capture students' interest in STEM, aiming at strengthening STEM literacy as well as encouraging careers in those fields. NASA pursues excellence in space education, thought it also faces challenges. The purpose of this research is to present and analyze, from a pedagogical perspective, impacts of selected NASA and Brazilian state-of-the-art practices in space science education in the teaching-learning process of STEM. The methodology elected for this purpose has been qualitative research, comprising exclusive interviews with NASA Education professionals, visits to NASA Education facilities as well as immersion in a sample of ongoing projects. The results presented provide pedagogical analyses in terms of benefits and constraints associated with these practices. The results here presented carry the added value of extracting the experience of NASA Education experts, along with corresponding pedagogical analyses of the enterprise

[1] This thesis has been carried out during a summer internship at NASA Goddard Space Flight Center, in the condition of a graduate International Space University (ISU) student, during the academic year of 2007-2008. The original content has been updated and adapted for the purpose of this publication.

[*] E-mail address: norma.reis@gmail.com

formulated by a non-U.S. educator. More comprehensive analyses would require immersion in schools and systematic interaction with participant teachers and students for a larger period of time. This investigation is not intended to present final conclusions, but to discuss and analyze the mentioned practices, to finally provide concrete recommendations and encourage further research in order to fill gaps, which inevitably remain – questions and provocations to be answered by new researchers.

Keywords: space science education, STEM education, NASA Education

INTRODUCTION TO PART I

Education for the 21st century features novel and complex challenges for all citizens. Schools are faced with requirements of working and interacting in a societal fabric characterized by globalization, social inequalities, violence, changes in the family structure, rapid scientific and technological advances, new demands of diversity and inclusiveness, new technologies, influence of media (television, internet), changes in the work market, just to mention a few. In this panorama, educators are faced with the question: *"Should we educate for work market or for life overall?"* Clearly, the second option is significantly more demanding in terms of time and teacher preparation to address humanistic questions. Further, schools are required to consider the increasing importance of scientific and technological knowledge in order to prepare citizens for effective inclusion in today's knowledge society, which requires from all individuals a novel, complex set of skills i.e., higher-order thinking, proficiency in technology utilization, teamwork and problem-solving skills, innovation capabilities, and intellectual autonomy. Though this investigation features an opening discussion regarding these pedagogical issues, its focus is to analyze space education possibilities for the betterment of Science, Technology, Engineering, and Mathematics (STEM) education at precollege level.

Society is permeated by countless benefits derived from space studies, from mobile phones to pacemakers. Astronautics has turned into reality humanity's dream of exploring further, stepping onto the Moon, and revolutionizing scientific fields such as astrophysics, heliophysics, and planetology with astonishing images and discoveries from space-based telescopes and probes. Through many advances in space sciences and technology (e.g., meteorology and remote sensing satellites), there have been uncountable benefits in global, high-resolution monitoring of Earth's natural resources - including forests, minerals, lands, and oceanic areas. Further, natural disasters mitigation is substantially more effective today thanks to satellites that we have 'up there.' Currently, there are also novel challenges and perspectives in the emerging "space society," including plans to send humans to Mars, to the Moon again, and other destinations i.e., asteroids and other planets' moons. Considering high costs and complexity involved in these missions, they should be carried out mainly via international cooperation. It is also intended the development of innovative, more cost-effective launching technologies, and continuation to microgravity scientific research in space-based laboratories, such as the International Space Station (ISS). There are also studies to develop propulsion systems that would enable humanity to someday reach other stars. The private sector is also making their place in space. There are space tourism companies already selling tickets for suborbital and orbital flights – though prices are currently 'stratospheric' for most citizens – some plans for space hotels, and other initiatives such as the Google Lunar X-Prize, which encourages engagement in space matters through competition. Although there is not yet sufficient awareness regarding space studies and exploration's mid- and long-term importance among populations, the combination of the abovementioned governmental and private enterprises demonstrate that humanity has been embracing space as part of its culture – the highest expression of Icarus' dream, *"boldly going where no one has gone before."*

Students learn about space matters in informal education environments i.e., their own house, the Web, museums, and planetariums. Bringing the space thematic inside the classroom contributes to enrich pedagogical practices and to motivate students to learn subjects such as Physics and Mathematics. Well-structured practices in space (science) education such as hands-on experiments, observations, projects, visits, and other activities using the space thematic as a learning vehicle strengthens STEM education, especially if these practices are articulated with school's pedagogical project. The strengthening of education in STEM, by its turn, is capable of contributing to a virtuous cycle of fostering careers in these strategic fields. Space education is also to the benefit of all students, as it can impact positively the teaching-learning process of all subjects, considering the interdisciplinary nature of space science. For this reason, a qualitative study and analysis of state-of-the-art practices in space (science) education at NASA – also recognized by its excellence in the field – and the Brazilian Space Agency case, is of substantial importance for an enhanced understanding of the possibilities and constraints of this historically recent phenomenon – space education. This development called "space education" remains quite scarcely investigated and documented. Indeed, it remains the "youngest child" of space exploration.

Aiming at studying and analyzing a sample of NASA state-of-the-art practices in space (science) education - especially those devoted to primary and secondary schools - this investigation has primarily two purposes: to analyze how educational practices in space (science) education are capable of contributing to the success of the teaching-learning process of STEM overall in precollege level; and to identify constraints deriving from this practice, from the viewpoint of NASA Education experts, and from my personal observations. The methodology chosen has been qualitative research, because education is a complex process influenced by subjectivities of the diverse array of individuals involved. Further, the use of interviews is capable of gathering substantial data based on experiences of professionals in their daily activities within a sample of educational endeavors. This investigation also intends to provide subsides for analyses of space (science) education efforts at NASA, Brazil, and other potential scenarios, based on selected examples and recommendations here presented. Finally, this research describes the complex structure and functioning of NASA Education, which may be of interest to STEM education managers, researchers, and decision-makers.

For organizational purposes, this chapter has been divided into the following sections: a) characterization of the research, represented respectively by the items: *Objectives*, *Justification* and *Methodology*; b) theoretical research, which was subdivided the following manner: the section "*Space and Education*," including the subsections "*Educating for the XXI century: Advances and Challenges*"; "*Space exploration and space education: what does that mean, anyway*"; "*Space education in the world: how far have we gone?*," aims at providing the reader with the necessary background to contextualize the following section, related to NASA and practices themselves, namely "*Space Education at NASA – A Model of Excellence?,*" including the subsections "*What NASA Is?*," *"NASA Education – A Brief*

Characterization," "NASA Education – A complex structure," "Resources for the Education Community," "NASA Education Technology," and *"State-of-the-art Practices in Space Science Education at NASA – Summary and Excerpts of Interviews with Specialists,"* the section *"The AEB School Program – A Brazilian initiative for the enhancement of STEM: Brief Summary and Excerpts of Interview with the Program Manager,"* and finally the section *"Pedagogical Analyses and Recommendations"*; c) conclusions, list of references and appendixes, including the paper "*Eclipses through the Centuries*" (adapted), an in-depth example of space science education research and content preparation conducted by the author as part of coverage of the total solar eclipse of August 1, 2008, in China, by the Sun-Earth Day team at NASA Goddard Space Flight Center.

Chapter 1

THE RESEARCH

In the coming sections we will see that NASA pursues several efforts for the betterment of STEM education at all levels. These enterprises are also aimed at preparing NASA's own workforce that will assure continuation to the agency's activities. NASA Goddard Space Flight Center is the largest NASA field center, and its missions are primarily focused on robotic exploration of the Earth, the Sun, the Moon, the Solar System, and beyond. Amongst other missions, NASA Goddard manages the Hubble Space Telescope (HST), and the Solar Dynamics Observatory (SDO). Virtually all NASA missions have associated education programs/projects, and one of these enterprises is focused on solar physics, or heliophysics, the scientific area in which I carried out my major internship research. The purpose of my ISU internship at NASA/GSFC has been to carry out research on "Eclipses through the Centuries," and to simultaneously write my space (science) education thesis, which originated this book. This chapter derived from qualitative research of NASA Education enterprise, with investigations of a sample of state-of-the-art NASA Education projects/programs. In the sequence, I provide information about this research's background and execution.

1.1. OBJECTIVE

To conduct pedagogical analyses of possibilities, challenges and constraints for a sample of NASA and Brazilian Space Agency state-of-the-art practices in space science education for elementary and secondary school levels.

1.1.1. Specific objectives

- To perform a survey on a sample of exemplary NASA space (science) education programs/projects/activities;
- To study the use of new technologies in the framework of space (science) education needs;
- To produce the paper "Eclipses through the Centuries" as a didactic resource for education and outreach purposes.

1.2. JUSTIFICATION

Scientific and technological development constitutes a fundamental pillar to national sustainable development [1], and can be translated into socioeconomic growth, sovereignty, and social welfare. Since the Industrial Revolution, scientific and technological advancement has been producing differentiated levels of development amongst nations [2]. For instance, products, processes and services, such as those belonging to the space sector, yield higher added value than most others. If a nation does not invest in this sector, its population will pay the price established by the ones pursuing that know-how, thus generating greater dependency. Actually, the 'scientific capital' – category used by Bordieu [3] in his approach to the societal uses of science – is unequally distributed in the world, so that it is necessary to foster mechanisms of cooperation in order to assure that all nations pursue at least a basic level of scientific and technological development to fulfill its essential societal needs. As for the social role of science and technology, in addition to achieving its intrinsic purposes such as discovery and exploration, it also need to be pursued for the service of social inclusiveness and democratization of access to technological goods, in the perspective of increasing social welfare [4]. Science and scientists do not exist separately from the societal fabric in which their practice is carried out. Finally, considering that science and technology is produced to a great extent with public resources, it is more than fair that promoting social welfare be taken into consideration as a high priority by scientists, technologists, and decision makers.

In order to assure higher levels of development in Science, Technology, Engineering, and Mathematics (STEM), federal governments of nations with diverse levels of socioeconomic development have been adopting several strategies in order to maintain and enhance their level of autonomy and competitiveness in these sectors. One such strategy refers to the elevation of citizens' proficiency in STEM subjects, by strengthening STEM taught in schools. This attitude aims to: a) promote understanding of the scientific efforts, considering that these enterprises are socially situated, and should be understood by all citizens; b) motivate young students to follow careers in STEM, and thus strengthen the sectors' development. In fact, STEM education plays an important role in the preparation of citizens considering that the current knowledge society is permeated by processes, products, and services requiring from all individuals a certain level of STEM literacy for effective social inclusiveness [4]. Nevertheless, education in STEM has been facing challenges as recent results of the International Programme of Student Evaluation (PISA) demonstrate that a considerable number of students of elementary and secondary levels in several countries with diverse socioeconomic profiles have not been learning sciences as they were expected to [5].

Certainly it is also correlated with both global literacy processes starting during or before elementary education, as well as with intra and extra-scholarly factors. As for intra-scholarly factors, it can be pointed out methodologies adopted by teachers, which sometimes are in dissonance with manners through which students learn. STEM education also faces challenges correlated to pre-service and in-service teacher training. Further, schools do not always offer proper conditions to the development of adequate pedagogical practices, and teachers do not always have enough time and didactic resources to prepare good and dynamic classes, such as using innovative educational technologies. Further, some pedagogical theories have misunderstood the sense of STEM teaching/learning, and some theorists claim that students may learn spontaneously through experience, without systematic, teacher-mediated contact with theoretical knowledge. However, practice shows that these two spheres of the scientific enterprise are intrinsically related. Another challenge consists of encouraging teachers to present scholar contents in an interdisciplinary manner, so that it is necessary that the Science teacher work along with the Mathematics, Geography and History teachers, for example. Knowledge should not be considered in compartments, but as a dynamic and dialogical whole. The artificial subdivision of knowledge in scholarly subjects plays an organizational role, but should not reduce knowledge to fragments, as this attitude contradicts the radical nature of knowledge, which is integral and interactive. Education in STEM is also challenging today because scientific knowledge changes extraordinarily fast. The internet has been revolutionizing the way individuals interact with information, produce and disseminate knowledge, and teachers are not always prepared to work with tech-savvy students, born in a world much more dynamic and demanding than the one of just some decades ago. It is critical to identify in this panorama innovative pathways to encourage STEM education for a nation's sustainable, long-term development and social welfare.

In order to illustrate challenges in STEM education, let us discuss the example of Physics teaching. Campbell [6] comments that this discipline is usually introduced at school in an a-historic manner, not connected with its roots in society and with its present and future application. For this reason, several students do not learn and do not like Physics. They have never understood what is Physics used for, anyway! According to Monteiro [7], some critical problems in the teaching/learning of sciences are: a) eminently theoretical approach; b) themes and problems a-historical and presented out of context; c) practical activities as appendices of theoretical classes; d) extremely linear programs; e) resistance to interdisciplinary work. In this framework, one factor that would contribute to the success of Physics education consists of having multidisciplinary teams translating technical speech into a language that is understood by the general public. Further, considering Physics as a curricular subject, it should be taken into account that the main purpose of the teaching/learning of this discipline in precollege level is not to prepare physicists, but to consider Physics as culture, and as possibility of understanding the world [8]. It means that we need a new design for the teaching/learning of this discipline both in theoretical/ conceptual, and methodological aspects. The advances of Physics, mainly in the two last centuries, as well as the history of science and the scientists, should be included. Likewise, the experimental approach of this science should be valued, as it makes possible the effective participation of students in knowledge construction, which will accompany students throughout their entire lives. In summary, it is necessary to bring students to the laboratory; to have them watch online videos and simulations; to make connections with real world and with other subjects; it is necessary to learn in context. A wide range of integrated methodological approaches would enable learners not to merely memorize and pass school exams, but to learn for life and for real world demands, as education is supposed to provoke long-term changes and preparation for citizenship, professional life, and continued learning.

Students learn better STEM concepts when they can put their hands on and construct things. Experimental or hands-on activities raise students' interest for STEM and their critical spirit, leading to better perceptions of relations between science and technology [9]. Indeed, experimentation is a strategy capable of generating discussions, requiring reflection, and hypothesis formulation. It teaches one how to analyze results, and offers students contact with objects, facts and tangible/observable phenomena present in daily life, thus making easier the effective understanding of principles, concepts, and phenomena. According to Ramos and

Ferreira [10], the utilization of experiments leads students to attempt to reproduce them and, from his/her interest, to understand what happened, and to explore theorization. Further, according to these authors, the use of this resource makes it possible to distinguish learning from memorization – a situation in which students understand the speech in the short-term memory, but not the observed phenomena. It is fundamental, however, that teachers understand that experiments play a key role and has pedagogical specificity in students' preparation, instead of considering them as mere appendixes. According to Soussan [11], it is also essential to distinguish which experiments students are capable of carrying out. This assumption should consider the meaningful experiences that students have been exposed to, their previous knowledge, historicity, as well as their means of accessing the world of knowledge in a given moment of their biological and psychosocial development.

Technologies also play a fundamental role in STEM education. Currently, students may access and explore virtual worlds, blogs, and social networks from their desktop or laptop computer, tablets or smartphones. They can explore STEM in these virtual spaces outside the classroom. Teachers should explore pedagogical benefits of integrating technologies to the curriculum. For example, students may not have access to a telescope or to Physics laboratories, but they could visit them at Web databases. Technology effectiveness in education depends upon several factors, and it is possible to affirm that they promote the betterment of the teaching/learning process in different curricular subjects. One example is sciences and mathematic, in which the introduction of educational technologies (i.e., software, simulations, and modeling) has shown to be producing positive results [12, 13, 14]. Further, it has also been observed in several contexts that students with special needs and belonging to low income communities learn better with technologies because these devices and virtual environments constitute for them a novelty, and a challenge [15]. Educators can take advantage of these tools that capture students' motivation to engage them in meaningful collaborative projects, and in innovative manners of constructing knowledge with technology.

Considering that the space thematic is capable of stimulating curiosity of youth and capturing their interest by STEM contents, it should be taken into high account the utilization of practices related to this thematic in STEM education. It also contributes to historical approaches of space activities as an important element of national culture. Brazil, for instance, has been developing all constituent elements of a complete space program. The country also has a professional astronaut, Marcos Pontes, who has been trained at NASA, and launched in 2006 aboard a Soyuz spacecraft [16]. It is important that Brazilian children learn about these facts, while they learn Physics and Geography, for instance. Achievements in space sector are inherently important for socioeconomic development, but 'space' should also become part of a people's culture, and in this sense space agencies should invest in education and public dissemination. It is important that students learn about history and recent accomplishments of their nation in the space sector, in order to promote awareness of what a given country has in terms of space infrastructure, and societal benefits derived of space activities. Space sector achievements are also an element of national pride, patriotism. It is important that educators teach students to care about the world, but it is difficult to prepare a 'citizen of the world' before preparing a good national citizen. Finally, as students become aware of their country's space program, it could raise their interest in following career paths in the space sector.

The object of this research, space science education for precollege students, constitutes a historically recent pedagogical practice and, as a distinct research topic, demands deeper investigations from interdisciplinary perspectives. Regarding the relevance of space science education for the development of nations, the United Nations Office for Outer Space Affairs,[1] by means of discussions carried out on occasion of the Third United Nations Conference on the Exploration and Peaceful Uses of Outer Space - UNISPACE III held in Vienna, in 1999, considered it as an important development factor and suggested that it should be an integral part of countries' global educational policy agendas. According to the United Nations' Committee on the Peaceful Uses of Outer Space, the introduction of elements of space science and technology into school science curricula could revitalize the education system, introduce the concept of high technology in a non-esoteric approach, and contribute do develop national capacities in STEM [17]. Further, all countries can take advantage of benefits of new technologies, which, in many cases, are spin-offs from space activities.

The relevance of research on NASA state-of-the-art pedagogical practices in space science education is that NASA Education pursues excellence in terms of qualified personnel, rich didactic resources, facilities, networks with scholarly communities, scope, and elements that can effectively contribute to elevate students' understanding of STEM. It is important to access a sample of these initiatives in order to investigate the elements that make these NASA efforts so successful. On the other hand, it is also important to identify constraints and propose channels by which these practices could be enhanced. Thus, this research also intends to contribute for analyses and eventual promotion of concrete changes in the scope of ongoing practices in space science education at NASA, Brazil, and beyond. Actually, from the perspective of educational *praxis*, theory does not exist separately from practice, and the final intent of research in education is not only to advance thinking in the field, but also to contribute to education betterment itself. It is also presented here a successful Brazilian effort in space science education, as a way of comparing two educational enterprises happening in two different countries, though with the same purposes – STEM education betterment, awareness about space activities, and enhanced citizenship preparation for a changing world.

1.3. Methodology

The present qualitative and participative investigation derives mostly from dialogues with NASA Education specialists and managers, and a manager of the so-called "AEB School Program," from the Brazilian Space Agency (AEB). It aims at providing useful perspectives and insights that contribute to analyze ongoing space (science) education precollege initiatives, from a pedagogical perspective. On account of the singularity of NASA's experience in space science education in terms of structure, implementation, quality, scope, and resources, it has been selected and analyzed a sample of education projects/programs from three NASA field centers, plus the Brazilian Space Agency's education program.

a) Institutions involved:

NASA Goddard Space Flight Center (internship host institution);
NASA Headquarters;
NASA Johnson Space Flight Center.
Brazilian Space Agency.

[1] United Nations Office for Outer Space Affairs Web site, http://www.oosa.unvienna.org/

Figure 1. NASA Goddard Space Flight Center, located in Greenbelt, Maryland. On the top, an aerial view of the largest NASA field center, located in an area of 1270 acres, part of which loaned by the U.S. Department of Agriculture. The center is responsible for robotic exploration of the Moon, the Sun, the solar system, and beyond. On the bottom left, there is an image of one of its main buildings, and on the bottom right, an image from inside a building. Credit: NASA.

b) Programs/ projects investigated:

From the institutions selected, professionals involved in NASA state-of-the-art education programs have been selected for participative interviews, along with a space education program manager of the Brazilian Space Agency (AEB), for comparative purposes.

Table 1. Projects and programs investigated through interviews

	Program/ project	Location	State/Country
1	Sun-Earth Connection Education Forum[2]	NASA/GSFC	MD/USA
2	Space Weather Action Center (SWAC)	NASA/GSFC	MD/USA
3	Teaching from Space (TFS)	NASA/JSC	TX/USA
4	Explorer Schools	NASA/GSFC	MD/USA
5	Radio Jove	NASA/GSFC	MD/USA
6	NASA Elementary and Secondary Education	NASA/HQ	DC/USA
7	Solar System Radio Explorer Kiosk	NASA/GSFC	MD/USA
8	GLOBE Program	NASA/GSFC	MD/USA
9	INSPIRE	NASA/GSFC	MD/USA
10	NASA Academy	NASA/GSFC	MD/USA
11	NASA TV Education	NASA/GSFC	MD/USA
12	AEB School Program	Brazilian Space Agency (AEB)	Federal District, Brazil

[2] Currently named Sun-Earth Day Team. Learn more on the Web site, http://sunearthday.nasa.gov/team.php

For logistical reasons, some interviews were conducted on spot (about 90 per cent), and others by phone - the following cases: a) Educator Astronaut Program, located in NASA Johnson Space Center, and b) AEB School Program, located in Brazil. A sample of complete and partial interviews is presented in this chapter, and all original and complete audio files are available under e-mail request to the author: norma.reis@gmail.com.

c) Professionals Interviewed:

Table 2. List of NASA Education professionals interviewed

Name	Program/Subject	Institution	Date
Robert E. Gabrys, Chief	State-of-the art programs in education at Goddard	NASA/GSFC	07/03
Shelley Canright, Manager	NASA Education policies	NASA/HQ	07/09
Ryan Fitzgibbonbs, TV Producer	NASA TV Education	NASA/GSFC	07/11
Carl E. Walz[3], Astronaut and Director	Astronaut selection and training	NASA/GSFC	08/06
Wanda Dias, Specialist	Radio Jove	NASA/GSFC	08/07
Igor Eberstein, Manager	INSPIRE	NASA/GSFC	08/08
Todd Toth, Manager	GLOBE Program	NASA/GSFC	08/08
Trena Ferrel, Manager Coordinator	NASA Explorer Schools	NASA/GSFC	08/08
Margareth J. Maher, Manager	NASA Education policies	NASA/GSFC	08/12
Timothy E. Eastman, Deputy Group Manager	Sun-Earth Connection Education Forum and Plasmas	NASA/GSFC	08/12
Troy Cline, Coordinator	Space Weather Action Center (SWAC)	NASA/GSFC	08/12
Ivette Maria Soares Rodrigues, Manager	AEB School Program	Brazilian Space Agency (AEB)	08/12
Sherrica Newsame, Manager	NASA Academy	NASA/GSFC	08/12
Cynthia L. McArthur, Manager	Educator Astronaut Program	NASA/JSC	08/12
James R. Thieman, Director	Radio Jove	NASA/GSFC	08/12
Leonard Garcia, Specialist	Solar System Radio Explorer Kiosk	NASA/GSFC	08/15

Considering the dynamism present in the scope of an interview, I had in mind the importance of raising fundamental questions that would help the educational community understand what each particular program is all about, how does that work, and what are its main benefits and constraints. Hence, in the course of an interview, in some cases different questions have been raised, in other circumstances, unexpected yet fruitful discussions have taken place. In some cases the researcher adopted a suggestive approach, even though this is not the primary goal of the research. When it is a research in education though, it is imperative that the individuals involved construct ideas together, and share similar or different viewpoints. This result indicates that the interviewing process was partially passive in the sense of collecting data, and partially active, in the sense of truly interacting with the interviewed persons, and in some cases suggesting different approaches within the initiatives investigated.

[3] Currently, Mr. Walz does not work at NASA Headquarters.

Chapter 2

SPACE AND EDUCATION – AN OPTIMAL MATCH FOR STEM BETTERMENT AT PRECOLLEGE LEVEL

If education alone is not capable of changing the world,
Without education the world does not change.
- Paulo Freire (1921-1997), Brazilian educator

This section discusses the engaging, motivational power of space science education exemplified by a set of NASA and Brazilian efforts. It provides fundamental concepts, discussions, and an overall contextualization of current trends and challenges in education; it also features a summarized narrative of the history, benefits, and perspectives of space exploration. Finally, it discusses how the 'space education phenomenon' – which in the format of education programs used to be privilege of professionals working on the pioneer space programs – has later entered precollege education walls in a growing number of countries. This background information facilitates understanding of this research's purpose, which consists of presenting pedagogical analyses of a sample of NASA and Brazilian state-of-the-art practices/projects in space science education for precollege level.

2.1. EDUCATING IN THE 21ST CENTURY - ADVANCES AND CHALLENGES

Education is theme of exhaustive debates and studies worldwide. Virtually everyone – government, private sector, and citizens – recognize that education access and betterment are critical. Federal and local governmental agendas foster investments in education to assure and/or maintain socioeconomic sustainable development. However, education purposes goes far beyond preparing for the work market, and assuring this development. The word 'education' comes from the Latin '*educare,*' which means 'to bring up, rear,' related to '*educere,*' meaning 'to lead forth,' or 'bring out.' Indeed, education is the process of bringing out in a particular individual the historically constructed culture. Let us explore some definitions of education. According to the Brazilian educator Paulo Freire[1] (1921-1997),

[1] Considered one of the most notable thinkers in the history of worldwide Pedagogy, Freire has influenced a movement called 'critical pedagogy,' and is famous for his work with popular education. He proposed a dialectical method for literacy and defended the dialogue with low-income populations not only as a teaching

education is an exclusively human process [18], through which individuals transform the world through intentional actions, and express reality with language. Freire also states that only humans are fully aware of themselves and the world. Similarly to Freire, the Brazilian educator Dermeval Saviani (1943-) states that the educational labor consists of producing intentionally in each person the humanity historically, collectively produced. According to Saviani [19], education is strongly correlated with culture – as societies become more complex, formal education (the one that takes place inside schools and universities) becomes essential to socialize culture. The author understands that scholarly knowledge should be elaborated, erudite, and scientific – not popular, fragmented, and spontaneous. In this scenario, formal education should socialize through curricular disciplines the systematic, elaborated knowledge, along with attitudes, habits, ideas, symbols, concepts, values for life, and identify effective means to have students learning them. Saviani states that there is currently a widespread misunderstanding about education's role in society as schools at times embrace secondary activities i.e., festivals, campaigns, and trips, in detriment of teaching students what they really need – elaborated knowledge to allow them to become professionally competent citizens in today's knowledge society. Saviani [19] adds that especially in low-income communities to deprive students of elaborated knowledge would mean to contribute to enlarge social inequalities.

There are several other definitions of education that could be presented, but the purpose here is just to provide a brief panorama of what education *is* and what *it is not*. The mentioned definitions assume that education consists of teaching socially valued knowledge, skills, values, and attitudes for citizenship and effective societal inclusion. Though preparing for work market is a priority in education, it should not be the exclusive one. Equally critical is to teach values and attitudes that promote cooperation, solidarity, environmental awareness, justice, and respect for the human person.

More than ever before, education in the 21st century is supposed to be for all and with quality. Through the history public schools have been established and expanded mainly to meet economic demands. It was with Industrial Revolution starting in the 18th century in England that education for 'masses,' or public schools flourished. In the beginning of the Revolution, formal education was aimed at elites only, as education for 'masses' would cause social disorder from the dominant classes' viewpoint. Workers if educated would supposedly 'rebel' against a prevailing inhuman system that explored young children, women, and men for exhaustive working hours, thus extinguishing any possibilities of learning and leisure. The current concept of infancy did not exist at that time. Actually, children were more exploited than adults because they were more malleable, fragile, and easier to control. In was only in the 19th century that public schools appeared in England. Although capitalists were afraid and reluctant that the 'mass' would 'learn so much' and create difficulties to their purposes, churches defended education for the 'mass' [20]. Further, later workers started fighting for their rights and against exploitation. Work hours have been increasingly reduced as a result, thus allocating time for formal education in public schools.

Currently, national governments have been working towards promoting equality through public policies in education. There is concern with human rights and education for all, though this entity called 'market' today requires complex skills and knowledge (not always

method, but also as a means of fostering political awareness. According to Paulo Freire, students should create their own learning pathway, instead of blindly following an established script, such as in technicist pedagogy.

accessible to everyone) for qualified inclusion. Otherwise, the individual may be trapped in the unfortunate paths of unemployment or sub employment. Instead of working in machines in repetitive tasks, the current worker is required to be productive in terms of innovation, information and communication technologies utilization, scientific knowledge, complex processes proficiency, and so forth. It is required from the new worker complex thinking, analytical, teamwork skills, because information, technology, and knowledge evolve extraordinarily fast, and working teams often cross national boundaries. The advent of the so-called knowledge society[2] generated an urgency towards educating all citizens not only in terms of number of school years, but also (or mainly) of quality to prepare workforce for new sectors, and to assure their effective participation in all spheres of a democratic and ethical society. According to Bell [21], the knowledge society is defined by the service sector, the predominance of theoretical knowledge, and intellectual technology. If in the industrial society the factory used to be the core of modern life, the knowledge society is organized around knowledge, generating innovation and change. In this also called information society, universities are the major institution. Humanity has also entered a new era called "digital age." The break is as revolutionary as it was the rise of the industrial society. In this society, knowledge is the 'passport' *per excellent* for social mobility [21]. One indicator for this fact is the growing access to increasingly higher levels of education by every citizen regardless of societal class, which is a tendency observed in an increasing number of industrial and developing countries.

However, we cannot be naïve and believe that the 'market,' which is still a strong influence in the world, aims at "reducing inequalities through education," simply to assure more welfare to people who were not educated before. Though there are political and economic powers that are socially committed to population welfare, the major motivator for this access to more years of qualified education for all is the urgency that societies today have for more competitive products and innovative processes to gather leadership or at least some level of economic independence to survive in today's global society. For this reason, more educated workers are required. In this scenario, formal education is considered a right of every citizen, and people are encouraged and sometimes even pressured to study more and more. As the world is currently seem in terms of a global community this concern of providing education for all crosses national boundaries. For instance, international organizations have been promoting initiatives to contribute to foster equality among different countries with regard to education provision and quality. The United Nations launched, for example, the so-called Millennium Development Goals[3], comprising among them the goal number two: "Achieve Universal Primary Education." The target would be that by 2015, children everywhere would be able to complete primary education. Updated data show that this goal will not be met as there are still nearly 69 million children outside school in the world, mainly in Africa and Asia. However, important progresses have been made through actions such as abolishing school fees, investing in teaching infrastructure, and providing school meals [22]. The fact is that, despite of social inequalities that still exist in the world,

[2] The term was coined in 1969, by the famous U.S. economist Peter Drucker (1909-2005), and started being used when the U.S. sociologist Daniel Bell (1919-2011) published the book *The Coming of Post-Industrial Society*, in 1973.

[3] The United Nations goals are (1) end poverty and hunger; (2) universal education; (3) gender equality; (4) child health; (5) maternal health; (6) combat HIV/AIDS; (7) environmental sustainability; (8) and global partnership. Extracted from the United Nations Web site, http://www.un.org/millenniumgoals/education.shtml.

and conflicting interests between the 'market' demands and citizens' interests, the scientific and technological development has contributed to create a world where the population is required to study more.

In this context, populations worldwide are increasingly aware of the importance of scholarly education as a 'passport' to better living conditions and professional success in today's world. It generates demands for access and quality at all levels, and leads to governmental concerns translated into educational policies, programs, and projects committed to provide access and quality at all educational levels. One significant difficulty observed in several nations is that though schools, universities and colleges have been assuring wider access, the education offered not always has the necessary quality. The result is a lack of qualified professionals required in the scope of complex of the knowledge society. Thus, at the same time that populations require education from governments, in many cases there is criticism to the quality of education offered.

As mentioned, scholarly education has been suffering several criticisms as it embraces several tasks sometimes to the detriment of its core mission. Indeed, schools have been facing difficulties in playing their major roles, which is to effectively educate in systematic, historically constructed, and socially valued knowledge, skills, and values. As a response, several pedagogical theories have been brought inside the classroom, comprising innovative methodologies and resources, claims for family involvement in scholarly affairs, interdisciplinary approaches, multiculturalism, and so forth. Some actions that have been producing positive impacts in education in several nations are: (a) investment in all educational levels; (b) provision of Information and Communication Technologies (ICT) to schools/students; (c) effective in-service and pre-service teacher training; (d) innovative teaching-learning methodologies, such as working through projects and hands-on activities. It is important to point out, though, that huge investments in education are positive, but not always sufficient to produce desired results. For example. international assessments such as PISA have shown that there are countries with several low-income, but high-achieving students, whilst there are countries with many high-income, but low performing students [5]. Factors presumably less impacting in learning outcomes such as the manner how education is perceived in a given culture has substantial influence on students' dedication to studies, and produce positive results in evaluations. For instance, in South Korea education is considered something very important, and virtually everyone wants to enter a good university. As a result, children study both at school in the morning, and have private lessons in the afternoon. They can alternatively further their education with a so-called cyber home learning system,[4] which provides them with customized learning and real-time interactive tools. The result of this positive perception of education can be accessed in PISA results for South Korea. Indeed, the problem with South Korean education is that children and teenagers study too much!

Let us explore some more elements of education, which will help further contextualize the 'space education phenomenon.' Education can be formal or informal. Formal education is systematic and takes place in specialized institutions such as schools and universities, providing students with some type of certification after completion. Informal education is not systematic and can occur at a wide range of spaces: home, in a church/temple, clubs, museums, theaters, on the disco, on the cyberspace, and so forth. In the framework of this research, I will concentrate on formal education, as the focus is to analyze how practices in

[4] Information on the Cyber Home Learning System at http://english.keris.or.kr/es_vod/1173950_539.html.

space (science) education can contribute to enhance precollege education. Formal education is systematic and committed with socializing knowledge historically and scientifically elaborated, and systematically organized in order to be assimilated by students in accordance with their specific ages and previous schooling. As mentioned, it also includes values, skills, and abilities valued by a given society in a given historical time. Some pedagogical theories disagree of the idea of knowledge transmission, because it would make more sense to see the educator as a mediator, a facilitator between knowledge and students. In this perspective, knowledge historically and scientifically elaborated would be in permanent construction and students would play a more active role in their own educative process.

Education is in service of individuals and society, thus it does not happen isolated, but interconnected to work market requirements, political, economic interests, and societal pressures. Although economy influences education, it is fallacious and somehow simplistic to infer that education is 'determined' by the work market. Educators interact dynamically with the societal tissue to construct their own realities in accordance or in conflict with that given panorama. In this interaction, sometimes the "economic system" wins, sometimes educators do; and sometimes both fail. There are several concepts and definitions of education constructed in the course of history arising from different interests and ruling classes. Some pedagogical theories regard education as a neutral endeavor. In other words, they discuss curriculum, methods and techniques, but not political and socioeconomic influence that impact the education process. According to the Brazilian author Libaneo [23], some of these theories are the traditional and the technicist pedagogies. Effective education is not only a result of intra-scholarly factors, though. There are extra-scholar factors with substantial impact on the educational process, such as students' material conditions, and parents' education background. One theory that considers these extra-scholarly factors, situating the education process within the realm of socioeconomic influence is named historical-critical pedagogy This theory situates student in society and history, and assumes that scholarly education specificity is to assure that each person learns the elaborated culture, without undervaluing or ignoring their own [19]. Simultaneously, it is assumed that education should be a pathway for reaching autonomy, liberty, and awareness. These historical-critical theorists regard education as influenced by (and influencing) the social and political scenario. Education would play an active, conscious, defined role in influencing the societal tissue, working towards social inclusion of underserved, underrepresented communities. This is accomplished basically by providing education with quality to low-income students, and promoting awareness by teaching them how to think critically and become agents of societal transformation in service of social justice.

Let us visit some more definitions. One states that education consists of promoting behavior change in a positive manner i.e., by bringing students from a certain level of understanding, interactions, and knowledge to a higher one. The Russian psychologist Lev Vygotsky (1896 – 1934), for instance, introduced the notion of zone of proximal development as "the distance between the actual developmental level as determined by independent problem solving and the level of potential development as determined through problem solving under adult guidance, or in collaboration with more capable peers" [24]. From another perspective, derived to a great extent from liberal theories of economy, education could be characterized as one's preparation to work market. This perspective is more prominent where and when the priority is to prepare the 'ideal employee,' and usually this type of education fails in preparing the 'integral citizen.' The concept of integral citizen refers not only to professional and academic knowledge and skills, but also the preparation of a critical thinker, a human being, a proactive member of a family, a society, a historical time, and a changing world. Preparing the 'integral citizen' requires actively studying themes such as ethics, human values and rights, politics, sexuality, and subjects such as literature, sociology, philosophy, psychology amongst others, which will assist the person not simply to get 'adjusted' to a given reality, but to better understand and interact with that in order to figure out how one could contribute to transform that by being creative, proactive, and constructively critical in the sense of identifying channels for promoting changes capable of bringing benefits for society. Education in this perspective is committed to contributing to transform society and promote equality among students and population in general. Hence, this category of education is not politically neutral, but committed. The mentioned Brazilian educator Paulo Freire regards education as a means of transforming society. In one of his several books he states that if education itself is not capable of changing society, without education society does not change [25]. Actually, education if taken isolated cannot promote societal transformation. It is capable of influencing subjects' minds, though, and affects the viewpoints and values that a person will carry in the course of life – and subjects do change society.

Education occurs in constant and dynamic interaction with *extrinsic* and *intrinsic* factors related to the education process. The *education process* has the main intent of assuring students' success in scholar activities and takes place in the scope of the actions carried out at the scholar institution, constituted by the diverse interactions involving the school pedagogical project, its curriculum, educators, students, community, principals, pedagogues, psychologists, and other professionals directly or indirectly involved with this process. The *extrinsic* factors stand for those not derived from the scholarly process, but that interact with scholarly reality, influencing positively or negatively the educational process. It can be mentioned some foremost ones such as social *inequalities*, changes in the traditional *family* structure, the influence of *media*, such as TV and internet, *psychological* issues, and so forth. One of the most influential of these factors is the so-called social inequality. Several types of inequalities still permeate society. The most noticeable is the economic one. Indeed, one of the major problems of education is that a considerable part of the world's population pursues scarce financial resources. These individuals in many cases do not have proper conditions to succeed in formal education, especially because their parents also did not receive enough education. However, in some contexts students succeed in spite of their socioeconomic disadvantages. They are called 'resilient' students. On average across OECD countries, 31 percent of students from disadvantaged backgrounds are resilient[5] [26].

I would like to highlight an important (though not unique) extrinsic factor to assure student success in scholarly process that is family involvement. The point here is the example and the encouragement provided by family environment, which will not be found elsewhere. The school environment is not capable of replacing family. Family is so important that it can compensate significantly low-income conditions. Studies show that parents with high educational attainment, combined with an emotionally stable and stimulating environment at home can minimize the negative effects of financial restrictions [27]. According to a Harvard

[5] A student is regarded 'resilient' here if he or she is in the bottom quarter of the *PISA index of economic, social and cultural status* (ESCS) in the assessed country and performs in the top quarter across students from all countries after accounting for socioeconomic condition. Source: *Education at a Glance 2011: OECD Indicators* © OECD, available at http://www.oecd.org/dataoecd/61/2/48631582.pdf.

study, family involvement in elementary school leads to higher reading scores, language growth and development, motivation to achieve, prosocial behavior, and quality work habits [28]. On the other hand, parents with low educational attainment and family fragmentation, contributes to reduce the level and quality of knowledge assimilation by students, as well as influence negatively the construction and consolidation of personality. The fact is that several children do not have family (or have families that do not offer emotional and educational support) and, in this case it is necessary to consider full-time schools in a special system comprising teachers, and other professionals such as psychologists, social assistants, pedagogues, and the entire school should be committed not only to providing academic knowledge, but also love and guidance to those young human beings. Students need to learn responsibility and confidence, but first they need at least an adult to encourage them, an adult to ask them with care how was their day, and how was school. In psychology, it is accepted that children need a model of woman and man. This model should be father and mother, but when they do not have them, they will end up adopting an alternative, usually the teacher. Without a good reference, children would be easily vulnerable to following not so acceptable and desirable behaviors in the societal tissue, such as drug addiction, early sexuality and pregnancy, criminality, and so forth.

This discussion of extrinsic factors does not constitute, though, a deterministic view of education that conceives students and teachers as "victims" of the system or *status quo*. In fact, there is always place and space for reaction, and education does contribute to enable persons to overcome barriers both academic and in life overall. In this sense, individuals are active; although in some cultures and historical moments there is a politically committed encouragement to continuity, passiveness, and adaptation to standards, reality is intrinsically dynamic – actually the only constant element of reality is change. Indeed, in a dialectic approach, the beginning of a process brings up the seeds of its destruction, and this destruction represents the beginning of a new reality, which will subsequently be destructed and replaced by a new one. In this destruction, though, not all elements of the former reality are discarded. The idea here is destruction by incorporation i.e. the elements considered useful are transferred to the new system under construction. Stated differently, one does not "throw out the baby out with the bathwater." In other words, many students facing unfavorable conditions will effectively overcome them, but it is obvious that they are not "competing" in conditions of equality with students who have access to refined goods and resources such as international travels, visits to museums, adequate housing and living conditions, family care, proper medical assistance, toys, leisure, and so forth. It is also predictable that the majority of those low-income students tend to face difficulties in their interactions with the educational process. Another important factor is pre-service and in-service teacher training, their working conditions, and remuneration. In countries that did not achieve a sense of worth regarding the teaching profession, usually educators need to work three periods per day, teaching in the morning, afternoon, and evening to have a reasonable income. In this panorama, these professionals usually do not have proper opportunities to adequately prepare good classes as well as do not count with opportunities of continuity of studies, in-service or at graduate level, so critical to enhance the quality of teaching.

In addition to the previously described extrinsic factors, there are factors *intrinsic* to the educative process. One of them refers to the student-teacher-knowledge interaction. It has to do with methodologies, methods and techniques targeted to assist students in their process of constructing knowledge. Although humans have stepped onto the Moon and internet has been revolutionizing the world of communication, though the concept of students as *tabula rasa*, and *alumnae*, which means a being "without light," have been left behind in the framework of pedagogical theories, it is a fact that still currently a considerable number of teachers worldwide use only chalkboard, voice and chalk to teach, as well as some yellow notebook containing the same lessons to be taught year after year. In other words, traditional pedagogical approaches still dominate several classrooms worldwide. Many teachers do not receive proper preparation to teach. Indeed, in many cases they have attended college, but do not have adequate pedagogical background, thus have not been exposed to effective teaching methodologies, sociological, psychological, philosophical subjects and so forth, intended to facilitate understanding of the education process. Another factor relates to the access to Information and Communication Technologies (ICT), which have already produced a significant impact on education worldwide, though have not sufficiently reached some developing countries, and even some underserved regions within industrialized nations. Another important factor relates to the managerial style of the scholarly institution. It is understood in Pedagogy that democratic management of schools is more effective, because when schools promote real engagement of parents, students, teachers, and community in the decision-making process of issues related to school and students, all these actors feel responsible for the results, and it generates some sort of synergy, and the main benefited is the student. On the other hand, when schools decide to adopt the traditional hierarchical, vertical managerial style restricting to a minimum the level of interaction and engagement of the actors involved in the educational process they end up alienated of the process and this lack of engagement contributes to failure of the overall educational process.

As for teaching, I still would like to highlight one important (though usually forgotten) *intrinsic factor* that is the human heart of the one who teaches. When teaching, for good or for bad, we teach ourselves. It is impossible to separate the practice from the one who practices; it is important to separate the piano of the pianist, as the melody depends on the interaction between them. However, we rarely think about these questions. *"Who is the self that teaches? How does the quality of my selfhood form – or deform – the way I relate to my students, my subject, my colleagues, my world? How can educational institutions sustain and deepen the selfhood from which good teaching comes?"* [29]. It is certain that my *self* influences the quality of my work. For instance, if I do not care about some human values, how will I teach them to my students? The quotidian would show my inconsistencies between speech and practice, and students would probably ignore my words. And it is not only in the realm of values and attitudes, but also knowledge. For example, if I am a teacher and I do not seek a variety of information about my subject to enrich my classes, how could I expect and require that my students do excellent research? They could eventually do that if their family environment favors that, but they would easily notice that I require from them things that contradict my practice. It may seem at first glance a very "subjective" concern, but our *persona* does influence our practices and impact students.

As discussed, education faces many challenges in the 21st century. Perhaps one of the most important is the urgency for valuing diversity and promoting inclusion of: students with special needs, students with learning difficulties, low-income students, illiterate adults, and other actors who have been historically excluded from the education system. It is necessary to mobilize academia, government, educators, and community to a common discussion and effort of constructing a new school aimed at reaching and serving every citizen, and identifying methodologies that address specific needs of each of these groups. Educators should also be aware of the fact that students start their education process in different levels of academic capacity. Schools need to foster discussions on how to break the cycle of simply encouraging the already encouraged and discouraging the discouraged. This process generates self-exclusion of the latter category of students, who perceive themselves as a 'failure' in the realm of that system. At a certain point, either at the first grades or after a precarious, low-quality college study, they decide to abandon formal education because they believe it was their fault not to have adapted to a system that does not meet their specific education demands, though we know that education is a basic right of every citizen in the world. I would not 'blame' the system here. Several national governments have been working seriously towards increasing both access and quality to education. However, there is a lot to be done in order to offer equal educational opportunities to all students, regardless of their starting conditions. As early childhood education produces positive impact in posterior levels, it is recommendable to invest in this area. Further, in the entire world there is an enormous population of persons with special needs – visually impaired, persons with locomotion impairment and so forth. These citizens are in many cases kept inside the house, as their families feel uncomfortable about having a child with some impairment, and to not want others to know. In other cases, families do not have financial conditions to equip their children with the apparatus they would need to learn properly and have better life quality. And even when their families are more conscious, well-off, and provide their children with special needs with proper education opportunities, school and university is not always prepared to effectively offer them education with quality, adapted to their specific needs and teaching-learning requirements (i.e., equipments and pedagogical approaches). Finally, even when these students succeed in all educational levels, there are barriers in the moment of finding a good job. It is evident that a number of them succeed in all phases, but public policies should be committed to assuring that every child with special needs will find proper education and employment opportunities, as they do have in most cases the same (at times higher) cognitive capabilities in comparison to their "normal" peers. The social prejudice towards individuals with special needs has to do with the lack of knowledge and awareness of those who think that an impairment, a "disability" is not something that can happen to everyone in any phase of one's live, following the same logic of those who have prejudice against elderly people, irrationally thinking that they will not get older as well…

Another challenge is the urgency to bring back to school adults who have been in the past excluded from the education system. This grown-up student needs to acquire knowledge in order to obtain a better job or even a promotion to improve their living conditions. Often, this student already has spouse and children, thus he or she does not have so much time to study. It is necessary to conceive a pedagogical project targeted to the interests and necessities of this group, a system capable of guaranteeing both their professional/academic progress and human development. However, it would be necessary to mobilize the entire society and to create mechanisms to stimulate every adult citizen in this condition to return to school as many of them would not do otherwise especially wives, who often have a profession, and also take care of family and house. Education open doors not only to better jobs, but also to citizenship. Further, educated people tend to be more aware of their rights and respect other people's rights; and they tend to be less tolerant with abuses such as domestic violence. Thus, it is urgent that governments establish public policies that assure that all citizens have access to education with quality at all levels, either at physical schools and universities, or at quality distance learning institutions.

Figure 2. Visually impaired students participating in a NASA's radio astronomy project. NASA carries out several programs and actions towards exploring the potential from all students in their missions. In this image, students with visual impairment learn how to build their own equipment in the scope of the Radio Jove project. Credit: NASA.

It is necessary to analyzing the scholarly education process and raise questions that would enable educators to write collectively the school's political pedagogical project, such as:

> *Are we educating for which society? With which political project are we committed?*
>
> *Are we educating for perpetuation of social inequalities or are we committed to a project of effective inclusiveness?*
>
> *Do we teach our students on real democracy and consequences of political manipulation?*
>
> *Are we educating for reproduction and passiveness or for intellectual autonomy and action?*
>
> *Are we preparing for life overall or for the work market only?*
>
> *Are we using evaluation as means of accessing students' progresses and difficulties and as a tool for finding out manners of assisting them in overcoming their difficulties, or are we only using exams to attribute marks, promoting some and excluding others, without any concern?*
>
> *Are we using continuous evaluation or only the 'end-of-semester' exam?*
>
> *How are we using standard evaluation applied by national governments to access overall students' learning? Are we using that as one parameter of*

analysis or taking that as an absolute reality? It is necessary to have in mind that national and international exams play an important role in providing some elements for analysis on how students are doing overall and provide comparisons, but that has also constraints in terms of accessing school specificities, thus they should not be taken as absolute and unique true.

Are we educating for intellectual autonomy, innovation and creativity or for passiveness and repetition of archaic ideas?

Does our school have a political-pedagogical project? Has it been constructed by the scholar collectivity or is that only a document on the principal's desk?

Do we teach ethics? How can we discuss ethics when it seems that the world outside – and sometimes inside school walls - seems to be permeated by examples that radically contradict the principle of ethics, honesty, and respect by the human being, such as bullying?

And what about sexuality? Is this discussion still a taboo in schools? And when these discussions do happen, how are they conducted? Only restricting themselves to teaching "contraceptive methods," or is it conducted an overall discussion on human relationships, respect, meanings and purposes of male/female personal relationships? It is not a matter of going back to the conservative school, but primarily not to vulgarize sexuality and relationships, as if they were banal, regarding human persons as objects, following the same logics of consumerism, as teenagers live in the era of shopping centers and must follow the "ritual of consumption" in order to be accepted in their "ghettos."

Moreover, do we use the knowledge students bring to classroom as a starting point to promote the education process or do we assume that their knowledge is unimportant, thus it does not need to be taken into consideration?

Do we consider that in a single classroom there are students with different levels of academic proficiency and background, and that all of them deserve the same respect, and customized learning opportunities?

Are we really committed to promoting inclusiveness in our schools? Or do we treat students with special needs as if they were "aliens," "poor guys," or "less capable" than the "normal" ones?

And what about diversity? Do we still see the "different" as "inferior"? Do we still conceive our model of world as the "standard" one, and our truths as the "right" ones?

What kind of discipline are we teaching? The one based on fear or on responsibility? Do we still feel the need to "scream" with our students?

Are we promoting meetings to discuss changes in the curriculum and regular planning in order to attend demands of each specific community as well as to incorporate the newest advances in science, technology, information and other realms of knowledge?

Do we intervene when we see students suffering injustices or when they are doing things that are not fair or socially desirable, or do we just "wash our hands," because it is easier to "let things the way they are," "it's their parents' responsibility" and "there is not so much we can do, anyway – we cannot solve all problems in the world"? In other words: do we care about our students' welfare? – Because educating is also an act of love.

Are we considering the benefits of informal education preparation, such as visits to museums, expositions, theaters, space centers, zoos, fields, beaches, and other spaces, allowing students to widen their horizons of reality?

Are we committed with an education in which students can be proactive by presenting works in science fairs, events, encouraging initiatives where they can organize their own governing systems such as meetings of students' representatives and other democratic decision-making spaces?

Are we committed to the logics of meritocracy of a few and indifference towards other students facing difficulties?

Or, on the opposite, are we somehow committed with the logics of mediocrity, in which if one does the minimum to "pass" the academic year that is fine, and even one should not go farther in order not to be regarded as "nerd" or other kind of jargon used to label children who "study too much?"

Are we encouraging competitiveness or cooperation in classrooms?

What kind of leadership are we teaching? A leadership in which all members of a team are encouraged to contribute with their know-how, or another kind that intends to centralize power due to the probable insecurity of leaders, who are afraid that if group members start to understand the overall managerial process, they could also identify failures and inconsistencies, start to figure out new possibilities, and threaten the leaders' "position?"

Are we concerned whether our students are effectively learning or are we just concerned about our teaching? Do we understand education as an interaction, that is, an action performed simultaneously by both teacher and student(s), transforming both parties, or are we just "throwing" information to be "swallowed" by our students the way they can – or cannot?

What is the importance of the topics we are teaching? Why did we choose these topics and not others? Which interests are driving us? Are these interests committed to a project of national state and society for all or to a project of perpetuation of interests of some particular group or societal class?

Are we contributing to the welfare of our nations and our citizens with our pedagogical practice?

Do we understand that education is different of teaching? That teaching is just one element of a much more complex process, which is education?

Do we have courage enough to be educators, not only teachers?

These questions are not delirium of a researcher who is peacefully sunbathing on some Caribbean beach; instead, they are provocations of someone who has effectively been a teacher of low-income communities in South Brazil, and faced many of these challenges. I faced difficulties and could not answer many of these questions. However, educators do need to discuss and look for answers for these questions – provisional answers, though, because reality will inevitably and dialectically change and raise new questions.

Education is a right of every child and teenager, stated both in international and national legislative documents, and it is the joint responsibility of national states, education systems, schools, and families to assure this right to every citizen. Indeed, societies as a whole needs to perform coordinated efforts to overcome current challenges in education. It is unacceptable to blame teachers, parents or students for many of our present educational problems. It is evident that, for instance, the teacher alone will not be able to transform their students' reality in all senses. This would be utopia. It is necessary to carry out this discussion in terms of real possibilities and constraints. It is also evident, though, that having teachers motivated and committed to their students will help substantially the achievement of better results. However, it is necessary to mobilize other sectors and actors, such as families, communities, non-governmental organizations (NGOs), and federal and local governments. It is necessary to work towards educating for life, not only for work market requirements, though it is highly necessary to provide students with tools to become a future employable citizen, a businessman, an actor, or whichever other career path the person chooses. In the current scenario of globalization, it is necessary to have students being capable of developing autonomy to go farther; to construct and change realities; to work towards promoting social justice and mutual respect among all citizens; to learn permanently along one's life; to reach higher skies. It is necessary to work together to construct a better and inclusive school for each and every children and teenager, regardless of their cultural, ethnic, socioeconomic, gender, or any other background. Education is a right of all, and for all, and it is our mission as educators to keep on the fight to make it happen, and with quality.

The previous discussion has been conducted to contextualize and signify a specific field research: Science, Technology, Engineering, and Mathematics (STEM) education, which has been essential in the knowledge society as STEM advancements have been provoking societal, cultural, and educational transformations by bringing up innovative and complex processes, products and services. This fact implies that all citizens should be literate in STEM to a certain extent in order to achieve better academic and professional proficiency in this contemporaneous panorama. In other words, STEM education plays a vital role in promoting effective inclusiveness of citizens in a societal fabric permeated by processes/products/services requiring of individuals a certain level of STEM literacy [30]. The latest PISA results (2009) demonstrate that in some parts of the world students are excelling in mathematics and science. Major examples are Eastern Asian countries such as China, Japan, and South Korea. Some common characteristics of these education systems are to encourage excellence of all students, and to select and retain the brightest professionals for the teaching career. On the other side of the spectrum, we have seen that there are several countries performing just average or below the expected. A considerable number of them belong to the so-called 'developed' world. These future professionals will probably lack the required skills to advance the frontiers of STEM knowledge and compete in the global economy, which would bring serious barriers for sustainable national development [5]. In this scenario it arises the urgency of constructing public policies to promote STEM education with quality, and revising elements such as curricula, methods, and teacher career.

It was in the 19th century that science was incorporated to school curriculum in Europe and in the United States. By the beginning of the 20th century in the U.S., curriculum reform movements began in science education. They were initially identified mainly with efforts of the U.S. philosopher and pedagogue John Dewey (1859-1952). He defended the importance of scientific education, and his studies became more meaningful in the 1950's when scientific knowledge started to be super valued in comparison to other areas of human knowledge. The theme became a great slogan, starting a worldwide movement in defense of scientific education [31]. The category *scientific literacy* defines the understanding of the main conceptual frameworks that constitute the foundations of science, the role they play in scientific efforts, the reason why they are widely accepted, as well as the understanding of the elements of the scientific method, of the analytical and deductive thinking, of the process of logical thinking, of trust in objective evidence, and the fundamental role of experiments in science. It also means that one can ask, find, or determine answers to questions derived from curiosity regarding everyday experiences. Further, scientific literacy enables citizens to distinguish between what science is and what it is not. According to a definition by PISA, *scientific literacy* is *"an individual's scientific knowledge, and use of that knowledge, to identify questions, acquire new knowledge, explain scientific phenomena and draw evidence based conclusions about science-related issues; their understanding of the characteristic features of science as a form of human knowledge and enquiry; their awareness of how science and technology shape our material, intellectual and cultural environments; and their willingness to engage in science-related issues, and with the ideas of science, as a reflective citizen"* [32]. Two principal purposes of STEM education are currently emphasized: a) to socialize STEM basic knowledge and skills for all in schools; and b) to encourage STEM careers. In order to reach those purposes, it is essential that scientists and educators conceive, discuss and implement collectively innovative approaches, methodologies, and resources for the teaching/learning process of scientific subjects. However, based on what we have discussed, science education will not perform a 'miracle,' as education belongs to a determined society, and there are other factors inside and outside school influencing learning.

Physics teaching can be used as a prime example in the realm of STEM education. Physics consists of a canonical science, the first of them. It is a natural science, comprising early efforts of philosophers such as Tales of Millet (624-547 BC) with his *Cosmogony*, Aristotle (384-322 BC) with its *Physics*, and names such as Galileo Galilei (1564-1642) and his astronomical findings, Isaac Newton (1642-1727) and his classical mechanics, and Albert Einstein (1879-1955) and his Special and General Theories of Relativity [33]. Physics has been facing revolutions mainly since the twentieth century through advances in relativity, quantum mechanics, particle physics, cosmology, chaos theory, and so forth. Considering precollege education, this subject is taught more deeply at secondary school level. As Physics deals with a wide range of scientific concepts and principles – mathematical, conceptual and (indirectly) even philosophical elements – allowing interdisciplinary approaches, it is essential that students at that level achieve effective proficiency in that subject. Schenberg [34] comments that the evolution of concepts in Physics is interesting because it is not a straightforward process, but a zigzag: even when it is reconsidered an idea which had existed previously, it is done differently of what have been formulated before. It is also interesting to observe that in regard to Physics theories it can be identified in the course of history an attempt to a global and unified interpretation of phenomena [33]. However, precollege and even college students face innumerous difficulties to understand this subject. According to Campbell, *"Physics is conventionally taught as a timeless and ahistorical (sic) subject, divorced from its roots in society and from any present or future use"* [6]. Indeed, this statement could not be generalized, and it is not a phenomenon restricted to this subject. It could be partially a result of the fragmentation of knowledge, which led to its artificial division – not always positive in scholarly education, and it can also have some relation with curriculum selection of contents, approaches, methodologies, and resources. This scenario claims for collective construction of innovative pedagogical practices.

Space (science) education is a historically recent education practice capable of contributing to enhance STEM education, because it shows how science and technology are used in the scope of real space-related researches and operations. In the following sections, we will explore data related to the space thematic such as the current status of space exploration, and concepts associated to the 'space education phenomenon', along with several examples of successful space (science) education national and international efforts worldwide, and also discussions on how these efforts motivate students and raise their curiosity towards a wide range of STEM topics, positively impacting learning. Space education also plays an important role in encouraging young students to consider STEM disciplines as possible career paths.

2.2. Space Exploration and Space Education: What Does That Mean, Anyway?

2.2.1. A brief characterization of space exploration

Space has been fascinating humanity since the first peoples started contemplating the skies. Even prior to the invention of writing, astronomical observations had already been conducted. However, in early times Astronomy was not the popular science we know today. At that time, the vast majority of the populace had no grasp of the nature and mechanism of what they observed in the night sky. Most ancient civilizations have viewed astronomical phenomena through the lenses of fear and apprehension. Comets, meteor showers, supernovae, lunar and solar eclipses have been regarded as bad omens by several societies. Solar eclipses were particularly feared because the first agricultural societies depended on the Sun in order to survive, and the supposed disappearance of the heavenly luminary in broad daylight due to some evil power would literally represent the end of their civilization! Sometimes, not even emperors pursued basic astronomical knowledge. Thus, the role of the Astronomer was very important for the government, as signs in the sky would supposedly indicate to emperors the decisions they should make, such as starting or interrupting a battle. For example, on May 28, 585 BC, it occurred the so-called "Battle of the Eclipse" between the Lydians and the Medes, at the Halys River. It was the final battle of a fifteen-year war, which was suddenly ceased thanks to an eclipse. The phenomenon was perceived as an indication that the gods wanted that battle to stop. The kings became impressed when they viewed the eclipse and stopped fighting. Further, governors perceived the impact that celestial phenomena exerted over populations, and started using them as manipulative tools. They used

Astronomical information to make the population believe that they had influence over what happened in the sky. For example, the ancient Chinese believed that during solar and lunar eclipses there was an invisible dragon in the sky that was angry with then, and would devour the Sun or the Moon. When an eclipse occurred, the emperor organized the population to make loud noises by banging drums and pots, and to shoot arrows in the air to frighten the supposed monster away. In this context, Astronomers should inform to emperors in advance the meanings of signs in the sky [35]. In summary, though in early times populations were already mesmerized by the starry sky, some celestial bodies such as the Sun and Moon were regarded as gods with enemies to fight, thus some astronomical events used to make most people fearful of some cosmic-caused tragedy…

Likewise, the dream of reaching outer space is as ancient as humanity. Mythological legends such as the well-known Icarus dream are fulfilled of episodes related to the conquest of the skies. A story tells that Icarus was a young man who has been imprisoned with his father Daedalos at a tower in the island of Crete, around 1325 BC. The reason was that Daedalos has helped the hero Theseus to kill the Minotaur and escape with Minos' daughter, Ariadne. Observing birds in the sky, Daedalos came up with the ingenious idea of constructing wings for his son and himself to fly for freedom. Then Daedalos built wings out of bird feathers and wax. Before leaving for what would supposedly be the first human flight in history, Deadalos warned Icarus that he should fly not so close to the Sun so that the wax on his wings would not melt, not so down close to the sea that he would not fall. Though initially Icarus promised to follow his father's advice, as a stubborn young man he got bored of just flying in straight line, and started having fun by venturing up and down. As expected, Icarus went too close to the Sun, his wings started melting and came apart. To his father's despair, Icarus fell into the sea near Samos [36, 37], and his body was washed ashore on a nearby Greek islet, which has been named Icaria[6] in his honor. Though this legend does not refer specifically to spaceflight, it can be associated to the human desire of exploring farther, going beyond unexplored frontiers, flying towards the freedom of the perhaps infinite Cosmos, thus reaching higher levels understanding of human identity and place in the world.

The 'Icarus dream' has become more sophisticated through the times. Similarly to the effort of a more recent age of explorations between the fifteenth and seventeenth centuries when Europeans were resolute to explore oceans to discover new lands and expand the world map, since early times there were individuals obstinate about inventing machines aimed at reaching high skies. Historically recent phenomenon, Astronautics is the science and practice of sending humans into space beyond the Earth's atmosphere in the inner of spacecrafts. According to the Brazilian scientist and writer R. Argentière[7] (1916-1995), the most ancient document referent to the primitive concept of Astronautics is dated of more than 5,000 years ago and describes King Ethan's supposed attempt of reaching the skies [38]. There is another interesting old legend of an official in Ancient China named Wan-Hu, who had supposedly tried to reach space by assembling a rocket-powered flying chair. On the date of the flight,

---

6 Icaria ranks among the top four regions worldwide with the longest living inhabitants (Carr, 2012). To learn more about Icaria, visit the 'For the Interim' Web site, http://fortheinterim.com/learngreekinicariagreece/

7 Argentière is a famous Brazilian scientist and an important science popularizer. He became a student of Madam Curie, met Werhner von Braun, and exchanged letters with Einstein. Proficient in eight languages, Angentière wrote more than 30 titles, and sold more than 30 million copies, amongst them *The Travel to the Moon* (1947), original "A Viagem a Lua." Extracted from 'Abril' Web site, http://super.abril.com.br/superarquivo/2002/conteudo_266266.shtml

Wan-Hu would seat himself on the chair and ask his assistants to light the rockets and... After a strong noise, nobody saw the courageous Chinese again... It is not known for sure what happened to Wan-Hu, but probably both his flying chair and himself were blown in pieces.[8] Another story is that of the Greek philosopher Luciano de Samosata (125-181), who said that after crossing the Hercules Columns – currently Strait of Gibraltar – the ship that he navigated was raised by an enormous waterspout, and after several days of celestial travel he docked in a big and bright island, which was the Moon! [39] There are other funny stories of individuals who supposedly tried to reach space without the required technology... Science fiction is another expression of the popular imaginary since long ago featuring stories related to space exploration. For instance, through the centuries numerous scientists and science fiction authors have written about a hypothetical travel to the Moon, such as the German scientist Johannes Kepler (1571-1630), with *Somnium* (Dream). He imagined that the Moon would have two kinds of inhabitants: the *subvolves*, who lived on the face of the Moon visible from the Earth, and the *privolves*, who lived on the far side, and would take long walks to the other side where they could contemplate the beautiful Earth [39]. Several other authors wrote stories about a travel to the Moon, amongst whom the Italian poet Ludovico Ariosto (1474-1533), with *Furious Orlando*, the English bishop Francis Godwin (1562-1633), with *The Man in the Moon*; the English cleric John Wilkins (1614-1672), with *The World of the Moon*; the English writer H. G. Wells (1866-1946), with *The First Men on the Moon*, and the notable French writer Jules Verne (1828-1905), with *From Earth To The Moon* [40]. Science fiction inspired the engineers and scientists who would become the pioneers of modern space rocketry that enabled humankind to finally turn our updated Icarus' dream into reality.

The foundations of Astronautics comprise inspiration, science, and technology. As for technology development, space exploration would not have become an ordinary reality today without rocketry. Records of the first rocketry developments are due to the Chinese, who invented gunpowder some centuries before Christ. The Chinese have always paid special attention to matters such as bombs and explosions. They used to love (and still do) fireworks, and created several kinds of them. Fireworks flourished in dynastic courts with colored lights and flame balls. They also used explosions and flames to supposedly scare away evil spirits. The Chinese invented several guns and fought their enemies with success. Gunpowder opened holes in the strongest fortifications, and signed the "death sentence" of the castle and the European aristocratic military feudalism [35]. Indeed, rockets have been extensively used in wars, and it has been also these combats that helped spread among nations the practice of rocketry. The combination of the Chinese developments, the Kepler's laws and the Newtonian Physics would have allowed humanity to reach outer space already in the 17th century. However, there was not a political motivation for that endeavor, and even in the most developed countries, only a minority of people would consider it important to invest a considerable amount of money to send artifacts with humans inside into space. In this scenario, although rockets already existed, space rockets[9] would arise much later only in the context of the cold war between the United States and the former Soviet Union.

[8]Extracted from the NASA Web site, http://www.grc.nasa.gov/WWW/k-12/TRC/Rockets/history_of_rockets.html

[9] A rocket is a device that burns fuel (propellant) to work. If chemical, this propellant can be solid, liquid or mixed, for example. As fuel mixes with oxygen inside a chamber escape gases are produced. These gases produce impulse in one direction, and the rocket is pulled in the opposite direction according to the Physics law of action and reaction, thus allowing spacecrafts to move in space. The first rockets used solid propellant. They had limitations that a Robert Goddard would find out later in his studies and experiments of space rocketry.

The first space rockets have been developed due to the effort of scientists, working independently in different countries, who reached approximately the same conclusions. These different personalities have been fundamental for the development and popularization of rocketry. The Russian schoolteacher Konstantin Eduardovich Tsiolkovsky (1857-1935) delineated the principles of spaceflight propulsion. Regarded as the father of Astronautics, he envisioned that space was a place for definite occupation. In the United States, the professor Robert Hutchings Goddard (1882-1945) also investigated rocketry. Goddard found out the limitations of solid rockets. He built, tested, and flew the first liquid rockets. At his time it was believed that rockets would only work with the presence of air. However, through testing, Goddard affirmed that rockets would work better in the vacuum of space. His ideas that rockets could carry people to the Moon have been regarded as ridiculous initially, until the government felt the need to invest in rocketry due to Cold War pressures. Goddard also developed the idea of multistage rockets to reach higher altitudes. Currently, Goddard is regarded the father of modern rocket propulsion, and the largest NASA field center was named after him, on May 1, 1959.[10] Another important rocket researcher was the German scientist Hermann Oberth (1894-1989). Inspired by Julius Verne, he was determined to find a way to make spaceflight possible. He independently reached the same conclusions as Tsiolkovsky and Goddard. However, he was the only researcher who was able to publish his ideas. In 1929, he published "The Rocket in the Interplanetary Space," a book that let the world know that rocketry was a serious topic [41]. In the book, he explained not only how rockets could fly in the vacuum of space or how to insert an artificial satellite in orbit, but also evaluated the effects of spaceflight on the organism of crewmembers. Oberth was also Wernher von Braun's professor, introducing him in the German program of rockets. Von Braun has become the central figure of the Germans' rocket program, and later the leader rocket engineer for the NASA's Apollo Program. In sum, in a favorable political scenario the rockets invented by the Chinese have been much later enhanced by Germans, Americans, and Russians to carry humans beyond the Earth's atmosphere, towards our closest celestial body.

Motivated by the Cold War and the International Geophysical Year[11] (IGY), the Space Age started in 1957, with the launch of the first artificial satellite, Sputnik I (the Russian word *sputnik* means "companion") into Earth's orbit, and human space exploration started in 1961, with the flight of the first cosmonaut Yuri Gagarin[12] (1934-1968). Some historians consider space exploration, and also exploration of the Antarctic and the oceans as the 'Third Age of Discovery.' We could define this age as the modern extension of a process of exploration

[10] Extracted from the NASA Web site, http://asd.gsfc.nasa.gov/archive/hubble/overview/glossary.html

[11] The IGY (*1957-1958*) was a United Nations international event to "*observe geophysical phenomena and to secure data from all parts of the world* (http://www.nas.edu/history/igy/)." The former Soviet Union and the U.S. launched satellites for this event. An important discovery of the time made with a satellite (the U.S. Vanguard) has been the Van Allen radiation belts. The IGY idea was born years before on April 5, 1950, in Silver Spring, Maryland, when a small group of scientists gathered to meet with the visiting English scientist Sydney Chapman. They were discussing Chapman's latest theories on Earth's atmosphere, and the conversation turned to geophysics, including topics such as the complexity of Earth's oceans and atmosphere, geomagnetism, geodesy, and the role of the Sun in Earth processes. Someone suggested that with all the new tools available, such as rockets, computers, and radar, it should be the right moment for a coordinated, worldwide study of Earth's systems (http://celebrating200years.noaa.gov/magazine/igy/welcome.htm).

[12] Gagarin wrote a book entitled "The First Soviet Cosmonaut Team: Their Lives and Legacies," which brings interesting information about the first cosmonauts, their training, lives, and other aspects of the early years of the Soviet human space program.

Figure 3. Robert Goddard and his liquid oxygen-gasoline rocket (left), and Wernher von Braun in front of Saturn V, at the Kennedy Space Center (right). At Auburn, Massachusetts, Dr. Goddard fired this rocket on March 16, 1926. He is credited for substantial advancements in modern rocketry. His researches have been essential to NASA's human space program, which enabled the Apollo missions to the Moon in the 1960's and 1970's. Dr. Wernher von Braun was the chief architect of the Saturn V launch vehicle, used in Apollo missions. Photo (left): Esther C. Goddard. Source: NASA.

beginning a thousand years ago with the Norse voyages to Greenland and North America [42]. However, this age has some singular characteristics in comparison with the previous ones. The First Age of Discovery (15th and 16th centuries), has been marked by Europe's market expansion and associated with Columbus, Prince Henry the Navigator, Magellan and other explorers; the Second Age of Discovery (18th and 19th centuries) has been characterized by geographic discoveries such as the voyages of Captain Cook, bonded with modern science [42]. In the 'Third Age of Discovery' or human space exploration, unless we eventually find intelligent life, interaction with other civilizations is absent; there are no 'natives' to exploit. According to Pyne, *"No one will live off the land on Deimos, go native on Titan, absorb the art of Venus, the mythology of Uranus, the religious precepts of Mars, or the literature of Ceres. There will be no one to talk to except ourselves. Discovery will be a colossal exercise of self-reference"* [42]. Another difference is that there will be no colonies, in the original sense of that being composed by foreign civilizations. Our colonies in space would comprise ourselves. These 'colonies' would refer to permanent outposts with rotating crews such as the International Space Station, and in the future there could be permanent colonies on the Moon, on Mars, and other celestial bodies. Another singularity of this Third Age of Discovery is that in the First and Second Ages of exploration the "crews" were essentially composed of explores with the same nationality; in contrast, space outposts will have crews comprising individuals of several different nationalities, but sharing the same goal. It has both advantages and disadvantages. It is certainly challenging to spend long periods of time living with people with so different cultural backgrounds, but on the other hand intercultural teams achieve higher performance, because they combine several skills and different approaches to solving problems – critical to mission success.

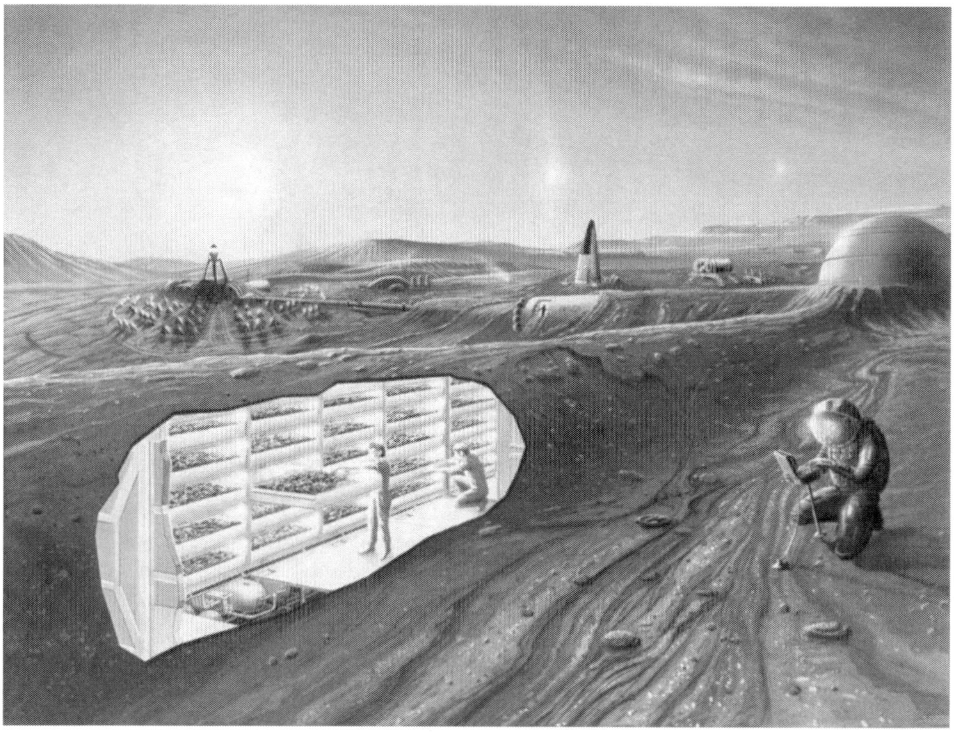

Figure 4. Artistic concept of a colony on Mars. A colony in another celestial body would be very different from the old colonies settled by our first explorers in other continents such as Africa and South America. The landscape on Mars would be a rocky desert without 'natives,' to learn from or "teach" our culture, any paradise landscapes, and probably without lots of treasures to explore. However, astronauts will need to learn how to take advantage of the planet's scarce resources for long-term missions, producing propellant with in-situ resources, resources, for example. Credit: NASA.

As previously discussed, the age of space exploration started in the context of the Cold War. It involved an expensive, obstinate space race between the United States and the former Soviet Union, lasting 11 years, and ending with the arrival of the first U.S astronauts who stepped onto the Moon on July 20, 1969. During and after that period, numerous satellites, spacecrafts, and probes have been launched into space, generating innovative processes and products that transformed our lives. Many other countries noticed the benefits of space activities, and started developing these activities in various scales. Some of them would focus on scientific discovery, whilst others on applications with socioeconomic return. Often, these developments have resulted in establishment of national space agencies with the principal role of coordinating their respective space programs. The scenario for space activities has changed a lot, though. The last decades faced dramatic reconfiguration in the space sector. The collapse of the former Soviet Union, and the end of the Cold War has transformed the political, social, and economical environment (…) space analysts point to a 'new techno-economic paradigm', concept referent to a closer relation between space activities, and economic and public returns [43]. The current arrangement marked by the participation of private spaceflight companies is also called New Space. The representatives of this new phase of space exploration already organize their conferences[13] involving private and public institutions, and have their publication venues.[14] For example, public expenditure of the U.S. and Russia in the space sector has been reduced. As a survival strategy perhaps, these space programs needed to identify public, private, and international cooperation channels, and prioritize space activities aimed at public and economic returns.

The largest expenditure in space still comes from national governments, but commercial players are steadily growing in number and influence. The arrival of these players into the space arena is called the "Personal Spaceflight Revolution," or the mentioned "New Space." It started with the Ansari X Prize, and has been followed by the Google Lunar X Prize, competitions that have individuals and private companies developing their own projects in space. For example, in 2004, the first private-developed spacecraft performed a suborbital flight in the scope of the Ansari X Prize. Indeed, an increasing number of private companies and individuals are investing in their ideas and implementing them [44]. As they do not depend upon governmental signatures execution may happen faster. This significant involvement of the private sector in space activities is positive also because it generates innovation. Systems of innovation comprise public and private actors, and are fundamental in the knowledge society. Further, private companies often face budgetary constraints, and deal constantly with the challenge of finding inexpensive solutions to survive in a competitive scenario; bringing their *modus operandis* and creativity to the space sector could mean to do more effective trade-offs in order to innovate and identify cheaper technologies and processes such as propulsion systems, launchers, and materials aimed at reducing costs of space travel.

Thus, I would affirm that a good articulation between government and private companies may lead to optimal, varied, and less expensive solutions in the scope of the portfolio of space enterprises. For instance, at NASA there is a large number of contractors from private companies [45], that strengthen the U.S. space program with their diverse problem solving skills, and knowledge. Further, there are several private actors carrying out well-structured space activities, from which governments could benefit in eventual partnerships. Private entrepreneurs in the space sector are investing in a wide range of efforts such as space tourism with suborbital and orbital trips to low Earth orbits (LEO), creating viable space habitats, developing commercial space vehicles to deliver cargo and people to LEO, and projecting space hotels. Today there are some companies that are selling tickets for suborbital and/or orbital flights such as Space Adventures,[15] Virgin Galactic,[16] and Rocketship Tours.[17] There is a private company[18] that currently develops a model of commercial space station that could be used as a backup to the ISS. The Russian space agency Roskosmos has been gathering resources for its missions through partnership in space tourism; the strategy is that sometimes the agency opens doors for private individuals to spend time as their "guests" in the International Space Station. For example. the first space tourist to visit the ISS was Dennis

[13] Example, New Space 2011, http://newspace2011.spacefrontier.org/
[14] Example New Space Journal, http://www.newspacejournal.com/
[15] Space Adventures Web site, http://www.spaceadventures.com/.
[16] Virgin Galactic Web site, http://www.virgingalactic.com/.
[17] Rocket Ship Tours Web site, http://www.rocketshiptours.com/.
[18] Orbital Technologies Web site, http://orbitaltechnologies.ru/.

Tito (1940-), an Italian-U.S. engineer and multimillionaire that has before worked at NASA Jet Propulsion Laboratory, and who had a childhood dream of flying into space. As he was reaching his sixties, he thought it would be his last chance to fly, and he found the way through Space Adventures, which articulated his mission with Roskosmos. At the time of his flight, many space agencies were afraid of problems that could arise from having a guest in the ISS, but Tito opened doors to turn this practice into something more usual.

Considering that national governments face budgetary, political constraints, and that both the public and private sectors have specific expertise and methodologies that instead of "competing" are complementary, current challenges in space exploration comprise creating synergy between public and private sector, in order to jointly reach strategic goals such as developing cheap and effective technologies such as space elevators to reach low Earth orbit without generating pollution; developing effective solutions to detect in advance and destroy "killer asteroids" or near-Earth objects; avoiding the creation of more space debris [44]; finding solutions to reach farthest destinations such as Mars and other habitable planets or moons, thus implementing the Tsiolkovsky concept of human race as a multi-planet species; and eventually creating a worldwide organization to coordinate international efforts in space and create synergy among space agencies to the benefit of space faring nations and humanity.

Figure 5. Dragon to the ISS in 2012. SpaceX was the first private company to send a spacecraft to resupply the International Space Station. On the left, the launch of the Dragon, and on the right its successful recovery, marking the completion of the mission. Source: NASA.

International cooperation is another characteristic of today's space exploration. National governments are investing less in the space sector, especially in colossal endeavors such as new trips to the Moon or the establishment of a colony on Mars. There is no political, economic, military urgency to accomplish these things, and governmental agendas have more urgent priorities, such as strengthening education and promoting equality. As seen, budgetary cuts have been encouraging space agencies to sum efforts to reach common goals. The International Space Station (ISS) is an emblematic symbol of international cooperation in the space sector, being the largest scientific and technical program in history, constructed with resources and expertise of 15 nations. It is designed basically for training astronauts in long-duration permanence in space, with the conduction of long-term research that take advantage of the microgravity environment, such as materials, life sciences, fluids, medicine, and polymers. The station has been also promoting breakthroughs in technology and engineering as well as partnerships with industries and universities worldwide. It is also used for educative purposes, such as radio amateur interactions with students on Earth, and educational scientific experiments. Another example of international cooperation is the European Space Agency (ESA), today composed of 19 countries. Individually, those countries could not carry out a very strong space program. However, through integration it has been established the space agency with the second largest budget in the world after the United States and before Russia. ESA started as a group of countries with maturing space programs. The cooperation did not prohibit the existence of national space agencies, and resulted in several successful programs – launchers, meteorological, observation, space science, and telecommunications satellites. Due to the increase in space activities benefits within social, political and economic European policies in recent years the European Union have been providing more support to ESA, promotes awareness about the strategic relevance of space activities, and contributes with 20 percent of the agency budget [46]. An agency that crosses national borders is an intelligent solution that optimizes resources, combines experience and knowledge, and allows more accomplishments in smaller periods of time. Also examples of international cooperation are missions carried out in partnership between two or more space agencies. The model of conception and development of missions in partnership is very common, such as the SOHO spacecraft, a joint effort by NASA and ESA. Further, international teams are more diverse, thus more effective in reaching the complex, multifaceted goals of space exploration. In summary, from an initial primarily military motivation in the beginning of the Space Age and a competitive approach, the current phase of space exploration is marked by commercial use of space, concern about reduction of mission costs [43]; "faster-cheaper-better" approaches for the development of technologies for spaceflight, international cooperation, private enterprises, and a serious concern about promoting public awareness through space education and outreach efforts.

The exploration of space has brought uncountable progresses to science and technology, to space sciences in particular, and has been enhancing everyone's lives. This activity leads to comprehensive knowledge of the Earth and space, providing a new perspective of human nature, our home planet and interactions among atmosphere, the oceans, surfaces, and human activities, and therefore is essential for economic growth and social wellbeing [47]. For example, knowledge of the 'Sun-Earth connections' due to satellites studying the Sun's produced space weather and its interactions with the Earth's magnetic field helps us better understand and predict space weather effects (coronal mass ejections, solar flares), helping us prevent satellite damages, blackouts, failures in communications with airplanes, and so forth. Actually, it is no longer possible to conceive human life without the direct and indirect contributions of space activities. There are several direct benefits of space exploration such as telecommunications, internet, weather forecasting, and television transmissions. For example, over 12,000 television channels are provided worldwide by telecommunications satellites, along with internet connections; knowledge about the climatic conditions of Venus and Mars provided by scientific satellites may help save us from ravages of global warming or from the next ice age; space systems allow monitoring the pollution of the oceans, the decreased vegetation, and the melting of glaciers and the ice cap; space systems have also shown that rain forests have been significantly depleted, thus lessening the amount of oxygen pumped into the atmosphere; space systems allow to monitor hurricanes and evacuate people from destructive paths. They have aided with search and rescue for lost boats and ships at sea, downed pilots, and explorers in the remotest parts of the world [44]. Earth observation satellites provide global measurements of key parameters in air quality management. Accurate air quality forecasts and informed regulation contribute to improve public health, crop health and tourism [48]. Tornados, hurricanes, tsunamis, volcanoes, drought, technological hazards, and so forth cause human and economic losses. For example, in the United States the annual direct loss due to drought only is estimated in $6-8 billion [49]. Data provided by satellites have also been helpful in saving countless lives through natural and technological disasters and hazards forecast and mitigation. Space activities also generate indirect benefits or spinoffs[19] in areas such as medicine, transportation, public safety, environmental resources, consumer goods, robotics, computer technology, industry, and recreation. Examples vary, comprising fire-resistant fabrics used by firefighters; water purification systems that brings clean water to people in remote areas; verification tools for secure online shopping; light sensors present in our cell phone cameras; environmentally sound methods for cleaning up oil spills; fuel cell technology to generate clean energy; and robotics originally designed for intricate repairs on the ISS, now used for a minimally invasive knee surgery procedure, where its precision control makes that ideal for inserting a very small implant [50].

Space activities produce the so-called high-tech products, those that generate the highest percentages of return on investment. In the United States, aerospace products and scientific instruments constitute the major high-tech products in the economy [51]. Though space activities generate high tech products capable of producing richness and elevating quality of life, it is common that the non-specialized public believes that money applied in space is waste. Interviews[20] with U.S. individuals in 2010, for instance, showed that 50 percent were favorable to cutting space exploration from the federal budget given the economic difficulties faced. Further, only 27 percent of U.S. citizens were favorable to sending humans to Mars. In developing countries these beliefs are even stronger, as in many cases a 'space culture' is still absent. These arguments disregard the fact that today most space expenditure is aimed at producing improvements in human life. Unfortunately, there are still people who have a reduced image of space activities as only astronauts floating in space. Although the future of space exploration is uncertain, and despite of all misunderstandings concerning the utility of space activities, considering all benefits derived from them and associated human needs, I believe it is likely that humanity through government and commercial partners will keep interested and expand this effort, which may lead to important national and global

[19] Spinoffs can be defined as products, processes or services that carries space-related technology, and that are made commercially available to a broader audience. While the original technology's purposes were space mission-related, it after fills quotidian life needs, thus enhancing people's standard of living (NASA, 2010).

[20] Access interview at http://www.rasmussenreports.com/public_content/business/econ_survey_questions/january_2010/toplines_space_program_january_13_14_2010.

achievements such as reaching Mars and farthest destinations with a multitude of innovative, unimagined technologies and processes, generating as a byproduct even more economic richness, social welfare, and public awareness and engagement.

Space exploration has brought benefits for numerous countries other than the United States and the former Soviet Union, which initiated the enterprise. An increasing number of countries in all continents have been carrying out space activities through established space agencies. While in the early 1960's only a few countries carried out space activities, in 2009 this number increased to more than fifty [52]. These nations recognize the importance of space activities such as telecommunications, environmental monitoring, resource management, infrastructure development and national security, along with national prestige [53]. For example, South Africa launched in 2009 its own Earth observation satellite called SumbandilaSat, the second homegrown South African satellite. The Indian program is one of the most cost effective in the world, prioritizing international partnerships and focusing on strategic actions. Amongst its main accomplishments, the Indian space agency has sent a polar orbiter for remote sensing of the Moon, called Chandrayaan, with the purpose of carrying out topographic, chemical, radioactive and mineral mapping of the Moon; India also has plans for future Moon human exploration by 2020 [54, 55, 56].

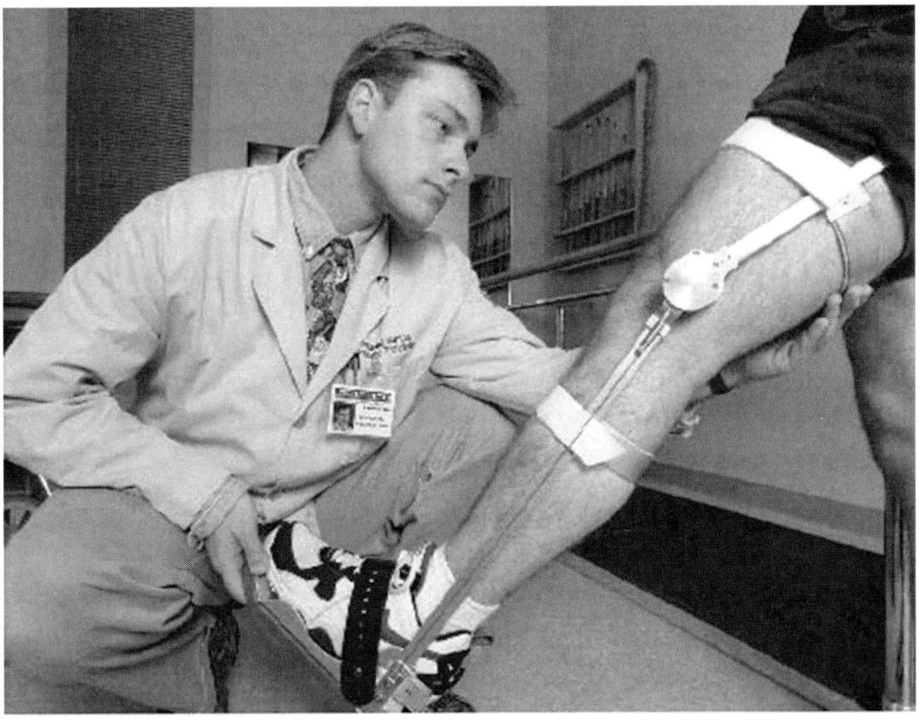

Figure 6. Lockable Knee Brace Speeds Rehabilitation. Inspired in the mechanical design of rocket engines and advanced propulsion at NASA Marshall Space Flight Center, a doctor worked along with NASA personnel to develop a very useful spinoff: an orthotic knee joint that is allowing people with weakened or absent quadriceps and varying degrees of knee instability to regain their mobility. Similar to the human knee, the selectively lockable joint works in three distinct, but complementary modes: free motion or automatic stance control; for walking; and manual lock, for standing. The device started being commercialized in 2002 as a very durable option for knee orthotics and has helped thousands of people ever since. Credit: NASA.

Another example is Brazil, which early in history identified in the space sector a unique opportunity to generate richness, social welfare, and to advance science and technology. The country carries out institutionalized space activities since the 1960s, with satellites, launch platforms – Alcantara Launch Center (CLA/DEPED),[21] Hell Barrier Launch Center (CLBI)[22] - rockets and launch vehicle (under development), center for satellite applications (such as Earth observation), along with an industrial complex targeted to meet demands of the sector. Brazil established educational centers of excellence internationally recognized such as the Technological Institute of Aeronautics (ITA),[23] and the National Institute of Space Research (INPE),[24] with outstanding graduate and undergraduate courses in several engineering fields, and more. The mentioned Brazilian Alcantara Launch Center has a privileged location in terms of propellant saving due to its proximity to the geographical and magnetic Earth equators, and offers a larger number of launch windows in comparison to other launch centers [40]. Currently, the Brazilian space sector counts with a workforce of nearly 3,000 professionals, and its priority are activities that generate socioeconomic benefits such as Earth observation, scientific and technological missions, telecommunications, and meteorology [57]. With a mission carried out by government, industry, academia and international partners the country has designed and integrated satellites such as the China-Brazil Earth-Resources Satellite (CBERS).[25] This remote sensing satellite provides frontiers control, environment monitoring and natural resources vigilance, monitoring rivers, the Amazon Region with the largest tropical forest in the world covering more than 5 million km^2 [57], and the Atlantic forests. Brazil has also been producing a large quantity of remote sensing satellite images and distributing them to national and international institutions.[26] Satellite technology is critical for a country such as Brazil, with the largest territorial extension of the southern part of the American continent: 8.5 million km^2 of continuous area, more than 7,400 kilometers of oceanic coasts and 15,700 kilometers of terrestrial borders, limiting with all South American countries, except for Chile and the Equator. A more recent achievement has been the launch of an astronaut into space. After eight years of training at NASA Johnson Space Center and some internal political definitions, in 2006 the first Brazilian astronaut Marcos Pontes was sent to the International Space Station aboard the Russian Soyuz spacecraft on occasion of the 100th anniversary of the first flight of the Brazilian inventor Alberto Santos Dumont (1873-1932) in his airplane 14 Bis [16]. Currently, Astronaut Pontes, waiting for assignment for his second space flight, has several other responsibilities in Brazil and other countries. He is an important disseminator of STEM among children and teenagers through lectures and media. The Brazilian space program also comprises the autonomous development and construction of a satellite launch vehicle, which should be launched in 2013. However, technology is not everything in the Brazilian space program. It also values inspiration and boldness. The Brazilian Space Agency carries out a space education program and outreach activities that assist schoolchildren, teachers, and the general public.

[21] Alcantara Launch Center Web site, http://www.cla.aer.mil.br/
[22] Hell Barrier Launch Center Web site, http://www.clbi.cta.br/new/index.php
[23] Technological Institute of Aeronautics Web site, http://www.ita.br/
[24] National Institute of Space Research Web site, http://www.inpe.br/
[25] CBERS satellite Web site, http://www.cbers.inpe.br/
[26] Accessing the Web site http://www.dgi.inpe.br/CDSR/ there is plenty of satellite images that can be downloaded for free. All that is needed is to subscribe. For instructions in English, French or Spanish on that Web site just select the desired language at the top of the page on the left.

In summary, several countries have been enjoying the benefits of space exploration today, and the tendency is that the knowledge society opens doors to more nations to "join the club." In some more decades perhaps, the awe-inspiring, incomparable experience of viewing the blue Earth against the dark background of space will not be a privilege of professional astronauts only; the ordinary person could pay an affordable price for a trip to suborbital or orbital space as launching systems costs reduce. The Icarus dream has not only became a reality, but it has been uniting the world in a sort of common aspiration for higher skies as a right of every citizen, increasingly fast scientific and technological discoveries, and cheapest and more effective solutions to betterment of our lives. In the sequence, let us visit some concepts and examples that show how the amazement of the space thematic has been used inside and outside schools to capture youth's imagination and curiosity, here on Earth.

2.2.2. Space education in the world to the betterment of STEM – An overview

The dramatic transformations that the world has been facing in terms of productive processes complexity increase and widespread access to information and technologies claim for a redesign in educational concepts, approaches, and methods. The current rapidly changing knowledge society requires the preparation of sufficient and well-prepared workforce in STEM; and it requires individuals capable of generating innovation. However, talking about education also (or mainly) means discussing issues such as attitude, values, and global awareness. As citizens of this planet, we need to open doors to children and teenagers to understand how fragile is our Earth environments; they also need to perceive the uniqueness of our planet – so far the only one we know in which intelligent life was able to flourish. This awareness would inspire adults, professionals responsible for a sustainably developed world. In the scope of STEM education, space education is a historically recent practice that plays the singular role of engaging students in fascinating themes in an interdisciplinary, interactive approach, in most cases also promoting meaningful uses of information and communication technologies. Space education plays the strategic roles of engaging students in STEM and space subjects at school, and of promoting STEM and the space sector as possible career paths. In addition to space education, outreach activities promote awareness of space themes for the general public, which is an important, socially-committed practice considering that one of the principal problems of science consists in not being part of people's culture [58]. I think that public outreach and engagement is not only a matter of dissemination, but also of fostering the concept of participative society. In other words, although most of the general public may not be specialist in space affairs, many of them are capable of thinking about the problems of science. At times, the best overall insights, solutions, or perceptions may come from individuals that are not immersed in the specific practices, thus we should not underestimate contributions coming from the general public. For example, recently I learned about an interesting effort by the Ecuadorian Civilian Space Agency, which is raising funds from the general public worldwide through a campaign[27] to launch the first satellite that will enable everyone who wants to join the effort, even children and teenagers, to have their own "webcam" in space. It will allow, for example,

[27] Learn more about the campaign and the effort on the "Your own Orbital Webcam," Indiegogo Web site, http://www.indiegogo.com/nee-01

more people worldwide to monitor near-Earth objects. The innovative thing about this endeavor is that for the first time in history, people could see what is "out there" directly, without intermediation from space agency specialists.

Prior to going deeper in the discussion about space education let us briefly re-visit some associated ideas and concepts. First of all, we are talking about science. Newell [59] defines the scientific work as a dynamic process through which scientists, individually and collectively work together to build up a commonly accepted explanation of the universe. That process involves observation, measurement, imagination, induction, hypothesis, generalization, deduction, test, communication, and mutual criticism. This process should be experienced by students at school, in order to allow them to understand how science is produced, develop higher-order thinking, and also to foster future careers in the field. As for the relation between science and technology, Morin [60] states that they are interdependent. Indeed, science generated several technological advances, and the latter made possible other scientific achievements. The classroom has been impacted by technology as students today live with internet, computers, cell phones, tablets, and so forth. In this research we explore and analyze examples of how space education contribute to STEM education. As we have discussed, STEM is critical for socioeconomic development in the knowledge society. However, STEM education has been facing several challenges such as: a) theoretical-methodological inadequacy; b) problems in pre- and in-service teacher preparation; c) quality didactic materials and technological equipment; d) fallacious pedagogical theories according to which students would learn spontaneously, without practical experience yet with overly-abstracted contact with theory; e) difficulties to work in an interdisciplinary approach; f) rapid changes in the world of knowledge.

Space education is situated inside Science, Technology, Engineering, and Mathematics (STEM) education, representing an important alternative for effective STEM teaching/learning. But what is space education anyway, and where can we find it? Space education stands for a wide range of pedagogical practices; didactic resources; subject matters; curricular and/or cross curricular topics; formal and/or informal; systematic and/or sporadic programs/projects/activities, which take advantage of the singular amazement that space matters exert on students' imagination, thus capturing early interest in STEM subjects, elucidating scientific knowledge in a non-abstract approach, spreading the benefits of space programs, and ultimately encouraging STEM careers [30]. Space education inspires and motivates students to learn scientific contents, therefore contributing to their integral scientific education. "*Space education is capable of offering the student an integrated comprehension of scientific and technological facts and phenomena. It consists in the awaken of the curiosity and interest of students with regard to processes, products and services derived from exploration of space environment, leading to an interdisciplinary understanding of science and technology and of the manner they affect daily life. Space exploration and its derivates can become the axis from which it is studied contents in subjects such as sciences, mathematics and technologies. They can be the starting and the final point of the pedagogical work*" [30]. Finally, it can be used to contextualize, illustrate contents and principles in STEM subjects. We will explore in this chapter examples of space science education efforts, in particular. But what is space science? It could be defined as the study of the Earth, the Sun, planets, and what lies around, between and beyond them, by means of disciplines such as Astronomy, solar and heliospheric physics, space plasma physics and planetary research [61]. The subject comprises scientific investigations made possible or significantly aided by the use of rockets, satellites, space probes, and so forth. These studies help us understand how our actions impact our near environment, and the overall cosmos. This 'space awareness' makes us feel insignificant in the shore of the cosmic ocean, and at the same time very important as we are made of the same material that stars are composed of.

Last but not least, when we talk about space education we are talking about human development. In my opinion, the importance of the 'space education phenomenon' goes far beyond 'helping STEM contents teaching/learning,' or 'encouraging STEM careers,' though these outcomes alone would be sufficient to motivate and justify this pedagogical practice in schools. Space education has to do with inspiring students to the beauty and mysteries of the Cosmos; it has to do with encouraging them to think less about their limitations and opening their eyes to their infinitude of possibilities; it has to do with helping students to be original, to be themselves. I strongly believe that space-related educational practices help students to be more confident. I think that studying space is something that also opens our eyes to possibilities that we have never considered previously; it has to do with going beyond apparent boundaries. The implicit idea that we transmit when we have students learning through space education is that they can accomplish everything that they want if they make consistent efforts toward their goals. The curiosity, motivation, and passion that arises from exploring space topics, especially when hands-on activities are used, help students not only to learn STEM contents, but also to incorporate innovative ways of doing things, widen their horizons of possibilities, and motivate them to fight for their dreams. I would also say that the space thematic has to do with STEM education, but also with poetry, arts, self-awareness, teamwork, cooperation, and planetary responsibility.

Space education is a historically recent phenomenon, born with the first national space programs. Along with the origins of the U.S. and former Soviet Union space programs, it started initially in these same countries a practice designated space education, primarily represented by space-related higher education programs targeted at preparing the staff of engineers and scientists necessary for implementation of those first space programs. In the 1980s, it started in those countries a practice dedicated to precollege students, educators, and the general public. Space education for this audience could be defined as a science education effort using space's singular source of excitement to capture students' interest in scientific and technological subjects. One of its main purposes has been to contribute for the enhancement of STEM education. Although the practice started with national space agencies' efforts, it has to be considered that Astronomy education, which for the purpose of this chapter I include inside the category 'space education,' has started much earlier. And why to categorize Astronomy education as space education as well? It is very difficult to dissociate these practices, because the ideas and ideals of these practices are very similar. Further, it has to be considered that after space agencies started space education efforts for schools, several other institutions and individuals came up with similar practices, though not necessarily connected and/or dependent on space agencies. Examples vary, comprising science museums, planetariums, universities, international organizations, foundations, and so forth. We can find astronomers who go to public places and set up telescopes for observations; teachers who have students building and launching their own water-propelled rockets; science centers that have students experiencing a spaceflight in simulated interplanetary flights, and so forth. Space education contributes to elucidate scientific knowledge in a non-abstract approach, to stimulate careers in scientific and technological fields – with special attention to the space sector – and to spread the benefits of space programs, obtaining sustainability for them.

Let us visit the cases of the first space agencies in the world that have started space education efforts for precollege level. Along with the former Soviet Union's space agency Roskosmos, NASA started space education activities for precollege level already in the 1980's. Indeed, education has also been considered as something very important, and strategic for NASA. For example, the 2011 NASA's Strategic Plan [62] identifies education as essential component that support the agency's goals and missions. According to Rosendhal et al. [63], *"Education and contributing to the public understanding of science have long been important components of NASA's mission."* Indeed, the agency pursues complex and integrated dissemination actions in education and outreach. NASA regards its investment in education – in all levels and modalities – strategic to the overall agency's enterprise. According to the White House, *"NASA's Education program is designed to attract future explorers to careers in NASA-related fields, allow university faculty to conduct NASA-related research, and teach the public about NASA's work. Specific programs target K-12, higher education, informal and e-education, and underrepresented and underserved populations"* [64]. Its educational effort is translated into printed and online resources distribution for teachers and students, teacher trainings, hands-on and interactive activities via TV channel and videoconference, technologies, events, visits for space centers, and so forth. NASA uses its high-quality resources such as inspiring missions, workforce, facilities, research and innovations to engage in STEM education and in the space thematic the educational community. The agency has education and outreach programs targeted to families, stakeholders, underserved and underrepresented communities. All of this also focusing on the development of their own space program, considering that education assures the preparation of the workforce required to give continuation to NASA missions. Finally, NASA contributes to the strengthening of U.S. education, as the agency encourages careers in STEM, strategic to national development. At the same time, the agency invests in marketing and its brand is the most well known among all space agencies. This is not only due to historical achievements such as the first manned Moon landing, or due to having the largest budget. NASA chose to be popular, to be among the general public, to inspire these people, and as far as possible to bring these people to their missions. In other words, NASA is popular and this makes a great difference. The other first nation that started carrying out space education activities for precollege level was the former Soviet Union, which started early efforts to provide space education to young students. Though they carried out several activities, it was inspired in NASA's Teacher in Space that Russia founded in 1988 an institution to better articulate and create synergy among these efforts, the All-Union Youth Aerospace Society "Soyuz," uniting all schools and universities working in space education and establishing contacts with aerospace societies in the United States, Japan and China. The idea of that society was to identify talented students interested in entering aerospace universities. The society has organized several competitions mainly for school children in several space-related topics. One example was one competition targeted to have students design and plan onboard experiments to Salyut Space Station. These competitions became a good student recruitment base for their leading aerospace universities. Some prizes granted to students comprised trips to space camps in the U.S. or other countries. Afterwards, it was established the Russian National Space Education program, to better organize all these separate efforts [65]. From both the Russian and the NASA cases, it is possible to state that capillarity and partnerships with universities and other institution/partners were critical to overall program success.

Further, it was fundamental to provide students and teachers with concrete perspectives, opportunities to develop their capabilities in space–related careers.

Currently, many other countries carry out space education initiatives aimed at primary, secondary (K-12), and graduate levels. In Brazil, for example, there are significant experiences in teacher training in several space subjects carried out by the AEB School program of the Brazilian Space Agency, national competitions such as the Brazilian Olympic of Astronomy and Astronautics (OBA),[28] and there are innovative isolated actions, such as space camp, and a public Brazilian middle school participating in the Tubesat Personal Satellite Kit[29] initiative. At that school, there are 120 fifth-grade students creating 22 full-scale TubeSat. The students who build the best mock-up will assemble the TubeSat that will go to orbit [66]. In Norway, the National Centre for Space Related Education - NAROM[30] enriches education by offering teacher training and projects with rockets for students [67]. There are efforts involving satellite construction and launch, such as an Indian initiative that has led to the creation of a community space education center in addition to the launch of an educational satellite (*Edusat*). Another example of satellite construction and launch is the ESA's Education Office in charge of the first European student mission to the Moon. Further, in 2010, high school students from different ESA Member States were able to watch their own satellites' fly aboard suborbital rockets in the first European *CanSat* competition. There is also the so-called International CanSat Competition[31] [68, 69, 70]. The Japanese space agency's JAXA Space Education Center invests in teaching materials. They consider that the first ten minutes of a class is when a teacher will (or not) capture students' heart, and develop introductory space education lessons in several topics to be used in school subjects such as sciences, arts, economics, social, and language. According to Hideo Nakamura Director of JAXA Space Education Center, *"space development can inspire children to have dreams and hope, enriching their hearts and minds"* [71]. The South African National Space Agency – SANSA, also offers space education initiatives such as workshops for science educators, outreach programs for schools and learners, and communication tools such as tele-education, thus improving STEM proficiency [72]. In Canada, Europe, Japan, and in the United Kingdom there are institutions such as the European Space Education Resource Office – ESERO, dedicated to training teachers in space education.[32] There are also cooperative efforts between countries, such as the Space Technology and Research Students program (*S*T*A*R*S)* launched to enable U.S. and Chinese students to collaborate online in science and space research, by interacting real-time with scientists, astronauts, and other students [73]. Another example of online collaboration space education initiative is an effort that NASA and Microsoft launched, which is an interactive Web site that allows users to become Mars explorers, giving real contributions. The "Be a Martian"[33] Web site invites members of the public to help scientists perform research tasks [74]. This kind of initiative generates higher levels of engagement, as participants are aware that their input is given consideration, and that they are offering actual contributions to out-of-this-world explorations. As for internship opportunities, NASA counts with a program called NASA Academy through

[28] Brazilian Astronomy and Astronautics Olympic Web site, http://www.oba.org.br/
[29] Interorbital Systems Web site, http://interorbital.com/TubeSat_1.htm, responsible for the effort.
[30] Norwegian Space Centre Web site, http://www.spacecentre.no
[31] European CanSat Competition Web site, http://www.esa.int/SPECIALS/CanSat/
[32] Extracted from the ESA Web site, http://www.esa.int/esaMI/ESERO_Project/SEM4KP4KXMF_0.html

which students from across the U.S., as well as from Australia, Canada, France, Italy, Japan, and many other countries worked directly with NASA scientists on cutting-edge research as part of a summer internship program. Just to name a few examples and implementers.

There are more successful space education enterprises that cross national borders and unite several countries in the amazement of space studies and exploration. One such example is the NASA's managed GLOBE program involving 112 countries, 54 thousand trained-teachers from 24 thousand schools worldwide [75]. This initiative that benefited more than 1,5 million students [75] consists of a hands-on science education program for primary and secondary students. It uses inquiry-based investigations of the environment and the Earth system, in which teachers undergo training and continuous support by means of didactic manuals and students perform scientifically valid measurements in different study areas of the environment and the Earth system (i.e., atmospheric, hydrological), and relate the data collected using the internet to collaborate with international colleagues and scientists [76]. Another outstanding example is the World Space Week,[34] an event carried out annually reaching space science education organisms and enthusiasts in more than fifty countries, consisting of the largest space public event in the world declared by the United Nations in 1999 [77], and aimed at spreading the benefits of space exploration to the educational community and the general public through a wide range of activities such as expositions, space parties, rocket launch competitions, art shows, TV shows, and several other traditional and innovative activities about the space thematic.

Figure 7. First European CanSat competition for high-school students. At 10 European countries, students have spent around 6 months to build their own small satellites, called CanSats because they have the shape of a soda can. The satellites were launched to an altitude of 1 km. The teams are composed by at least four students and one teacher, and should carry out experiments, such as calculating the variation of Earth's magnetic field, performing GPS measurements, tracking the Sun or measuring radiation. Credit: © Andoya Rocket Range, http://www.rocketrange.no/?p=695 Image in public domain.

[33]At this interactive Web site you can effectively contribute in Mars exploration, http://beamartian.jpl.nasa.gov/welcome

[34] World Space Week Web site, http://www.spaceweek.org

Figure 8. Activities at World Space Week 2010 engage students at Ekiti, Nigeria, in Sun-earth themes. The World Space Week is celebrated worldwide and comprises several types of activities and events. In order to promote awareness about the Sun and space weather, a solar system scale model along with NASA materials are used with students. Credit: Romeo Durscher. Image in public domain.

Some initiatives promote the use of information and communication technologies to have students interacting with space education contents. The following two initiatives enable students to somehow have an "eye" in space, as they can select spots from above to photograph. The NASA-sponsored ISS GRAIL EarthKAM program (Earth Knowledge Acquired by Middle School Students) is a program in which students of the U.S. and other 41 countries[35] obtain pictures from specific spots on the surface of the Earth with a camera onboard the International Space Station [78]. In 2011, NASA launched the GRAIL (Gravity Recovery and Interior Laboratory spacecraft. It is the first planetary robotic mission including an imaging experiment devoted totally to education and outreach, called the MoonKAM (Moon Knowledge Acquired by Middle school students) system [79], through which students can target interesting areas of the Moon that they want to investigate, using cameras onboard those twin probes. GRAIL is also the first probe to capture a video of the Moon's far side. Robotics is another successful approach to teach scientific research and to foster innovation in schools. An example is the GigaPan project, international effort managed by the Carnegie Mellon University and NASA Ames Research Center. In this project students from several countries use a robotic platform developed based on technology from rovers sent to Mars. An ordinary digital camera is assembled to that platform, and students configure the equipment to obtain several pictures of a specific location. The purpose is to obtain high-resolution panoramic images that result from the combination of several pictures taken during the exposition and that are edited with specific software. Students picture landscapes and local

[35] Information extracted from the NASA Web site, "Students Photograph Earth From Space via EarthKAM Program," http://www.nasa.gov/mission_pages/station/research/benefits/earth_kam.html.

events that signify their culture and communities. These images are grouped as projects on a dedicated Web site, promoting cultural interchange among communities from several parts of the globe [80]. Another example is the Moonbuggy race, a successful NASA hands-on initiative that started national and became international. It consists of having secondary and college students' categories of competitors building and driving their own vehicle in a simulated lunar terrain with "craters," rocks, "lava" ridges, inclines and "lunar" soil. The competition joins international teams of students to design a vehicle that needs to address a series of engineering problems such as those faced in the Apollo missions. Each Moonbuggy vehicle is human-powered and carries two students, one female and one male, over a half-mile simulated lunar terrain course, with lunar-like obstacles such as craters and rocks. The top three winning teams in each division (one high school division and one college division) will be those having the shortest total times in assembling their vehicles and traversing the terrain course. The competition lasts two days [81].

Some efforts are thematic and explore online interaction. For example, NASA organizes an international celebration called Sun-Earth day with events and programs that are held throughout the year culminating with a celebration on/near the Spring Equinox. Social media networking related groups are formed on Facebook and Twitter [82]. Another interesting event is the International Observe the Moon Night (InOMN), which occurs at science centers, museums, schools, and backyards worldwide. The goal of InOMN is to engage the public in annual lunar observation campaigns sharing the excitement of lunar science and exploration [83]. There is another effort called "Global Friendship Through Space Education," which aims at promoting friendship among youth at different countries through the study of space-related topics and involvement in activities i.e., international space camps, collaborative research on NASA's distance learning Web site, videoconferencing between partner schools and NASA, and e-mailing between students about space-related projects and for sharing information regarding their quotidian [84]. There are also outreach efforts by NASA, the Japanese JAXA and other space agencies through which people from all around the world could send their names into space [74] in robotic missions to the solar system and beyond.

An indicator of the relevance of the space thematic for education betterment and socioeconomic development is the support from international organizations for human development such as the United Nations and UNESCO. The latter not only support space education efforts around the world, but also counts with a space education program launched in 2002 based on recommendations of two of conferences:

- World Conference on Science[36] organized by UNESCO and the International Council for Science – ICSU, Budapest 1999.
- Third United Nations Conference on the Peaceful Uses of Outer Space (UNISPACE-III)[37] organized by the UN Committee on the Peaceful Uses of Outer Space (COPUOS) and the UN Office for Outer Space Affairs, Vienna 1999.

The UNESCO space education program aims at enhancing education in space subjects in schools, particularly in developing countries, promoting integration of these subjects in national curricula, and contributing to prepare future space sector workforce. The

[36] Conference Web site, http://www.unesco.org/science/wcs/
[37] Conference Web site, http://www.un.org/events/unispace3/

implementation occurs through collaboration of national ministries and/or space agencies, NGOs, and so forth [85]. Strategies comprise workshops and training for secondary teachers and students [86], conferences, fellowships, grants, competitions, events, space camps, and multimedia promotional materials. The program also provides a platform to implement common activities and establish regional and international cooperation. According to the UNESCO Web site, "*The programme puts space in the forefront and not only brings a new dimension to science education but also introduces new knowledge, values and perspectives related to the study of the Earth and its system. Moreover, space studies develop critical thinking and the problem-solving and decision-making skills which are central to quality education - the priority goal of the UN Decade on Education for Sustainable Development (2005-2014).*" Surely space education is related to sustainable development *"a constantly evolving concept, is thus the will to improve everyone's quality of life, including that of future generations, by reconciling economic growth, social development and environmental protection."*[38]

An increasing number of countries with space capability have been carrying out space education initiatives for all educational levels. However, it is difficult to evaluate the current status of space educational practices in the world. Let us make a few considerations, though. Space education activities are in many cases dependent on national space agencies budgets and workforce. Although it is positive in the sense that these agencies pursue the adequate know-how to properly share exciting discoveries from their missions, there are some constraints such as insufficient preparation of agency's professionals to establish effective linkages with the educational community through proper communication, and insufficient number of professionals available inside space agencies to carry out this work. Further, in several space agencies investments still target mainly the preparation of graduate and undergraduate students, and activities for youth and the general public are regarded as non-priority expenditure. Usually when some teacher training is offered at precollege level, they are short courses, less than 1-week-long, which do not comprise online, telephone or in situ mentoring after the course is concluded. Thus, in many cases teachers return to schools with unanswered questions, and if they find difficulties in implementing the proposed curricular space education projects, they would not find proper support.

Further, it is not all countries that have developed the desired capillarity among space agency, universities, schools, science centers and so forth, in order to establish an articulate, mature space education strategy. And even when this happens, implementation difficulties arise due to incompatibilities with national or local education system goals. The result is that in many cases teachers are discouraged from adopting space education strategies because they would supposedly be a 'waste of time,' considering other educational priorities related to immediate outcomes. And… the sad result of this understanding is that when governments prioritize short-term results, long-term gains may be forgotten, and this generates losses in terms of preparing innovative minds, and aware citizens for a changing world.

Let us briefly analyze the example of the United States, which pursues the space agency with the largest space education effort in the world. Even in the U.S., space education faces challenges. Although there was an increase in science teaching with the National Defense Education Act of 1958 (the reaction to Sputnik), with the end of the Cold War between the

---

[38] Extracted from "UN Decade of Education for Sustainable Development 2005 – 2014. The DESD at a glance," available at http://unesdoc.unesco.org/images/0014/001416/141629e.pdf

U.S. and former Soviet Union and the consequent reduction in space sector expenditure, the science teaching emphasis have been reduced in the long run. Some factors of the current education system makes it difficult to schools to have a significant quantity and quality of space science teaching, such as the movement 'less is more' that assumes that it is better to teach fewer topics in depth and with hands-on activities than to teach more science superficially, and the fact that the large amount of elementary and middle school teachers do not have a major or a minor in any science or science education field. It makes difficult for these educators to know how to integrate space education to the curriculum, or to teach the curriculum through space education interdisciplinary practices. Finally, the *No Child Left Behind* policies and its emphasis in national standard examinations have teachers focusing on the content that will be required in the exams. It means that even when science is evaluated emphasis placed in other disciplines such as Biology, Chemistry, and Physics [87]. Due to this concern in having students performing well in national exams, many teachers think there would be no time to teach space sciences or through space education.

Despite of these difficulties, it is also true that space sciences are now part of our cultural landscape. Missions to the solar system such as the Hubble space telescope, Voyager and Galileo, and rovers sent to Mars such as Spirit and Opportunity have been producing information, videos, and images that are now widely accessed. Further, space education unites the world around themes that cross national frontiers such as exoplanets and global warming. Perhaps one of the main challenges still consists of bringing space education to classroom with proper communication. It is critical to find education specialists able to help translate knowledge produced from thousands of hundreds of space scientists to teachers, and vice versa, through practices such as having university students working with teachers to help them implement innovative practices with due guidance [87].

In the coming section, I present and discuss selected NASA Education didactic resources, technologies and innovative efforts, which can be used by educators in the U.S. and beyond as engaging tools to enrich STEM classes and to motivate students to become better learners and explorers, in the scope of innovative pedagogical approaches.

Chapter 3

Space Education at NASA – A Model of Excellence?

Prior to exploring NASA Education, let us discuss meanings of this emblematic space agency that has been capturing sympathy, fascination, and polemic of the entire world through its more than fifty years of existence, especially after sending the first humans to the Moon.

3.1. What NASA Is?

The U.S. National Aeronautics and Space Administration (NASA), stands for the largest and most popular space agency in the world. Its insignia is recognized all around the planet. For example, in 1998 a telephone survey with respondents from the European Space Agency (ESA) Member States[1] showed that only 12% of the general public spontaneously knew ESA (with significant variations among countries) while 54% spontaneously knew NASA [88]. And if you ask children and teenagers in different countries, they will have at least heard something about NASA! This incontestable popularity (though not the same as during the time of the first manned missions to the Moon) derives not only from the fact that NASA has the biggest budget among all space agencies in the world[2], but also because the agency utilizes several combined marketing strategies in the scope of its education and outreach initiatives, aimed at engaging students and educators, and sharing NASA missions with the general public, respectively. NASA continuously shares its discoveries with society, and regards education a strategic element of its mission. Actually, in the publication *2011 NASA Strategic Plan* [62], the agency's Strategic Goal 6 is to "Share NASA with the public,

---

[1] Current ESA Member States: Austria, Belgium, Czech Republic, Denmark, Finland, France, Germany, Greece, Ireland, Italy, Luxembourg, The Netherlands, Norway, Portugal, Romania, Spain, Sweden, Switzerland and the United Kingdom. Extracted from the European Space Agency's Web site, http://www.esa.int/esaMI/About_ESA/SEMP936LARE_0.html.

[2] In 2010, NASA received 18.7 billion dollars. Extracted from the White House Web site, http://www.whitehouse.gov/omb/factsheet_department_nasa/. Although this seems to be a high amount, NASA budget used to be substantially higher during the Apollo era. Lately, there have been cuts in important areas, i.e. education and human exploration. These cuts have been causing difficulties to NASA Education managers who need to keep innovating with smaller amounts of financial resources, and human resource reductions.

educators, and students to provide opportunities to participate in our Mission, foster innovation, and contribute to a strong national economy." To achieve this, the agency communicates also with stakeholders comprising the media and policymakers. Every time NASA launches a spacecraft we can accompany the fact on TV, on the radio, on the internet. For example, in the fourth quarter of 2010 alone, NASA sent more than 100-million digital messages to its 244 thousand subscribers; and more than 315 million messages during the entire year. Through its Web site, NASA offers forty subscription topics, such as the popular 'Image of the Day' [89]. Efforts like these build continued sustainability for NASA explorations, as the agency fulfills its social role of generating educational benefits, and promoting public awareness of the activities carried out inside (and outside!) its walls.

The political motivation that brought NASA into existence was the reaction to Sputnik, best illustrated by the U.S. President John F. Kennedy (1917-1963)'s mission statement: "I believe that this nation should commit itself to achieving the goal, before this decade is out, of landing a man on the Moon and returning him safely to Earth.[3]" This has been accomplished mainly due to expertise and concentrated work of a group of rocket engineers led by the brilliant German scientist Wernher von Braun (1912-1977)[4]. Originated from the National Advisory Committee for Aeronautics (NACA), which was primarily a research-oriented organization, NASA was established on July 29, 1958 by the National Aeronautics and Space Act, "to provide for research into problems of flight within and outside Earth's atmosphere" and to ensure that the United States conducts activities in space devoted to peaceful purposes for the benefit of humanity [62].

Figure 9. The official NASA seal. Different from the known NASA insignia (the "meatball"), this seal is used in more formal agency's events. This picture was taken from a building wall at NASA Goddard Space Flight Center. Credit: Norma T. O. Reis.

[3] Kennedy's "Special Message to Congress on Urgent National Needs," delivered on May 25, 1961, three weeks after Alan B. Shepard (1923-1998) became the first U.S. astronaut in space. Extracted from the NASA Web site, http://www.nasa.gov/vision/space/features/jfk_speech.html.

[4] In that early NASA culture, painstaking attention to detail, exceptional performance, long work-hours, flexibility, control of the entire process, and testing were essential to mission success. Von Braun said in 1964 that "A good engineer gets stale very fast if he doesn't keep his hands dirty." Contrarily to what common sense would suppose, most of those first hard-working NASA engineers came from low-income families [45].

The agency's work is oriented by a vision and a mission. The NASA's current vision is "To reach for new heights and reveal the unknown, so that what we do and learn will benefit all humankind," and its mission is to "Drive advances in science, technology, and exploration to enhance knowledge, education, innovation, economic vitality, and stewardship of Earth." Let us visit some elements associated with the achievement of the agency's vision and mission. In order to accomplish its vision and mission, NASA recruits talented professionals, building a diverse workforce inclusive of all, regardless of race, color, national origin, sex, religion, age, disability, genetic information, sexual orientation, status as a parent, or gender identity [62]. During my ISU internship at NASA Goddard, I was able to observe several dedicated professionals with visual disability working in areas such as radio astronomy and education. It is known at NASA that a diverse workforce offers a wider range of solutions to the complex, multifaceted space-related challenges. In my opinion, NASA's vision and mission that embraces the planet by comprising ambitious goals such as colonization of other celestial bodies, searching for life on other planets, and monitoring of Earth's climate change, requires more international cooperation – because these goals are broad, complex, and benefit the entire planet. I think that NASA, with its 'open-doors' policy, could create even more synergy and achieve even higher by expanding, optimizing, and valuing international partnerships in science, exploration, and education, in the scope of 'win-win' negotiations. We can do many good things by ourselves, but we can accomplish more together. I would add that NASA should also invest on comparative studies of how private companies and other space agencies have been achieving their goals. These comparative studies help organizations to learn lessons, choose strategic investments, and identify areas of cooperation. And what is the future of NASA? I believe in the vision of Tsiolkovsky that the future of humanity is outer space; that humans will leave the cradle of Earth and become a space species, colonizing other celestial bodies. In this scenario, I believe that NASA will keep on playing a key role at creating a "space culture" among the public. I trust that with time, space will be a very popular topic, and common citizens will want to spend their vacation in space paying an accessible price. I believe that with technology advances spaceflight will become as safe as aircraft travel, and that parents could bring their children and perhaps even pets into space! Thus, I believe in a bright future for NASA, working cooperatively with other space agencies and the private sector, reaching farther destinations in the solar system and beyond, developing cheaper-better propulsion systems and spacecrafts, and keeping on fueling dreams of children, teenagers, and 'grown-up space-kids' in the U.S. and in the entire world!

And what about NASA's culture? It is impossible to define one single "NASA culture." There are several different cultures inside NASA! The agency's missions are accomplished by ten space centers, along with some facilities (see Appendix 1). These centers are geographically distant, and have evolved mostly independently. Each center has its own history and culture. For example, I would say that NASA Goddard is the 'NASA academia;' professionals are very friendly and it has somehow an informal atmosphere; NASA Johnson Space Center, on the other hand, is more formal, and its history relates with the first military astronauts. However, there are characteristics and values shared by all NASA centers – the 'NASA DNA' – such as teamwork as key to mission success, excellence, and integrity[5]. Further, NASA is not all about science and technology; it is also about values for life. For

[5] Extracted from the NASA's Employee Orientation Web site, http://employeeorientation.nasa.gov/main/CoreValues.htm.

example, I remember one sentence shared by NASA on its Web site several years ago when I was a child, which I found on a poster while an intern in 2008: "If something is bad, say no!" Indeed, NASA serves children and teenagers all around the world, inspiring them to make good choices for themselves and for the planet. Let us explore a little more the NASA culture. Especially with the success of its Apollo missions, the agency acquired a reputation of high-performance government organization, and it became a habit to evaluate other U.S. government programs using space agency standards. However, every organization matures and is affected by conditions they have to deal with, such as political demands. According to McCurdy, [45], through the decades NASA has grown more bureaucratic, conservative, and risk-averse, as it needs to assure its survival after the end of huge investments of the Apollo years. Indeed, NASA was born and grew applying mostly the culture of rocket scientists, at a time when government in a different geopolitical scenario provided more financial resources and freedom to those professionals. The 'mature' NASA exists inside a political tissue, thus also needs experts to dialogue with the U.S. Congress, to negotiate budgets, and to identify projects that would satisfy government needs. However, NASA keeps on innovating and attracting exceptional professionals, because of NASA's excellence reputation, and because of its work atmosphere. From my experience as an intern, I affirm that NASA is an excellent place to work. Indeed, NASA is regarded the 5th best place to work in the U.S. government.[6]

Figure 10. Building at NASA Goddard main entrance. This space center, which is the largest NASA center, has 34 main buildings, all of them numbered as the one here shown. This center is responsible for robotic missions such as the Hubble Space Telescope, and the Solar Dynamics Observatory. At the background, the inactive rocket belongs to the Goddard Visitor Center, opened to the public. Credit: Norma T. O. Reis.

[6] Data from 2011, extracted from the Partnership for Public Service Web site, http://bestplacestowork.org/BPTW/rankings/.

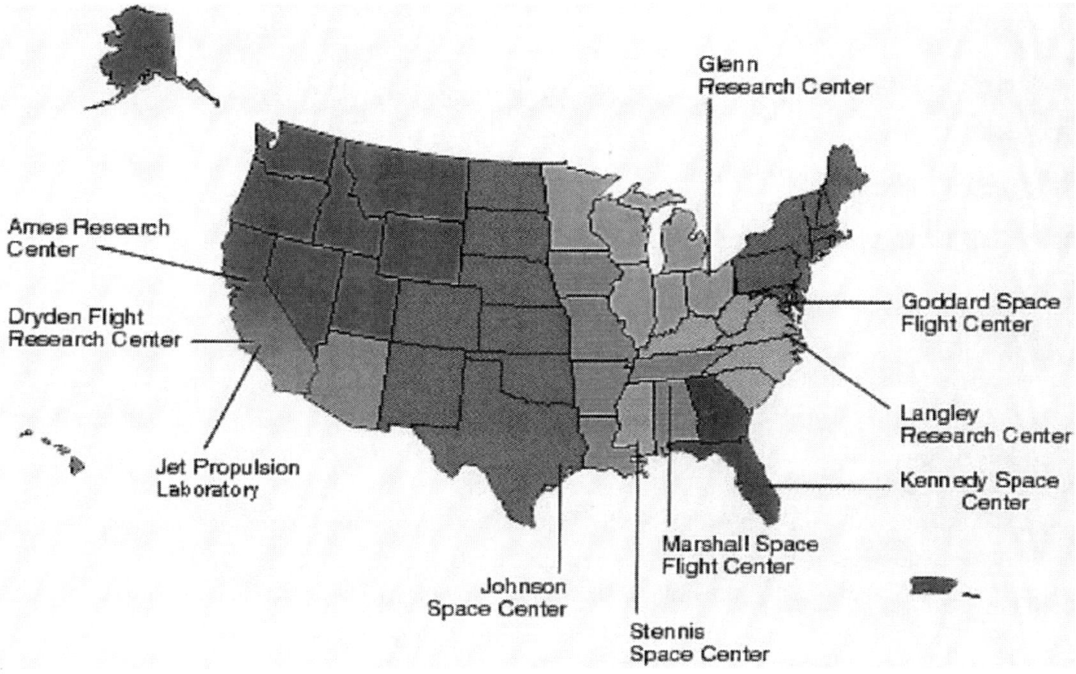

Figure 11. The NASA centers and their corresponding areas of responsibility in education. Each area is shown here in different tones of gray. Source: NASA. Credit: Norma T. O. Reis.

The NASA's culture of excellence, diversified portfolio of activities, and friendly work atmosphere has led to outstanding achievements. And how has NASA contributed to society's well being? Currently, a multitude of systems, processes or services incorporate NASA-inspired technologies. NASA develops technologies that have led to the invention of many household goods, such as solar water heaters, reflective insulation, water purification, coolers, home security systems, enriched baby food, satellite transmission, trash compacting, and so forth. In the medical field, it can be accounted heart pumps, invisible braces, ultrasound skin damage assessment, heart pumps, cool suits, eye surgery, breast biopsy system, cancer diagnosis imaging, and so forth. In sports and recreation, NASA technology is incorporated in fabric structures, stadium material, tennis rackets, ice rackets, ice skates, golf equipment materials, joysticks, virtual reality, cool suits, and so forth. In the transportation field, it can be mentioned fabric structures, vehicle controller for the disabled, quality control system, cleaner burning cars, metalized materials, advanced lubricants, crash analysis, flexible circuits, and so forth. With these contributions, NASA has been generating innovation in several areas of research and development. Indeed, "many of NASA's missions and activities power innovation that creates new jobs, new markets, and new technologies [90]." It is important to mention, however, that these products, services and processes usually are not fully developed based on space technology. Usually, professionals from different areas work along with NASA experts, and combine skills, resources, and institutions to develop these benefits, and make them available to society.

And how does NASA operate to execute its explorations and science? Considering the complexity of these missions, the agency's structure is also complex. The NASA missions are divided in three mission directorates (Appendix 02):

a) Science Mission Directorate (SMD);
b) Human Exploration and Operations (HEO); and
c) Aeronautics Research Mission Directorate (ARMD).

The lead offices for these directorates is located in Washington, D.C., at NASA Headquarters, and related divisions are present in each space field center, provided that the center carries out activities correlated to a given directorate. For instance, NASA Goddard does not develop activities in Aeronautics, thus it does not have an ARMD division. Along NASA history, several field centers have been constituted to serve specific, complementary purposes. These centers are autonomous, though they can collaborate with other NASA centers, and all of them report to NASA Headquarters, from which global policies and guidelines emanate. As for physical structure, NASA has 10 centers, 01 headquarters, 01 institute and 05 facilities (Appendix 03). Some centers focus on human exploration i.e., astronaut selection and training, while others, on robotic missions.

Figure 12. The Lunar Reconnaissance Orbiter (LRO), sent to identify safe landing sites, lunar resources, and to study how lunar radiation could affect humans. The probe was still being tested when this image was obtained in 2008, at NASA Goddard Space Flight Center, where it was developed. Credit: Norma T. O. Reis.

As for reaching new heights and revealing the unknown, let us consider that NASA explores to expand the human presence in space and to better protect the Earth, by means of manned and robotic missions. NASA made it possible for humanity to step onto the Moon in the 1960's and 1970's, and afterwards the focus was the development of a reusable spacecraft, the space shuttle, a vehicle that lifted off like a rocket and landed like an airplane, and also the International Space Station (ISS), along with satellites to study the Earth, and probes to study the solar system and beyond. An example is the Hubble Space Telescope (HST) that changed human views of the universe and promoted substantial advancements in astronomy. The agency acquired maturity through the accomplishment of many missions, and now exploration plans are even more ambitious, though with some economic constraints. In 2004, the agency came up with its *Vision for Exploration*, a plan aimed at continuing manned exploration by returning to the Moon by 2020, enabling future exploration of Mars and other destinations. In order to implement the Vision, NASA established the program *Constellation*, which included a robotic presence on the Moon that would pave the way for human presence, with establishment of a lunar outpost, so that crews would be able to live on the lunar surface up to six months. Crew would learn how to use lunar resources and eventually, the new system could rotate crews to and from a lunar outpost every six months. Although part of this plan has been accomplished, such as sending the Lunar Reconnaissance Orbiter (LRO) to study the Moon and identify some possible landing sites, the overall plan had to be replaced by a new one, considered more cost-effective. Some U.S. government analysts regarded it would not be a necessary step to return to the Moon as a prerequisite to travel to Mars.

3.1.1. NASA's New Vision for Human Space Exploration

Through decades of exploration, NASA has learned lessons from successes and difficulties. Surely, the agency's first main challenge has been the first human travel to the Moon with a safe return. In my opinion, the second principal challenge for NASA has been the assembly of the International Space Station (ISS), combining technology and expertise of several countries. NASA robotic missions such as space telescopes and rovers studied from our neighbor Moon up to much farther destinations, such as exoplanets, and distant galaxies. Currently, the agency operates more than 50 science missions and has more than 25 others in development [62]. These missions have generated valuable data with rapid, open access. An example of such is the Solar Dynamics Observatory (SDO), launched in 2010, and designed to study the causes of solar variability and its impacts on Earth and human technological systems. It has been contributing to enhance our understanding of the Sun's influence on Earth and near-Earth space by studying the solar atmosphere (corona) on small scales of space and time, and in many wavelengths simultaneously. This mission studies in more detail the Sun's behavior [91], and how solar activity causes space weather[7]. With missions studying the Earth, the Moon, the Sun, the solar system, and farther into the Cosmos, we learn more not only about the Universe, but also about how our home planet has formed, how it should evolve in the future, and how we should work towards preserving Earth's resources.

[7] Space weather could be defined as the combination of "eruptions from the Sun, disturbances in the solar wind, and the twisting and stretching of Earth's magnetosphere" resulting of the interaction with Sun's plasma emissions. Extracted from a NASA Educational printed material named "The Forecast? Look for Storms, Gale-Force Winds, and Plasma Blobs."

Currently, NASA's major exploration goal is to expand human presence in the solar system through manned missions to an asteroid, Mars, and other destinations. Let us see what official U.S. documents state about the agency goals for the decades to come. According to the White House [92], by 2025, NASA should send manned missions beyond the Moon, including an asteroid. A mission like that will bring unique challenges, such as those derived of the fact that there is no gravity on an asteroid to attract an astronaut to its surface. Further, it would help answer scientific questions about solar system formation, and origin of water and organics on Earth [93]. In my opinion, the prime motivation for this trip should be that NASA (and other institutions around the world) has efforts to detect, track, catalog, and characterize near-Earth objects[8] (NEOs), especially those who could pose major threat for Earth. For the eventuality of an asteroid coming in a collision trajectory[9], many experts support that the best measure to avoid collision would be deflection or nuclear explosion[10], though there are other possibilities considered. In any case, having astronauts traveling to an asteroid would be an important know-how to acquire. NASA also plans to explore the LaGrange points, which represent opportunities for construction, fueling and repairing of complex in-space systems [93]. As for the Moon, there are more nations such as China, Japan, and India interested in a human visit within the coming decades. Later, by the mid-2030's, NASA plans to send humans to orbit Mars, and return them safely. This would be a remarkable historic accomplishment after the first human flight to the Moon, to have humans stepping on another planet. In order to accomplish that, NASA plans to use scientific data from robotic satellites that study conditions in remote atmospheres and seek resources like water- or oxygen-rich soil, which may be used [62] in long-term explorations. For example, NASA sent in 2011 a rover called 'Curiosity' to study Mars' habitability.

International collaboration is essential for Mars exploration plans. There are more nations interested in this effort. A notable international enterprise is the Mars500, the first full duration simulation of a manned flight to Mars. The study, conducted by Russia's Institute of Biomedical Problems (IBMP), with extensive participation by ESA, was concluded on November 2011. During 520 days[11], six volunteers lived and worked like astronauts, hermetically isolated in four sealed interconnected habitat modules, and one external module, the 'Martian surface.' They lived isolated from external contact, with limited consumables, delay in communications, and other harsh conditions of a real interplanetary travel. The purpose of the experiment was to prepare, learn lessons, and collect data for future human missions to the Moon and Mars [95]. According to project's leaders, the next step could be to replicate the experiment in the ISS, considering that it will be operational only until 2020. One element absent in the terrestrial experiment was galactic radiation, thus the experiment in the ISS should be more complete [96]. Further, there are arguments supporting that we would

[8] Comets and (mostly) asteroids passing close to Earth's neighborhood. Not all NEOs are potentially hazardous. Learn more about NEOs at the NASA's Near Earth Object Program Web site, http://neo.jpl.nasa.gov/.

[9] The dinosaurs' extinction was probably due to asteroid collision with the Earth. These collisions with large asteroids are rare, but smaller rocks could also pose threats if they fall in populated areas. If the asteroid that reached Siberia in 1908 had come into a city (learn more about the Tunguska impact in http://science.nasa.gov/science-news/science-at-nasa/2008/30jun_tunguska/), the result would have been a disaster. For this reason, space agencies are establishing programs to track and catalog these objects.

[10] For this purpose, it would be better to detect the asteroid several orbital periods in advance, and have the explosion to happen the farthest possible from Earth, because even a small deflection at a more distant point would produce a large deviation from Earth [94].

[11] This is the period estimated for a complete voyage to Mars, comprising the 'go' and 'return' trips, along with the permanence on the surface.

be ready to explore Mars with human crews by 2016, and that a long-term plan such as NASA's would not succeed, because in a long time frame, political changes would make the effort always delayed [97]. Indeed, according to a NASA document "Human spaceflight (HSF) programs are complex and can occur on decadal timescales, yet funding is annual and political cycles occur on 2, 4, and 6-yearintervals." [98]

Although there were initial plans for stopping activities at the International Space Station, NASA plans to continue using the ISS until 2020 or beyond [99], to maximize the potential of the laboratory. The ISS also provides a stable destination to ease the growth and evolution of new commercial space transportation opportunities. Further, it has been lots of cooperative work and time to assemble the international outpost. With the retirement of the space shuttle in 2011, another important component of current NASA strategy for exploration is to foster development of private sector to provide reliable, cost-effective commercial spaceflight of crew and cargo to and from the ISS. NASA has been helping develop the private sector by providing it with advices from experts, access to agency's facilities, and development funding. The idea is that by having these services offered nationally, NASA will not need to purchase them from other countries, and perhaps the cost of access to space could be reduced. Further, it allows that NASA focuses efforts and money in more cutting-edge areas, such as developing more efficient launch systems, and its multi-purpose crew vehicle Orion in order to enable missions beyond low Earth orbits, reaching cislunar space, near-Earth asteroids, Mars, and other celestial bodies (92, 62, 93].

Finally, it should be said that NASA is aware that exploration beyond low Earth orbit is a long-term and expensive effort, thus it is uncertain if NASA funds only would be enough to guarantee the overall execution of such an ambitious exploration strategy. In this scenario, international cooperation is very important. U.S. policies state that government should identify areas of potential international cooperation that may include space science; space exploration, including human space flight activities; space transportation; space surveillance for debris monitoring and awareness; Earth science and observation; environmental monitoring; satellite communications; geospatial information products and services; disaster mitigation and relief; search and rescue; long-term preservation of the space environment for human activity and use; among others [92]. NASA should also continue a program of space science observation that comprises studying the Sun, and searching *Ea*rth-like planets orbiting other stars, identify potentially resource-rich planets, among other goals. These studies are also supportive of human space exploration. In the realm of environmental Earth observation and weather, NASA should enhance climate change research and sustained monitoring capabilities. For that, one of the strategies should be to use international partnerships to help enhance weather, climate, ocean, and coastal space observation.

If this vision becomes a reality in the coming decades, children and teenagers of tomorrow will have even more inspiration, and their horizons would broad even more. They perhaps will dream about interstellar travel, and even about exploring other galaxies. NASA already develops advanced studies about plasma propulsion, space elevator, solar sails, and other possibilities, which may lead humanity farther, and revolutionize our way of living in the next generations. I would end by inferring that in my opinion the NASA vision "To reach for new heights and reveal the unknown, so that what we do and learn will benefit all humankind," comprises a core commitment with the U.S., but also with the world. Surely, NASA is a United States' space agency, thus its primary goal is to generate benefits to the U.S population. However, considering the international nature of space exploration and the prominence of the agency in the international scenario, NASA may be also regarded as of service of humanity - though national. Indeed, countless children and teenagers from all around the planet are passionate about NASA missions! I am one of those children who some time ago was exploring the NASA Web site in South Brazil. And NASA has inspired me to research the 'space education phenomenon!'

3.2. NASA Education at a Glance

This research is particularly committed to analyzing the component "education" of NASA's mission, which is translated into several programs, projects and educative resources. One of the most important elements that make NASA Education a model of excellence is that the agency works towards being open doors to society. The purpose of NASA is to create channels to reach the entire society and make people aware of what NASA is and of what NASA does. This attitude has been part of 'NASA's DNA,' since the agency was born. According to Quinn et al. [100] "public outreach and science education have been important components of the mission of NASA since the Space Act created NASA in 1958." This Act, in its Sec. 203, states that NASA should "provide for the widest practicable and appropriate dissemination of information concerning its activities and the results thereof." There is a strong correlation between transparency, dissemination of information, and the commercial uses of space. NASA has created policies and processes to accomplish this, such as making available raw science data archived by all NASA missions, for open use, as previously mentioned. The main channel of opening NASA to society is the NASA.gov. Further, NASA use social media to reach out the public [62]. The NASA's Strategic Plan (2011) confirms this commitment to promoting 'NASA awareness' through education and outreach: "At NASA, sharing information is a mandate within our founding legislation. Throughout our history, it has been a priority to make data from science missions, research, and other discoveries available (...) Our [NASA's] missions are a natural means of interacting with the public and supporting students and teachers." The publication mentions some channels through which NASA accomplishes this:

a) traditional: print, television, live events, and
b) online: virtual events, live streaming video, online chats, social media.

An online presence fosters transparency in NASA's operations and management practices [62]. As for the importance of a public institution being open doors to society? It is not only 'important,' it is an obligation. Public organizations exist to the benefit of society. It is the taxes that the 'ordinary person' pays that sustain our cutting-edge science. Thus, NASA fulfills its societal obligations with the population both by producing advancements that benefit society, and by sharing and engaging every citizen, every student, and every teacher in the agency's wonders, missions, discoveries, and challenges. However, I notice that NASA does not see this as merely an institutional obligation; for NASA people, it is a pleasure, and it is fun to be with the public, and share its mission with everyone.

Education plays an especially important role in the scope of the overall NASA mission, and a strong indication is that education is part of NASA's Strategic Plan. "For nearly 50 years, NASA's journeys into air and space have developed humankind's understanding of the universe, advanced technology breakthroughs, enhanced air travel safety and security, and expanded the frontiers of scientific research. These accomplishments share a common genesis: education [101]." Further, the 2011 NASA's Strategic Plan highlights the urgency to improve STEM education to assure economic growth, and the role of NASA's efforts to inspire the passion and creativity of students. Indeed, NASA works towards contributing to strengthen the U.S. education system, especially with regard to STEM education. The disciplines that NASA investigates can be easily related to several curricular subjects. For example, activities that incorporate NASA themes like meteorology, oceanography, and technology, help to build analytical skills and encourage interest in STEM, and can be integrated in the curriculum of subjects such as Physics and Mathematics. And how does NASA benefit from its investment in education? It is the inspiration for NASA's mission that captures the interest of young students who will become the talented and creative professionals that will give continuity and expand NASA studies and explorations. In other words, by investing in education, NASA is sowing the seeds of its own future existence! However, in order to accomplish this goal of improving STEM education it is mandatory to engage and prepare educators as well. For that, NASA has ten NASA field centers, and one strategy to provide quality space education is to employ education specialists at each of these centers to work with local and regional partners to best identify resources and opportunities that meet the needs of the education community. This system allows NASA to be responsive to national initiatives such as "Educate to Innovate[12]," while maintaining flexibility in delivering products and services useful to educators. NASA specialists work directly with precollege educators through local and national education organizations. In those interactions, elementary, secondary, and informal educators learn how to translate NASA's current research and technology advances into meaningful education experiences. The agency also offers several opportunities as in-service and pre-service teacher training. However, the federal precollege STEM education lead is the National Science Foundation and the U.S. Department of Education. NASA, as a discoverer and technology producer, is an agency with an important, though complementary role in STEM education, oriented by its core scientific, engineering, and exploration missions [100].

In order to strengthen NASA mission by attracting and retaining students to STEM, the agency carries out numerous, diversified, and synergic initiatives to inspire, engage, and educate students, educators, families, and communities in the passion and discovery of space studies and exploration. One of these efforts is informal education, which comprises experiences such as after-school programs, activities in libraries and zoos, youth groups, community groups, media, community-based actions, exhibits, and activities for parents [62]. As for resource distribution, NASA's Web site pursues links dedicated hyperlinks to scientists, policy-makers, general public, educators, and students. Through an education dedicated Web site, the agency shares several rich online resources to the educational community at all educational levels – from kindergarten to post doctorate, thus benefitting the U.S., and beyond. On the NASA Education Web site, teachers and students will find innovative projects that use technology; high-quality didactic resources; podcasts; webcasts;

[12] Learn more on the White House Web site, http://www.whitehouse.gov/issues/education/educate-innovate.

an educational TV channel; blogs; games; animations; videos; visualizations; and so forth. As for printed materials, U.S. teachers can easily obtain them at some of the ten NASA field centers (including NASA JPL), or they can even send letters to those centers requesting didactic resources. At times, even teachers/students from other countries may receive printed materials from NASA field centers, by letter request. And what kinds of printed material does NASA produce? It is an incredible variety. Through its several missions and space field centers, the agency produces lithographs; set of images of the solar system bodies containing updated discoveries; information about current and past missions; information about astronaut selection and training; teacher guides on hot topics such as microgravity and rocketry; books about interdisciplinary themes such as space mathematics; missions stamps; bookmarks with space-related information; and so forth.

In order to align the agency's efforts in education with the goals of the U.S. education system, and to assure that the agency will prepare its future workforce, NASA also sustains a wide range of programs and projects to audiences from early childhood education to post doctorate. Examples of initiatives are summer internships opportunities at NASA; radio astronomy program; space weather-related actions; environment-related programs; videoconference actions; inclusive programs serving underserved and underrepresented populations, and so forth. Pre-service and in-service teacher training are also highly valued by NASA, considering that the teacher is the key person to assure success to the agency's education efforts. NASA concern with education and educators reach higher. For example, there is an effort for K-12 education named Albert Einstein Distinguished Educator Fellows Program, aimed at providing STEM education policy-making learning opportunities in federal agencies, including Congress. Further, the agency offers visits to its field centers; events such as conferences; and face-to-face interactions with NASA's personnel. The 2011 NASA's Strategic Plan mentions some different levels of engagement of NASA Education efforts, as follows: "Our [NASA's] participatory engagement activities span the communications spectrum ranging from passive activities—like watching online NASA videos—to highly interactive activities that use NASA-related social media tools or provide hands-on experiences. We also use these activities to collaborate with the public on interpretation of data and discoveries. We foster prize-based competitions, offering opportunities for organizations and private individuals to propose innovative solutions to specific challenges we have identified. By increasing the mechanisms through which the public can directly and specifically contribute to our missions, we can bring additional creativity and capability to some of our biggest challenges, and leverage our resources to accomplish even more toward our goals" [62]. An interesting example of challenger-oriented effort through which students and the general public can collaborate with NASA in important endeavors is through the U.S. federal government Challenger.gov Web site. As the name suggests, it brings challenges that a third party should solve. These challenges are aimed at developing solutions that address specific governmental needs, and often there are prizes associated, monetary or non-monetary.[13]

Another factor that generates success to NASA Education efforts is the intentional integration between research and education. According to Timothy E. Eastman [102], "(...) research and education as two sides of the same coin. One cannot do effective research or reach deep understanding of the physical systems of the world unless at the same time (...)

[13] More information on the Challenger.gov Web site, http://challenge.gov/NASA.

commit to constantly pass that information on to the next generation and be effective at training the next generation for adapting to new technology, scientific methods and understanding." Indeed, all NASA missions have the responsibility of carrying out education and outreach actions; and we can easily observe in NASA Education initiatives that the component 'research' is present. Students are required not only to access NASA data and learn from them, but also to produce research following the scientific method, sharing findings with their peers, and at times even with members of the scientific community. Indeed, NASA understands that knowledge and innovation can be produced by graduated scientists and engineers, but also by young students with bright ideas. The twenty-first century youth is significantly different from their peers of decades ago. As they have access to up-to-date, unlimited information through new technologies, and can communicate much faster by using several types of media to mean what they say, it becomes easier for them to face new challenges. Recently, for instance, a schoolboy from Canada sent a toy to Earth's orbit[14] through a not so high money investment. Current youth is increasingly less aware of the meaning of the word 'limit' in terms of accomplishments. And part of NASA's role is to help students understand and apply the scientific method in the scope of engaging projects that would enable them to develop unique, advanced academic, technological, scientific, cooperative skills for proper participation as aware, responsible and prepared citizens of a rapidly changing world that needs environmental care, technological innovation, and human values as solid foundation for sustained prosperity and welfare.

The education efforts at NASA are characterized by valuing and encouraging diversity and inclusiveness. They target all students regardless of specific conditions such as belonging to low-income families, having special needs, or being part of whichever minority group. In my opinion, the agency understands that all groups of people pursue singular, important contributions to offer in the scope of the wide range of NASA missions, and the countless types of tasks each of these missions require; and for this reason, the agency recruits talented professionals with diverse backgrounds. According to a publication of the NASA's Office of Education, one of the agency's strategic goals in education is the outcome "Identify, cultivate, and sustain a diverse workforce and inclusive work environment that is needed to conduct NASA missions. [103]"

NASA recognizes that cultivating diversity and inclusiveness strengthen the overall agency's missions because people with different backgrounds deal with problems from different perspectives and may identify creative solutions. Examples vary. Visually impaired professionals are more skilled at listening than people who see, and can conduct very good researches in radio astronomy. It has been noticed that multi-gender astronaut crews perform better than single-gender ones. Thus, the agency carries out education programs aimed at encouraging all students to pursue careers in STEM, and possibly become part of NASA's workforce. In other words, the educational NASA enterprise works towards serving the entire population with balance, equity and inclusiveness by engaging underrepresented and underserved minorities such as women and persons with special needs [103, 104]. The agency has several programs that focus these populations, such as:

[14] More information on the Mail Online Web site, http://www.dailymail.co.uk/news/article-2092501/Lego-astronaut-makes-debut-space-help-Mathew-Ho-Asad-Muhammad.html.

- NASA Minority University Research and Education Programs: http://www.nasa.gov/offices/education/programs/national/murep/home/index.html
- NASA Explorer Schools: http://explorerschools.nasa.gov/portal/site/nes/
- The Science, Engineering, Mathematics and Aerospace Academy (SEMAA): http://www.nasa.gov/offices/education/programs/national/semaa/home/index.html

Figure 13. Minorities inside NASA Education. The agency has programs that target the so-called minorities such as women and black, in order to assure the that the agency's future workforce will be diversified, and thus more effective in the complex, multifaceted tasks of space exploration. Source: NASA. Credit: Norma T. O. Reis.

And how does NASA reach the educational community? We have seen that the core physical infrastructure of the agency consists of 10 field centers, plus headquarters. The personnel and resources available at these places would be insufficient to serve the entire U.S educational community. Further, the number of experts in education at NASA is limited, considering the agency's primary focus on science, technology, and engineering [100]. However, NASA is excellent at establishing strategic linkages and partnerships at local, regional, national, and sometimes at international level. As previously mentioned, after the Cold War period space faring nations demanded socioeconomic benefits from their space programs, as political motivations weakened. In this framework, NASA's Office of Space Science hired education leaders and scientists to establish its own education and outreach programs (EPOs). In 1993, after getting advice from national education experts, the professor Jeffrey Rosenthal set up an "ecosystem" for EPO. Instead of each scientist, institution, and mission pursuing EPO on its own, the new EPO programs would coordinate and amplify individual efforts so that the global effect would be greater than the sum of its individual parts. The system was composed of forums to coordinate EPO activities among like missions, and regional brokers/facilitators to link missions/scientists to school systems, museums, educational publishers and so forth [87]. In other words, this structure amplified dialogue between NASA missions and the education community. In 1994, a mandate established that all NASA's space science missions committed 1-2 percent of their resources to EPO activities, which must be carried out in partnership with professional educators, and reach the maximum audience possible [100]. This complex structure added by this obligation increased scope and impact of NASA Educational efforts.

As for partnerships, it is foreseen in NASA's official documents that the agency will "build strategic partnerships and linkages between STEM formal and informal education providers" in order to promote STEM subjects and awareness of NASA mission [104, 105]. And who are these providers? One of them is NASA itself, with its several space education enterprises, offered through its physical and/or virtual infrastructure. Other providers and partners are schools, universities, museums, planetariums, science centers, aerospace companies, and so forth. These institutions combine efforts with NASA to reach a larger number of educators and students; they can use their facilities to conduct activities, and also help by designing and executing new programs. Indeed, NASA seeks partnerships with different institutions to engage new or broader audiences, and works collaboratively to leverage the agency's partners' resources. For that, institutions with wide range areas of expertise are invited on the NASA Web site to submit proposals,[15] such as grants or

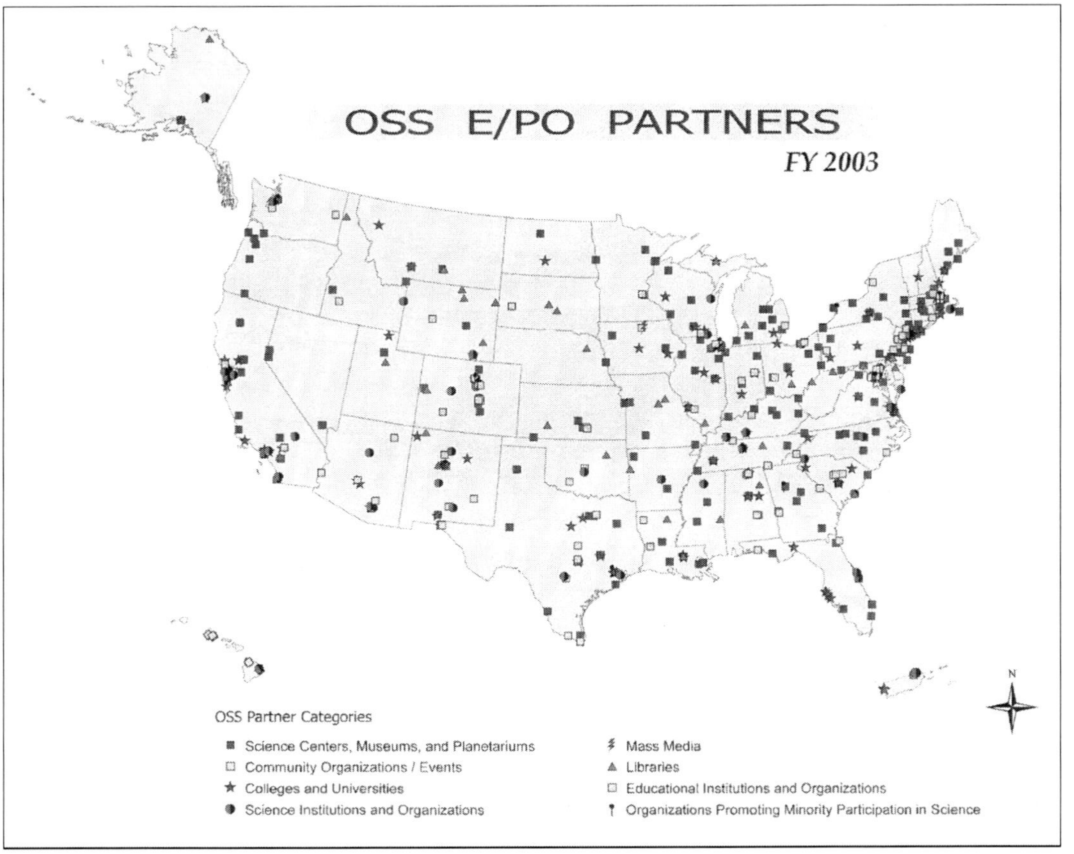

Figure 14. NASA science E/PO and its several partners. It is the large network of partners, and adoption of effective communication strategies that enables the agency to reach so many students and educators in the U.S., and beyond. Source: NASA. Credit: Norma T. O. Reis.

[15] More information on the NASA's Press Release Archives Web site, http://www.nasa.gov/home/hqnews/2011/apr/HQ_11-120_Education.html.

cooperative agreements, which may be funded [104]. As an example of linkage and partnership, it should be mentioned the National Space Grant College and Fellowship Program, which comprises 52 consortia comprising academic institutions, industry groups, and educational organizations. They represent all fifty U.S. states, the District of Columbia, and Puerto Rico. More than 850 university affiliates provide internships, mentoring for competitions, and formal NASA coursework. State-based consortia, by its turn, are responsive to specific needs of their regions, bringing NASA content and education support to K–12 school districts, and informal learning centers [62]. Several higher education who wish to pursue an educational/professional experience at NASA may find here excellent opportunities.

In the scope of overall NASA's mission, it should be considered that international collaboration is present, and very effective at producing knowledge in more comprehensive, innovative manners, regardless of some eventual communication difficulties due to cultural differences between team members. Indeed, "Several NASA Education projects are implemented in collaboration with international organizations" [103]. One example is the mentioned GLOBE Program, devoted to environmental studies in which students conduct experiments along with scientists all around the world. When students have opportunities to interact with peers from other countries, they become more motivated to the things they are learning, and they also broaden their understanding of other cultures, societal differences and similarities, and learning styles. Another example of NASA effort that comprises international participation is the One Stop Shopping Initiative (OSSI), where students from the U.S. and other countries may send proposals to NASA internship, fellowship, and scholarship opportunities. Some benefits for NASA deriving from this approach are the broadening of STEM education perspectives, to provide U.S. educational community experience in working with international peers, and to expand the agency's appreciation of diversity in multicultural scenarios. Further, NASA is part of the International Space Education Board[16] (ISEB), which facilitates discussions with other participating space agencies. Considering the above, I infer that NASA is open doors to the world. In my opinion, this attitude generates benefits both to international students and for the agency, as different cultures' profiles may bring innovative manners of solving problems inside the multitude of NASA's science, aeronautics, and human exploration missions.

3.2.1. NASA Education – A complex structure

There are some structural and conceptual elements that must be known in order to provide a reasonable understanding of what the NASA Education enterprise is all about, and how it is implemented. In the sequence, I present some of these elements in the format of topics. The information presented here has been extracted mostly from NASA official publications, and interviews with NASA Education managers and specialists. This information is essential, though not at all sufficient to characterize the entire NASA Education structure and functioning, whose level of complexity cannot be fully described in a single thesis.

[16] More information on the International Astronautical Federation Web site, http://www.iafastro.com/index.html?title=International_Space_Education_Board.

Goals:

- The three main goals or outcomes that align NASA Education activities.
- Contribute to develop the Science, Technology, Engineering, and Mathematics (STEM) workforce in disciplines needed to achieve NASA's strategic goals through a portfolio of investments;
- Attract and retain students in STEM disciplines by means of a progression of educational opportunities for students, teachers, and faculty; and
- Build between STEM formal and informal providers strategic partnerships and linkages that promote STEM literacy and awareness of NASA's mission [105].

b) Actors

The agency implements its education and outreach activities by means of integrated and synergic actions and/or interactions from/between the following agents:

Office of Education (OE): coordinates and integrates NASA's education framework, implementation approach and policies. They are responsible for planning and implementing national programs, as well as for communicating their policies. They lead the development of education programs that are unique to their Centers. There is one OE at NASA Headquarters and one in each NASA Center [105]. Each center is responsible for offering education programs, projects and services for a given region of the United States.

Mission Directorates *Education and Public Outreach Offices* (EPOs): responsible for incorporating education components into their specific missions and research programs. They provide discipline-specific content, funding, and human resources to design and implement education programs/projects/products/services [105]. The main office at NASA Headquarters coordinates the EPOs of each space center, but the centers have autonomy to conduct their own activities. Likewise, the Mission Directorate EPOs at each space center coordinate the EPOs of each specific mission present in a center and subordinated to its area. In this case, as each mission has a given peculiarity, the level of interference is even smaller in comparison to the higher hierarchical level, according to J.M. Maher [106]. It means that the mission EPOs have substantial autonomy to conduct their activities. Budgets are debated on a yearly basis at headquarters (due to the yearly cycle of budget authority given by the U.S. Congress) so that each mission division attempts to present the best proposals in order to obtain more conditions to develop their EPO activities. Almost all NASA missions have an EPO plan, and all of them develop education and outreach activities, according to J. M. Maher [106]. For example, it can be mentioned the spacecraft Solar Dynamics Observatory – SDO, designed to research the Sun's influence on Earth and Near-Earth space by studying the solar atmosphere on small scales of space and time and in many wavelengths, simultaneously.[17] The mission Web site has several education and outreach links.

Center Education Offices: responsible for implementing education programs, projects, and activities for the Mission Directorates and the Office of Education, and also for planning and implementing education programs peculiar to and funded by their Centers, in their area of geographic responsibility for K-12 education [105].

[17] Learn more on the NASA's Solar Dynamics Observatory Web site, http://sdo.gsfc.nasa.gov/.

Program/project managers: responsible for making decisions about their authority. They have authority and autonomy with regard to schedules, budgets, human and capital assets for their programs or projects. These education programs/projects belong to a specific mission [105].

External Education Implementation Partners: organizations that implement education activities for the Agency, such as contractors, academic institutions, museums, science centers, and so forth [105].

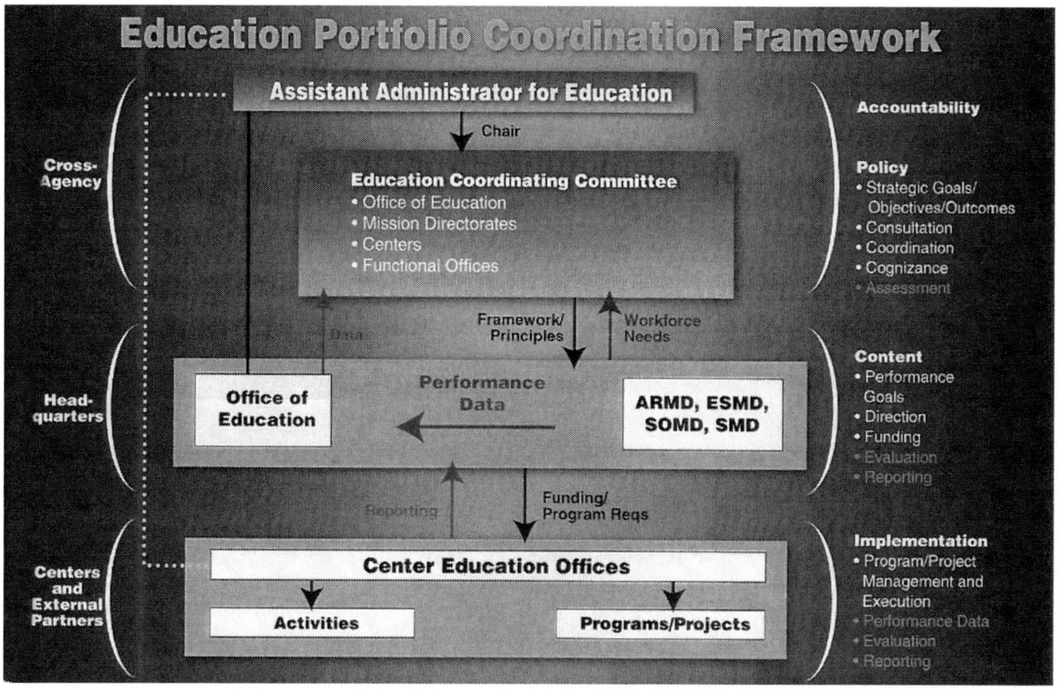

Figure 15. NASA Education actors – responsibilities and connections. This graphic shows who does what and connections among the diverse actors, in a scheme involving from external partners (universities, museums etc.); the Offices of Education and Mission Directorates Offices in each NASA center; until the upper level which is the agency's assistant administrator for education, who reports directly to the NASA's administrator. Note the division of tasks among the different actors. For instance, according to the graphic, funding matters are a responsibility of NASA Headquarters, and implementation are under responsibility of NASA centers and external partners. Source: NASA. Credit: Norma T. O. Reis.

c) Areas

A portfolio of investments is adopted in five strategic areas:

Higher Education;
Minority University Research and Education;
Elementary and Secondary education;
Informal education;
Education Technology and Products [107].

d) Categories of involvement

Educational activities are targeted to four categories of involvement: inspire, engage, educate and employ the youth. Each category should lead to the next. First, students are inspired and engaged by means of informal and formal education activities. Afterwards, they can choose careers in STEM, which may lead to employment at NASA, so that providing sustainability to the achievement of the agency's mid and long-term strategic goals.

Inspire: activities that promote awareness of NASA's mission, mainly by informal education and outreach activities, targeted both to general public and educational community. Examples: expositions and NASA TV programs.

Engage: activities that somehow add in participant interaction with NASA content for the development of a deeper understanding. Examples: robotic projects, interaction between students and NASA's content and experts through videoconferencing (NASA's Digital Learning Network).

Educate: activities targeted to students or pre- and in-service educators, using NASA resources to develop/enhance STEM knowledge and skills. Activities in elementary and secondary levels are supplementary to formal classroom education, and at higher education level NASA may develop specific university curricula in support of its mission. Examples: in-service teacher training actions, scientific research projects in topics such as space weather, radio astronomy, and Earth's environment.

Employ: development of personnel for employment in disciplines needed to achieve NASA mission and strategic goals. Examples: internships, fellowships, and other professional trainings [107].

e) Principles

The principles operating the education portfolio aim to ensure programs, projects, products and activities alignment and excellence, forming the basis for their evaluation.

Table 1. Principles guiding and operating the NASA Education portfolio

Principle	Description
Relevance	Usefulness to the education community
Content	Use content, people and facilities to involve educators, students, and/or the public in STEM.
Diversity	Promotion of inclusiveness of underrepresented, underserved individuals, such as women and impaired people, encouraging them to careers in STEM.
Evaluation	Use of metrics to measure if the program/project achieved its outcomes, as well as evaluation of annual performance goals. Models and techniques are appropriate to content and scale of targeted activity/product/program.
Continuity	Projects and activities draw from audience who has already demonstrated interest in NASA, connecting participants to the next level of involvement.
Partnerships/ Sustainability	Education investments leverage and achieve sustainability by their intrinsic design and the involvement of local, regional, national partners in their design, development, or dissemination. As suitable, key aspects of projects/activities are replicable, scalable, and show potential for continuance after period of direct NASA funding.

Source: Elaborated based on the NASA Strategic Coordination Framework: A Portfolio Approach, 2007.

Evaluation, by its turn, is a very strong component of NASA Education strategy. The intention is to measure the intended impact in proportion to the investment. Different parameters are used for different initiatives, because they can vary tremendously and one single set of parameters would not succeed in evaluating all of them. In the sequence, a detailed illustration of how this process occurs.

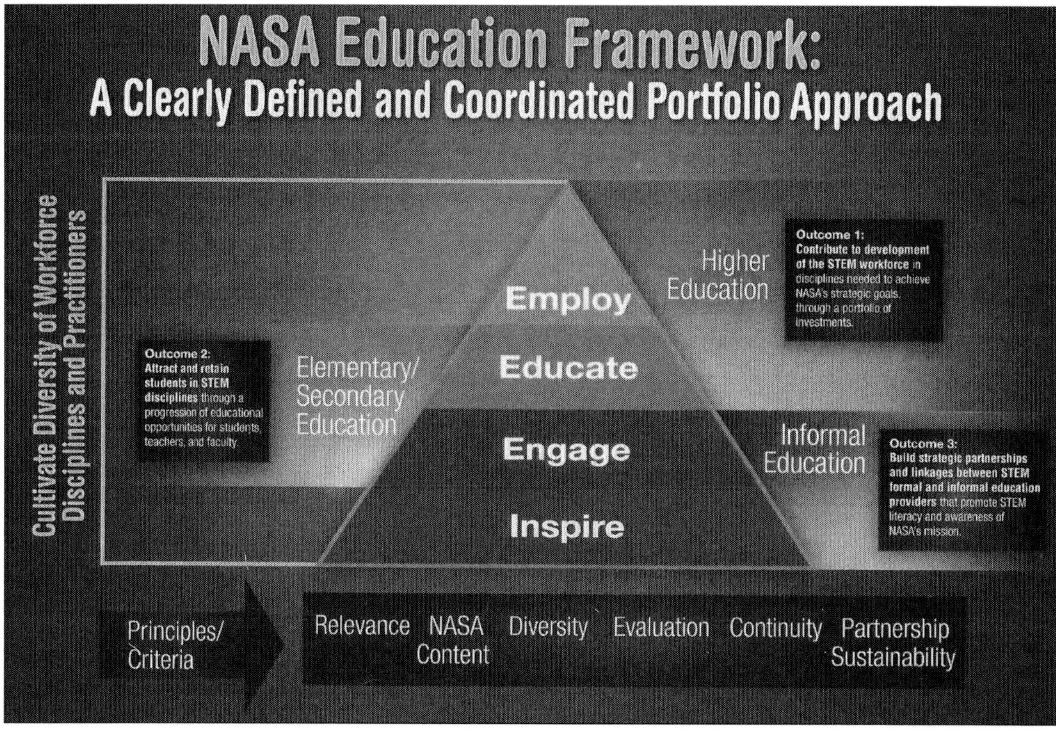

Figure 16. Relationships among NASA Education expected outcomes, levels of involvement, areas, and principles. Note that different areas related to specific outcomes and levels of involvement. For instance, elementary and secondary programs' priority is to promote activities of engagement and education, targeting at attracting and retaining students to STEM. However, they could also carry out inspiration actions. They could not carry out employment actions, since these are mission of the higher education area, which offers opportunities such as fellowships, internships, and scholarships. All areas work based on shared principles such as diversity, continuity, and relevance. Source: NASA. Credit: Norma T. O. Reis.

f) The partnerships

NASA actuates in regime of partnership with several national and international actors, because this strategy strengthens the scope and effectiveness of actions promoted.

Strategic partnerships strengthen the programs and increase the scope of the actions promoted. NASA has been working towards establishing new partnerships with different societal actors.

Inside NASA, education teams meet regularly in order to accompany, conceive improvement possibilities, and evaluate their projects. This institutional commitment with quality educational efforts is translated in several innovative, creative, content-rich products, services, and solutions for the education community. Let us see some of them in the sequence.

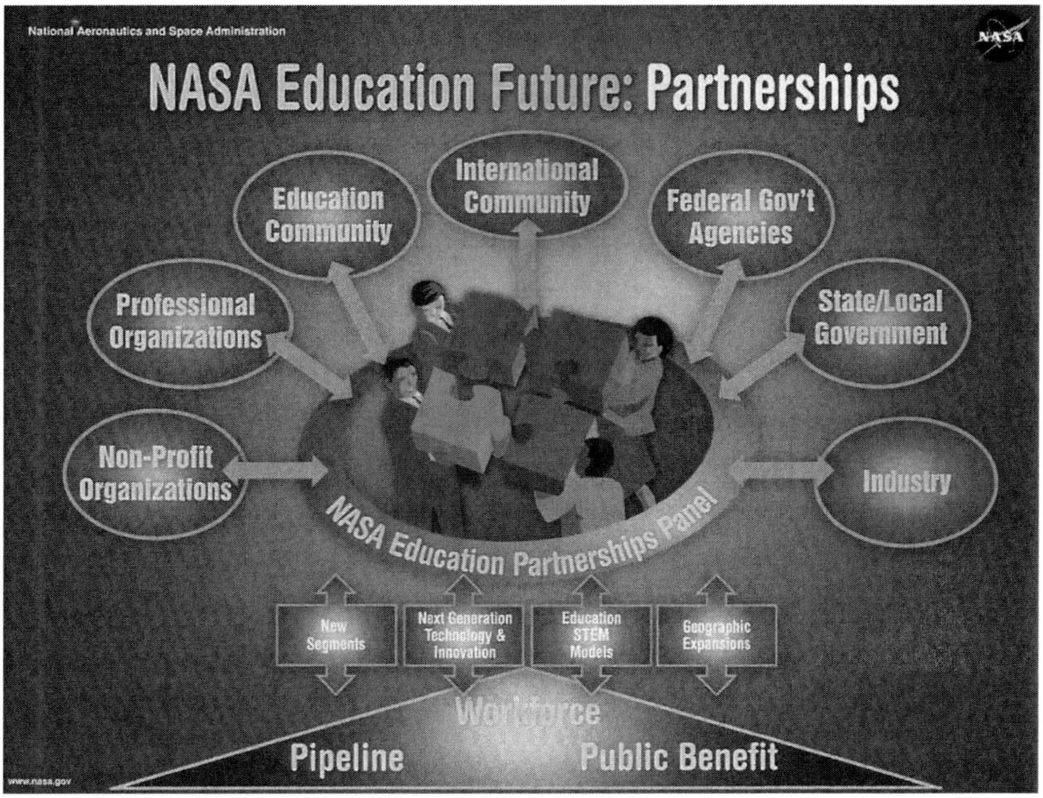

Figure 17. Partnerships to achieve more with quality. NASA's Education excellence and scope owns a lot to its strong system of partnerships that comprise the private initiative, other federal agencies, and international community, for instance. Several NASA programs such as the Space Weather Action Center and Radio Jove are implemented in many countries, which allow participating students to share findings with international colleagues, incorporate new learning strategies, become more creative and innovative, and learn about different cultures. Source: NASA. Credit: Norma T. O. Reis.

I present in the sequence a very comprehensive interview about NASA education structure and functioning.

Interview with a NASA education specialist (summer 2008), with permission

Interviewed: Margareth (Peggy) Maher
Position: Education Specialist, Office of Education
Employer: NASA Goddard Space Flight Center

What is your position here at GSFC?

I am an aerospace education specialist.

Could you tell me a little about NASA education structure?

Yes, at NASA there are four [currently, three] mission directorates. They have education and public outreach (EPO). There is the aeronautics directorate; the space operations directorate that has the Hubble Space Telescope, the Cassini spacecraft, the SOHO and so forth; and then exploration systems directorate is human spaceflight. They always produce materials for each astronaut mission, space shuttle or International Space Station (ISS) missions. And I would say that most of the education efforts are carried out in the sciences mission directorate.

Where are the directorates physically located?

The Headquarters is the main office for all of them, and is located in Washington, D.C., but they have branches in all the ten NASA centers. Every mission directorate is present in all centers. I am not absolutely sure if there is an aeronautics directorate office here at Goddard, because we do very little on aeronautics. And there is an EPO education lead at NASA Headquarters supposed to organize the EPO, but that do not necessarily decide for what everyone will be doing. Each mission competes for the funds, and part of the competition is a plan for EPO. This plan usually hires somebody outside of NASA to do the work – people from universities, former teachers; these people are hired by the mission. They are usually contractors, but sometimes they work for the government. For example, the SOHO mission has an education and public outreach component plan, and they hired Steel Hill [a specialist at NASA Goddard], but NASA did not hire him directly. Instead, NASA chose people for the mission SOHO, and the SOHO team chose Steel. They can also choose a group, i.e. a university. For example, Arizona is doing, there is a university doing EPO for the Swift satellite. People who competed for Swift chose that university to do the education component, and the university selected the people. They are all responsible for meeting NASA's education objectives, and they all have to report on information such as how many educators they have reached, how many are using their materials, if they think students are learning from their resources, and they have to provide all the evaluation measures. Yes, all EPO groups have to report to NASA on their efforts.

And does every center have an EPO?

Each center has several EPOs. At NASA Goddard we have 30 missions [in 2008], if we talk about EPO, in this center there is SOHO, STEREO, GLAST, Messenger, LandSat, Solar Dynamics Observatory (SDO), Lunar Reconnaissance Orbiter (LRO), New Horizons, and probably some other missions that I forgot. NASA Goddard is a large center. We employ a lot of people, and it is the biggest science center among all NASA centers, thus there are more scientists here. However, I know there are EPO projects at all NASA centers, and in some universities such as the University of Arizona, and the University of Colorado. By the way, Colorado State is where the GLOBE program is located.

So these EPO groups are related to NASA missions…

Yes, they are related to missions or at times to smaller projects. Sometimes you get funding for a project that is related to an instrument that will fly on a mission someday. You have only the instrument, but not have the satellite to put it on. So you are hoping to find a satellite for that purpose, then you develop an EPO plan for that instrument. Yes, there are little pockets as well. Further, there is grant money, so right now there is a solicitation for NASA funds for two different projects. One is for global climate change education, for high school. NASA occasionally funds things like that.

What else should we know about this structure?

All the missions are under a director, and the director has an EPO lead who should know all the EPO groups within his directorate. There is an EPO lead at NASA Headquarters, who is just coordinating, not controlling the EPO. They know all the EPO groups. It can be very difficult to know all of the education efforts at NASA!

And what about the offices of education?

They are not mission-related, they are separate from the mission directorates, and they are under the administration. Each center's education office is responsible for serving certain part of the country. Some initiatives carried out by the education offices for higher education comprise internships; they have partnerships with professors in universities to do research that include undergraduate or graduate students; and K-12 programs such as the NASA Explorer Schools or workshops for teachers within the center's region of responsibility. Yes, the education offices are very much focused on a particular region. They work with EPO, though not very closely. In my opinion, they should work more closely with EPO groups. Yes, because you are doing education, but it does not mean that you have to talk to the education office; I am not sure why.

And what is the difference between what the education office does, and the EPO groups do in a same NASA center?

The EPO is very much more concentrated on their mission and the science around it. The education office does reach out to the EPO. For example, this summer we had workshops for teachers here. When we offer these workshops, we do not teach them all ourselves; we organize them and ask EPO groups if they want to share their resources with those educators. We bring the teachers and EPO groups train the teachers, though we help with that as well. Actually, the education offices are not terribly large, there is somebody who runs higher education, somebody who runs K-12 education, someone in charge of the Goddard education page, and we have somebody here who does e-learning technologies, so that is all of the distance learning things, videoconferencing; not every center has that, it is a contract for all of NASA.

If the education offices are not managing the EPOs, it means that the EPO groups are autonomous; so what does the education office means?

At the Headquarters level the office of education does have some responsibility for EPO, since all of the evaluation goes through the office of education, but in our office we work on agency wide initiatives. For example, Hubble servicing mission is coming up. There is a lot of that going through the EPO group, but because it is such a big push, all of NASA is interested in the public being educated on Hubble, not just the Hubble mission. Thus, we have directions from the headquarters education that we [office of education] should also be teaching Hubble, not just the EPO group. Further, the International Polar Year, which is right now, which would involve EPO groups from a lot of earth sciences missions and so, we do work with those EPO groups. We get direction from headquarters for their specific kind of campaigns that they want us to work on. And then we have the Interdisciplinary National Science Project Incorporating Research and Education Experience – INSPIRE. It is not only the INSPIRE team who work with that, but also people in the education office; the Intergovernmental Personnel Act – IPA; we have three IPAs, they also call teachers (...)

because NASA pays their school, and then their school continues to pay their teacher, so they have the same salary, the same benefits, the same retirement, life status for salary, and those people do sometimes produce education materials, and work with EPO groups and scientists, because not all of the EPO groups have a lot of teaching background. For example, Steel Hill's [one expert at NASA Goddard) background is in English, and there is another one with background in Political Sciences. Thus, sometimes they come to us for help, and we want this to happen more. Sometimes they come to us for teaching expertise, or for help about which states' education standards they have to meet, so they come to us sometimes for expertise. We try to establish a relationship where they understand that this expertise is something that we have, which they do not, so we can help them with that. There are lots of projects that come from Headquarters, and from the education office; my group has a very different function, the Aerospace Education Services Project (AESP). We travel, thus we work closely with EPO groups, because they may not have any money to go places, little money to travel to conferences and so forth, but we are on the road all the time, we go to school districts, and we develop relationships with schools and the states' department of education, and universities and regional education services center, so we can connect with teachers, and connect EPO with teachers, and we can go there and teach their materials. For example, right now I am talking to the Space Weather Action Center (SWAC) team about providing training for all of the specialists by my group, so we can teach the SWAC to teachers in their states; not to give a conference with all teachers together, but it is more direct way of delivering to the teachers. And we can also follow up, go back to them later and say *'do you still need help, did you understand everything, do you want more training.'* Thus, my group has a very specific function, as we are out in the field, and everybody else is here on the center. Another thing we do is that we have someone go to the Heliophysics and Planetary Sciences, Astrophysics, and Earth Sciences. All have group meetings, not everybody, but EPO, and some of the science leads, and one person managing that, and we have a representative from the education office who is always there, so we are supposed to know when things happen at Goddard. We do not know when things happen at JPL and Glenn; it is a little hard to communicate between NASA centers, unless it comes from Headquarters. Thus, I would say that the education office picks up everything else, the EPO very focus on the mission, we work with EPO sometimes, and we do everything else that has to do with education.

Can you tell me a little bit more about the AESP activities?

Yes, I know the state science director or supervisor for all states. I know a lot of teachers in those states, and people in universities. In Pennsylvania, for example, they have intermediate units for original services, so I know people in all intermediate units. I know the space grant people. For example, in May we had a meeting with superintendents, when they said that the districts are very small, so they have decided it would be more cost effective if they combine big areas of the state into one district, with one superintendent, and then reducing from 200 and something to 80; and so there were certain things that may not interest right now because they have all these other things going on, and until they decide (...) they could not choose to work with you (...) That is the kind of thing the AESP does with the states, and we report on that to the office of education. The EPO groups are efforts focused on their science and their mission. It is an interesting system, and I suppose it has been like that since the NASA's early days, but it does not really make a lot of sense to me that you have an education office, but then most of the education projects are not in the education office, I do not know why. I think it is because it is the way it was designed long ago, so it is difficult to go back and change that.

Sometimes you work with the EPO groups, sometimes you do not...

Yes, in my group we try to work with the EPO very much because we do not produce our didactic resources, we use our people's resources when we go out train teachers, so we have to work with them. And we try to share things from our center with other people.

What are the main tools that you use to evaluate NASA education programs?

There is an online evaluation system. Every time you do any kind of education service and event, like a workshop, conference, museum nights, anything, you enter into the system and inform how many people were there, the products you talked about, and so forth. The system also has mechanisms to allow the teacher who does an event, to set up a page for the participants to go to and provide feedback. However, what has been missing here, and is often missing in education evaluation is to find out whether participants took those things and used them, and if the students learned from that. How do you really say that my product improved students' academic performance? Thus, we do not really have a good measure for that, because we are not in the classroom. We have people training teachers, but how do you go from that to saying that students learned more because of this? However, I think that it is common in education – how do you really measure that? That is one of the reasons they have instituted so much testing in the United States. It is very hard to say if my program made those students grades better. A lot of states do not have the budget to break that down. The tests are head by objectives, the use of the objectives that are needed to meet. Thus, I am going to ask questions for each objective to see if students met those. In Texas, for example, which is a large state with a lot of money, there have been testing for a long time. There, you can go and see that the students in your classroom made this average score on objective 2.5, but in most states you cannot. For example, I am teaching about the sun, and there is an objective of having students learn about the sun, then I should be able to go and see in those classes with teachers we have trained if students did better than in classes we did not. However, you really cannot use that in many states, and even if you could, how do you know if your program really made the difference, or if there were other things going on in those schools that lead to the improvement. It is very hard to figure out. We just started a new evaluation process, especially trying to focus on the follow-up: if the teachers use that, if they are still using that one year later. This sort of reporting is important because it does not focus on if teachers came, if they like that, or learned that, but if they effectively used that. If students learned from that is still something difficult to be accessed.

Do you have some study on whether NASA education efforts have increased the number of students interested in STEM careers?

I do not know, the follow up part is very difficult in education because there is all these privacy laws, so it is very difficult to get especially children's contact information. For example, this summer we had students who came here for an education workshop. It was targeted for 12-13 years old girls. All of the presenters were scientists, engineers, and administrators. Everything for women, because it was a crew awareness initiative, in which we show the female crews of NASA who had been successful in her careers. We conducted activities, tours, and things like that. We would like to have followed up with those girls and see if after 10 years they would have graduated from college, if they pursued a STEM career, but it is very hard to keep up with them. We have their information because their parents provided that to us, but if it ever changes, and it often does, then... And that is the same with the other programs we conduct. It is very hard to maintain a database of personal information because. For one thing, if that information is lost or stolen – especially stolen – then we are liable for that. We do not like keeping personal information of people who do not work with us, thus I think there has not been in a record to do such a study, but I could be wrong, because they could actually go out to the general population and ask them if they have ever been to this kind of presentation, or done an internship or whatever; they could do a large survey, ask if they graduated in seniors from college, but I do not know if they have done something like that.

How is education carried out by the mission directorates?

For example, there is the aeronautics directorate; everything that NASA does in aeronautics is run through this mission directorate. There are many levels, but all of them would report back to the head of that mission directorate. They have EPO in the mission directorates; there is a person in each directorate for that. That person is ultimately in charge, if an EPO project is chosen or is evaluated, or if something is going wrong. We [AESP] communicate with a lot of EPO groups, and we also dialogue with the mission directorates, but we know that a lot of actions do not take place there; it is just administration paper, but we do want to know what they do and how; we want our group, and actually the education office also, to be more integrated with the EPO, for the mission people come to us and say *'can you help us with an EPO plan, or can you suggest somebody?'* Most of EPO people are very good, but there are few without education background, they really should not be doing that.

Do you education office have meetings with Headquarters?

Our chief of education does. He regularly attends meetings at Headquarters. The other centers do not have meetings as often as he does, because NASA Goddard is close to Headquarters. However, the other centers do have meetings twice a year, and they may have meetings by phone. More than that, they do not have the education office meet with the mission directors. Occasionally, they have meeting with everybody, but this is not a regular thing, because mission directors do not see *'why they need us.'* However, actually, education has been a part of NASA since it was created; education was a big piece of NASA's original mission, so it has been always part of it. In the beginning, in the Apollo years and during all of the early exploration, there was so much interest on what NASA was doing that many people came into careers in aerospace. However, I think there are other great things going on currently. On one occasion, someone said that NASA was going down the hill after Apollo. I think this person only sees human exploration - astronauts. He is not considering all observatories – SOHO, Hubble, Chandra, Spitzer, all those space telescopes, all the planetary exploration - Voyager, Cassini, Galileo, all the Vikings to Mars. When we look at those, and then at our earth missions, we see that it is tremendous the things that we are doing now. In the Apollo era, we NASA had about 4 percent of the entire budget of the United States. And all they were really working towards putting a man on the moon. There was their one goal. Thus, it was 4 percent of the budget with one goal, and of course lots of things to learn to get to that goal. Currently, we have probably a little over of one half of one percent, so from 4 percent, we are now about 0.6 percent, and we are looking at the distant universe, looking at the solar system, we are looking at the sun, looking back at earth, we have all these missions with this budget. How can one say we are not doing a lot? However, many people are more excited with human experience.

And what are some lessons learned from the past in NASA education?

From my experience, I have heard more than one person say that they were unsure that attending teacher conferences really was worthy the effort, because you see lecturers for such a short time, usually one hour, maybe 90 minutes, and teachers have a lot of other things that they are going to see, during 03 days or a week. They have lots of different things - not only a small part of it; you do not see them for very long. We still attend a lot of conferences, but I hear more and more EPO people saying they are not sure these events bring the biggest benefit. It is better to have the teacher for a whole day or a whole week, or even longer, and to develop a relationship with the teacher. If the educators only see your materials and go back to classroom, they are much more likely to use that, than if they see 10 or 20 different things, and then return to classroom. There is still a lot of people who attend education conferences, but they are beginning to say, *'I do not know if this is really worthy.'* Further, there have been some individual actions - not a big program - that did not do very well because they did not do a good job at matching the teacher's needs. For example, there was a program in which teachers worked as if they were real scientists, doing experiments. It was related to ice, but how would students experience that in places where there was no ice. Teachers were unhappy because they were learning things they could not use.

And what about NASA education greatest success?

There have been a few programs in which we bring teachers for an extended period, usually 02 weeks, and then we follow up with them, and make them sort of experts in their area. They are interested enough, and they go back as NASA experts. Sometimes they get paid a little bit, but often it is just their own interest to share what they know. I think those have been the best efforts, because the teachers commit to that longer time, can become very interested, and those teachers have gone on, some of those came work to NASA, many of them work as teacher leaders, for example, as curriculum specialists, sometimes as administrators, or have gone to universities, but they have become real leaders and have been very focused on NASA as leaders in STEM.

About education inside NASA's overall mission…

The NASA mission is *to understand and protect our home planet, to explore the universe and search for life* – this is that everyone knows that we do – *and inspire the next generation of explorers*, and this is the education part of it. We want to be prepared to continue our missions, when people who are doing that now retire. The part of the mission statement ends with *"as only NASA can"* because, for example, NOAA, they are also protecting our home planet, along with the environmental protection agency that monitors pollution, and things like that; so the *"as only NASA can"* means that we still do those things, but we do not copy what other institutions are doing, we do things that they cannot do, like NOAA, the satellite for weather (...) we study hurricanes and other topics in different ways. This is what we understand for National Aeronautics and Space Administration. And many people forget the "Aeronautics" in it. NASA developed the air traffic control system used by all airports, for example. Goddard is the biggest science center. We do more science here. Langley Research Center in Virginia does a lot of aeronautics. They do a lot of testing; for example, they have a big structure where they can drop things from. Kennedy Space Center mostly does launches, and all of the sciences around launches. Marshall Space Flight Center in Huntsville, Alabama is most famous for developing rockets. In the 1950's especially, in the early NASA, it was where the rockets were tested and Huntsville got very big; those rockets were huge, and it was like an earthquake every time they set up a rocket, so they moved the testing facilities to Stennis, which is in Mississippi, a place where there are not many people around. Johnson Space Center is where the astronauts live, and where the mission control room is located. So every time there is a human in space the control of that is in Johnson. The Jet Propulsion Laboratory is home for the Mars rovers, the Cassini probe to Saturn, and so forth. Dryden does almost all flight research; they do a lot of experiments on planes. And the Ames does a little bit of many things, they are more famous for astrobiology, so life on other planets, how is that like, how we can detect that, things like that. Thus, you know, in Goddard the most famous mission is the Hubble, but Goddard is also the biggest earth science center, with more earth science studies in the entire world.

For NASA missions…

NASA has space operations that run the Hubble, the satellites, exploration systems with humans, the shuttle, space station, the Constellation program, then aeronautics and the science mission directorate, and within sciences, it is Heliophysics, Earth sciences, Planetary sciences. (...) So Goddard has all of those areas, but we do not have very much aeronautics. Further, you may have learned that education is a little strange at NASA, because we have an education office, but we do not control all education initiatives because each mission – such as the Hubble Space Telescope, and the Solar Dynamics Observatory – has hundreds of millions of dollars usually, and a section of it is for education and outreach, and the mission hires someone to do that. A very good articulated effort is in Heliophysics, with missions like SOHO, STEREO, SDO, TRACE. All of them have education, and the Sun-Earth Connection tries to connect all of those and help them keep up with what is going on. But, the strange thing to me is that we have an education office, though the education office is not consulted on whom they hire and what they are doing for EPO, it is kind of disconnected. For example, in my program, we do not develop a lot of educational resources, and we bring other materials like the Sun-Earth Connection educational resources to teachers. The Space Weather Action Center team asked our group to teach their program in our [AESP] training, so they want to train us to teach that when we go out. We try to inform teachers on the areas of knowledge involved in a particular NASA activity, and we ask what they need. If in some case they mention they would need something that the SWAC would supply, then the SWAC team asks us to teach that. All of the missions know what our program is, and that we can help them by taking them away from the centers out to the states. Even that has not been very well organized. It is a big agency, and still a lot of communication to do.

How are the education efforts divided?

It is divided in higher education and pre-college. Now within pre-college we also divided (elementary, middle, secondary) because we cannot talk about all of those levels as if they were the same. However, there is not one group just working with high school, we all work with all of them. I think it is probably because when NASA was formed, educating people was an important part of it, and they organized these categories. The way that NASA works is: we want to go to Saturn, so we need to go to Saturn, and we need to look at these things like the atmosphere, moons or whatever. Thus, we have people sending proposals on how they think they can do that, and within that proposal it is included the education for it, from those they choose the best and go more in-depth. All of that happens before it becomes an official program. Thus basically, they develop their education plan before talking to us. The Sun-Earth Connection group has been very good at making understand the services they can provide so they have some of these groups come to us, and say *'we need an education plan for this mission.' 'Do you have ideas, can you point us someone who can help us?'* That is where we want to be; we want these people writing proposals including educations plans to know who we are, and consulting us. And the Sun-Earth Connection is one of the groups that are getting that done. As for working as an education specialist at NASA, we have access to first-hand data from missions, it is a great place to work. You see things that no one else has seen yet!

3.2.2. Resources for the Education Community

NASA offers a wide range of high-quality didactic resources for educators and students, both online by means of its Education home page, and printed materials distributed by its ten field centers for their respective geographical area of responsibility. In the sequence, you will find a summary of those that I consider the most helpful. I provide the following descriptions mostly based on information from the NASA Education home page [108].

a) Educator Resource Center Network (ERCN): each NASA field center has a space called Visitor Center, open to the educational community and the general public. The mission of the Educator Resource Center (ERC) is to identify teachers' needs, distribute didactic materials, help educators use NASA's resources, offer in-service and pre-service teacher training, and promote awareness about NASA. The combination of all ERCs forms the ERC Network, comprising all NASA centers (each center is responsible for specific states). In order to benefit more educators, NASA has partnered with universities, museums, and other institutions that serve as regional ERCs in many states [109]. Prior to visiting, it is recommended to access a specific NASA center's ERC Web site for details regarding resources, directions, hours of operation, and so forth. A list of contacts is available in Appendix 2.

b) NASA's Education home page (education.nasa.gov): main gateway for NASA Education contents, didactic resources, information, programs, services, events, opportunities, and points of presence within states. It serves students, educators, families, and other members of the educational community and the general public. An advantage of making available online resources and opportunities is that not all teachers and students can travel to visit a NASA center. How would them experience NASA otherwise? Further, it is more cost-effective to provide on the cyberspace inspiration and engagement activities that do not need to be carried out on a NASA center. Finally, the NASA Education home page outreaches also internationally. From my experience, I know that education professionals from many countries use NASA resources to enrich their classes. The NASA Education home page receives between 3.5 and 4.5 million visits per month [110], and some of its resources are presented in the sequence.

- *Teaching materials* (http://search.nasa.gov/search/edFilterSearch.jsp?empty=true): curriculum support materials, such as publications and indexed Web sites. Educators can search by grade level, subject (careers, Earth sciences, life science, mathematics, physical science, space science, and technology), keywords, language (English and Spanish), and type of material (educator guides, classroom activities, lithographs, brochures, and so forth). It is one of the best sources for finding NASA's teaching resources.

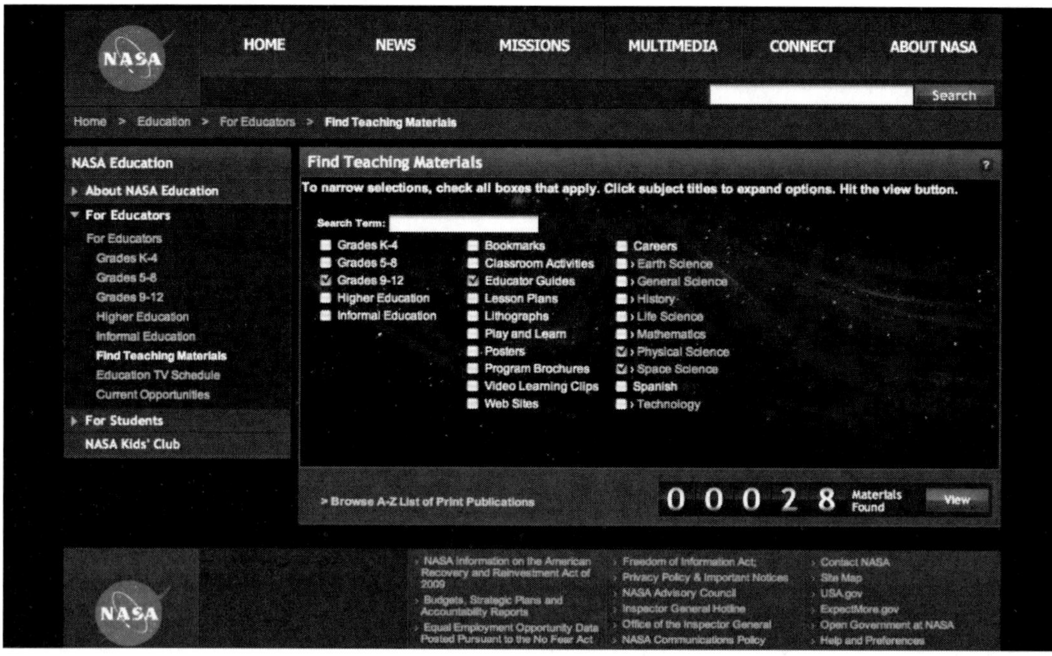

Figure 18. Finding teaching materials. There is an area on the right column on the NASA Education Web site, called "For Educators," where teachers will find a subcategory called "Find Teaching Materials," from which they could choose parameters to find the right didactic resources to enrich classroom activities. There are resources available from early education to higher education, in several areas such as technology and history, and in two languages: English and Spanish, though the quantity of resources in Spanish is not so large. Source: NASA. Credit: Norma T. O. Reis.

- *Express e-mail sign-up* (http://www.nasa.gov/audience/foreducators/Express_Landing.html): mailing list that enables educators to receive messages related to educational materials, activities, projects, events, and so forth.
- *Education news* (http://www.nasa.gov/audience/forstudents/k-4/index.html): news releases highlighting missions, discoveries, and activities related to education. This area provides information about the agency's work with schools, sponsored competitions, partnerships, opportunities, and other information of interest to the educational community. Visit that on the right column of the page.
- *Education calendar* (http://www.nasa.gov/audience/foreducators/Calendar.html): teachers, professors, and students find out about webcasts, workshops, contests, events, and other opportunities. There is also a link where educators can find the same information listed in order of expiration at http://www.nasa.gov/audience/foreducators/current-opps-index.html

- *Education programs* (http://www.nasa.gov/offices/education/programs/index.html): opportunities for students and educators of all scholarly levels. There is the Alphabetical List of All NASA Education Projects, an A-Z list containing all NASA programs. (http://www.nasa.gov/offices/education/programs/descriptions/All_Alpha.html) where students and educators identify interesting opportunities such as workshops, internships and fellowships, design challenges, Space Grant information, professional development, research announcements, or even apply for a job at NASA. Considering that a teacher, school or student would like to engage in some NASA specific activity, on this Web site one can find all information to get started. For example, if a teacher wants to involve her/his students in some NASA robotic initiative, she/he can visit that Web site. From the list, he/she could choose a robotic project (such as the FIRST Robotics Competition),[18] and contact the program manager at NASA,[19] to obtain more information on participation requirements. Some programs' Web sites are also available in Spanish.
- *One Stop Shopping Initiative* (http://intern.nasa.gov/): recent NASA Education solution that enable students from the U.S., and also from other countries to find in a single place a portfolio of rich professional/academic opportunities i.e., internships, scholarships, and fellowships, to which they may apply. Prior to this initiative, it was more difficult for students to identify opportunities, because they were spread among several places through countless NASA links.

Comments and questions http://www.nasa.gov/offices/education/contacts/comments_suggestions.html: students and teachers can send questions about NASA Education programs and the NASA Education Web site.

c) Central Operation of Resources for Educators – CORE (http://education.nasa.gov/edprograms/core/home/index.html): distributes NASA-produced supplemental curriculum materials in multimedia format (CD-ROM, DVD programs etc.) for a minimal fee to national and international educators via mail order, thus serving educators that cannot visit a NASA ERC. *CORE* is a nonprofit organization sponsored jointly by NASA and the Lorain County Joint Vocational School in Oberlin-OH.

By exploring the NASA Education Web site, educators and students will also find links to education offices in all NASA field centers (http://www.nasa.gov/offices/education/contacts/cdirect.html), for precollege education (Contact Information - http://www.nasa.gov/offices/education/contacts/precoll.html), information about the NASA mission directorates (http://www.nasa.gov/offices/education/missions/index.html), data regarding different NASA Education areas such as higher education and informal education, announcements, NASA Education official publications, informal education activities (http://www.nasa.gov/audience/foreducators/informal/index.html), the NASA Kids' Club (http://www.nasa.gov/audience/forkids/kidsclub/flash/index.html), and even to learn how to submit a proposal to NASA

[18] Extracted from the NASA's Web site, http://www.nasa.gov/audience/forstudents/9-12/features/first-robotics-2008.html.

[19] Extracted from the NASA's FIRST Robotics Web site, http://www.nasa.gov/offices/education/programs/descriptions/FIRST_Robotics.html.

(Business Opportunities - http://www.nasa.gov/about/business/ index.html). Students will find plenty of interesting resources such as videos, games, dictionaries, search engines, informative texts, and topics to help in homework. These resources can be found by grade level, i.e. K-4 (http://www.nasa.gov/audience/forstudents/k-4/index.html). Further, students will find in this link "Current Opportunities": http://www.nasa.gov/audience/forstudents /current-opps-index.html opportunities especially for them such as chats, lectures, webcasts, challenges, and competitions, divided by grade level. There is another interesting feature on the NASA Web site of interest not only to educators, but also to the general public, which is the so-called 'My NASA,' where one can customize to visit items of personal interest, every time one visits the NASA Web site (http://mynasa.nasa.gov/); in order to use that, one just needs to register. And whether a teacher or student only wants to illustrate the lesson or homework with awe-inspiring NASA images, videos or audios, one can visit NASA Images, http://www.nasaimages.org/. Finally, there is an interesting space for research in a wide range of space exploration topics on technical NASA reports at the NASA Technical Report Server Web site, http://ntrs.nasa.gov/?method=aboutntrs.

In the sequence, I will present some examples of how NASA has been exploring the potential of technologies, media, and digital content to innovate in education, and to foster broader interaction, engagement, and collaboration between NASA, students, teachers, and the general public. One interesting feature of these initiatives is their increasing level of involvement of the "ordinary citizen" with NASA, not only to "be aware of," but also to truly collaborate in the realm of science and exploration missions in tasks such as analyzing data, sharing results, and presenting solutions in the format of proposals to solve governmental challenges. If I were to summarize this NASA's attitude in one word, I would choose "democratic." Instead of adopting the standard behavior of some science institutions that work mostly with "closed-doors," and promote timid outreach actions, regarding outreach and education as "second-class activities," NASA, on the other side of the spectrum, truly considers and executes education and outreach efforts as one of its strategic goals, and with humbleness (something essential to an individual or organization learn more and develop) adopts a politically adequate and socially committed attitude of respecting, involving, helping, listening, and bringing the public "inside NASA." A tangible result of this attitude has been the bright in the eyes of many low-income children, further translated into space careers, new scientific discoveries and education projects by many grown-up "NASA kids!"

3.2.3 Education Technology and Innovative Programs

In several parts of the world, it has been noticed an increasing concern about the utilization of Information and Communication Technologies (ICT), media and digital content in schools and universities. Governments, private initiative, and citizens from nations with different levels of socioeconomic development have been working towards promoting the utilization of ICT and other technological resources integrated to curriculum, in order to promote education betterment and innovation. The motivation for these efforts relies mainly in the fact that we live in the so-called knowledge society, in which simple workers are no longer required.

People should develop sophisticated thinking skills, problem-solving capabilities, and other competencies directly or indirectly associated to new technologies.

NASA Education efforts are aligned with the U.S. education demands, thus the agency has been investing strongly in media production, online interactive and collaborative tools, and programs that use ICT and educational technologies in creative, innovative manners. The NASA mission directorates, centers and missions produce contents in multimedia format for a wide range of audiences, and aimed at enhancing formal and informal education [107] Technologies provide an important delivery mechanism for NASA instructional and outreach products [111]. By visiting the NASA Education home page, educators will find resources such as videos, games, simulations, interactive features, chats, webcasts, immersive environments, learner networking, and so forth. This type of product represents an effective channel through which students and teachers can learn on demand, focusing on topics they need or have more interest. It fosters interaction and makes it possible that teachers and students participate in enriching educational activities inside their classrooms or houses.

I consider these opportunities especially important for low-income students, or individuals with disabilities, who otherwise perhaps could not access NASA opportunities on spot. For example, even if a student cannot travel to a NASA center to participate of an event, he or she could access and send questions in real times during webcasts or chats with NASA scientists. "Education technologies (i.e., online games and professional development) enable scaled up delivery of NASA's materials to educator audiences with differing backgrounds, abilities, and interests [112]." As for professional development, educators can also find opportunities to enhance teaching skills by attending high-quality courses and events online.

There is a portfolio of e-Education opportunities and products developed by NASA in the scope of its several national and regional education programs and services, and in the sequence you could learn more about some of them, such as the Digital Learning Network (DLN), learning technologies projects, podcasting, and so forth. The purpose of developing these technologies is to combine "*advanced educational technologies with NASA content*, producing learning tools that engage and inspire today's tech-savvy students," producing learning technologies for the classroom that facilitate learning, is engaging as well as educational [113]. Some examples are listed in the sequence. I have provided the following short descriptions based on information available at each respective NASA Web site visited. All of these Web sites are informed in the sequence.

a) NASA Multimedia

Multimedia (www.nasa.gov/multimedia): the main space for NASA's multimedia resources. It features ISS mission coverage, live special events, interactive educational live shows, electronic field trips, aviation and space news, and historical NASA footage. Links to a variety of NASA resources can be found, such as the NASA image of the day, videos, podcasts, and interactive features. The World Book @ NASA link leads to articles about space phenomena, astronauts, and more. NASA television live events, news, and special programs are listed, along with the regular daily schedule (public and media channels), the NASA TV education channel, and NASA TV video file information.

NASA Apps (http://www.nasa.gov/centers/ames/iphone/index.html): there are plenty of applications that NASA has been making available for users of tablets and smartphones. You can select the apps according to the operational system of your mobile device.

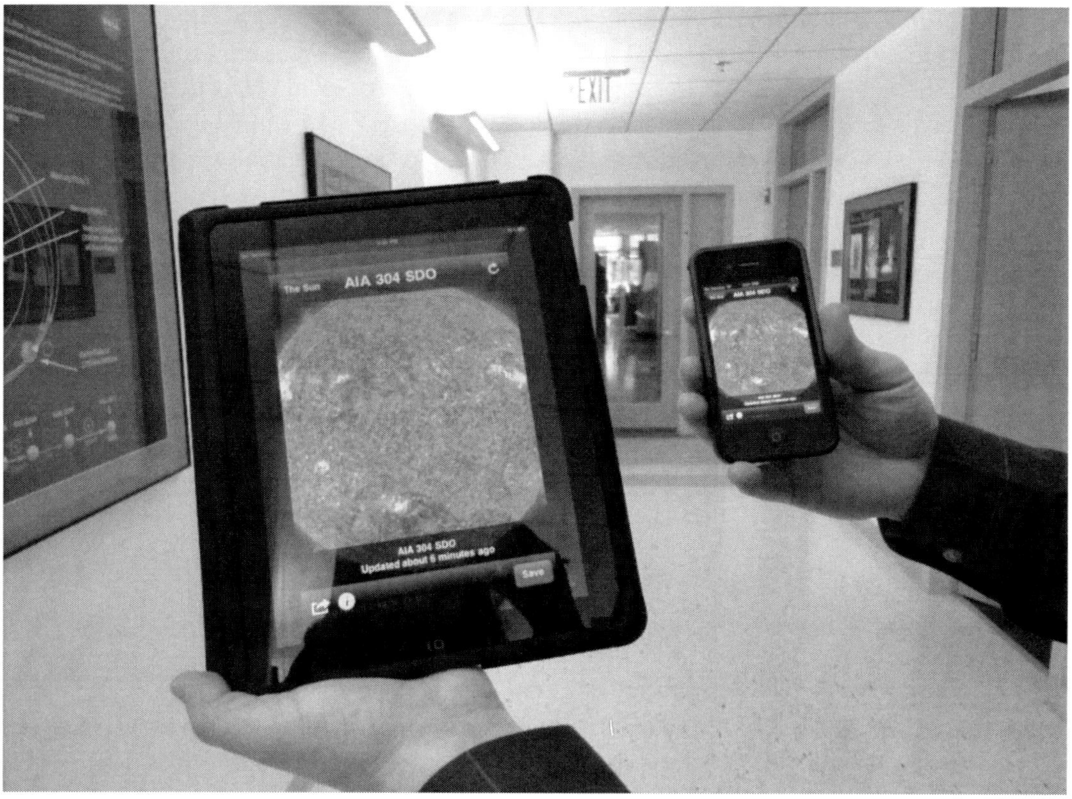

Figure 19. Study the Sun on your tablet or smartphone. There is application of the Space Weather Media Viewer that allows users to explore Sun's and learn more about space weather on demand. Credit: Norma T. O. Reis.

Podcasting (http://www.nasa.gov/multimedia/podcasting/): a very interesting learning tool in video or audio formats, which can be accessed both over the internet from the application I-Tunes (or similar), and transferred to portable devices, such as MP3 players, cell phones or tablets. The user just needs to subscribe to a podcast series so that new episodes are automatically delivered to his/her computer for playback at any time [107]. By visiting the 'podcast' area on the NASA Web site, users could select podcast series such as Solar System, Earth, and Station. Periodically, teams responsible for those series at NASA produce new podcast editions, which are delivered directly to the subscribed user's computer. There is another interesting series called 'This Week at NASA' where one can find out, weekly, important recent things happening at the different NASA field centers. More information can be fount at: http://www.nasa.gov/nso. In addition to this main Web site, there are also podcasts delivered through specific NASA missions' educational home pages such as the NASA's Sun-Earth Connection http://sunearthday.nasa.gov/2008/multimedia/podcasts.php.

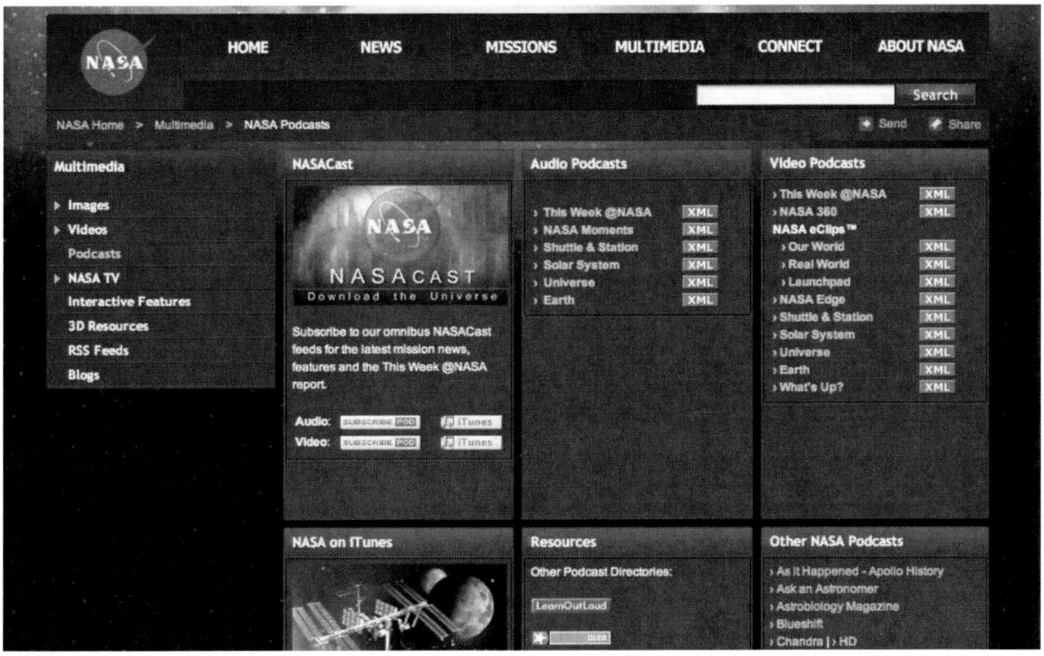

Figure 20. Podcast for learning on demand. NASA produces podcasts on several topics of interest. Everyone can subscribe to receive new editions periodically and automatically on the computer, and listen to NASA content everywhere, at any time with a mobile device. These audio or video podcasts utilize several approaches such as storytelling and interviews, for presenting educational content, news, and recent mission discoveries. For examples, visit also the NASA's Sun-Earth Day podcasts, http://venustransit.nasa.gov/2012/multimedia/podcasts.php, which invites users to send suggestions of topics they would like to hear in future editions. Source: NASA. Credit: Norma T. O. Reis.

*NASA TV (*www.nasa.gov/ntv*)*: it open doors to people all around the planet to watch NASA missions, news conferences, and activities in real time, especially those related to astronauts working at the International Space Station. Its main purpose is to assure enhanced communications between ground people (engineers, scientists) and astronauts in space, in order to guarantee crew safety and well being, and safe vehicle operation. This TV serves the public and the media, sharing content for reports and documentaries. NASA TV content is distributed by cable, by satellite TV providers, and online. Recently, it was launched the so-called NASA HDTV (http://www.ustream.tv/nasahdtv), with superior image quality.

b) NASA TV Education Channel
(http://www.nasa.gov/audience/foreducators/topnav/schedule/index.html)*:* the agency counts with an area for educational TV programs, comprising interesting programs such as:

- NASA eCLIPS (http://www.nasa.gov/audience/foreducators/nasaeclips/index.html)– short educational videos produced weekly to inspire, engage the general public and precollege students, exhibiting real world connections by exploring current applications of STEM topics. Educational material for eCLIPS programs is selected based on national curriculum standards;

- NASA Connect (http://connect.larc.nasa.gov) – a series of instructional programs aimed at improving STEM teaching in grades 5-8. The program connects STEM concepts taught in classroom to NASA research. Programs support national educational standards; includes teacher guide, and use a classroom experiment and web-based activity to complement concepts explored. They are free to educators, and available in public domain. You need to visit the Web site and register.
- NASA SCIence Files (http://nasa.ibiblio.org/program.php?program=NASA-SCI-Files), standards-based, technology focused, video and web-based, distance learning effort aimed at providing a series of programs to introduce students in grades 3-5 to NASA; integrate STEM through the use of problem-based learning, scientific inquiry, and the scientific method; and to motivate students to become critical thinkers. Educators can stream or download full programs. The NASA SCIence Files includes a teacher guide and several other instructional resources.

c) NASA Social Networking

Twitter: the first NASA Web sites appeared in the early 1990's and through the time the agency's primary Web site www.nasa.gov has become much more interactive. One example is Twitter. More than 320 thousand people follow NASA, and many agency's astronauts and other employees tweet on their personal accounts. It provides the public a personal insight of the agency's activities [114]. For example, scientific discoveries such as water ice on Mars are shared with the public very fast, in an informal, though professional way. There are also events called 'Tweet-ups' [114] (also called NASA Social) in which the public (U.S. and non-U.S. citizens) has encounters with NASA experts and facilities. An interested person can register, and there is random selection to events such as spacecraft launches. Registration opens at http://www.nasa.gov/connect/social/index.html. NASA Social is aimed at social media users who follow @NASA on several social media platforms (connection possibilities at the 'Connect and Collaborate with NASA' page). The goal of NASA Social is to allow individuals who regularly interact with each other via the mentioned online platforms to meet in person and discuss about NASA. When selected to participate of an event, participants are supposed to cover their travel expenses to the specified NASA center. At the event, participants tweet about their impressions of the event, sharing pictures, videos and so forth. More information http://www.nasa.gov/connect/social/index.html.

Facebook: NASA has several communities on Facebook, to which users can subscribe and receive updated information about subjects and/or missions of interest.

YouTube: NASA videos have become one the most popular contents on YouTube. NASA has a channel http://www.youtube.com/user/nasatelevision?ob=4&feature=results_main, with more than 31 million views (05/12/2012), and there are also channels for each NASA field center, such as NASA Goddard channel http://www.youtube.com/user/nasaexplorer?feature=results_main, with more than 19 million views (05/12/2012). Several NASA missions have their own channels on YouTube, such as the popular Little SDO, available at http://www.youtube.com/user/littlesdohmi?feature=results_main, sharing astonishing videos of the Sun in several wavelengths (Little SDO is also on Twitter http://twitter.com/NASA_SDO) and Facebook (http://www.facebook.com/NASA.Little.SDO); and there are channels for education programs, such as the 'Be a Martian,' at http://www.youtube.com/user/beamartian?feature=results_main.

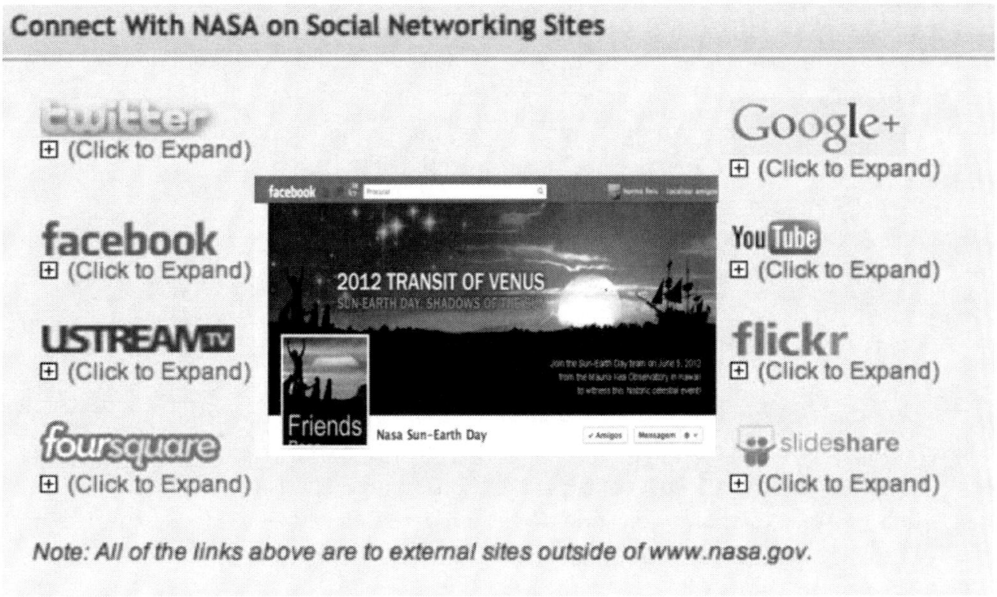

Figure 21. NASA on social media Web sites. The agency is present in an increasing number of social networking Web sites, which strengthens its institutional commitment of openness to the public, and provides a personal insight of NASA's missions. At the center, an example of NASA's community available on Facebook. People with the same interest can subscribe and receive updated information about missions and discoveries. Source: NASA. Credit: Norma T. O. Reis.

Blogging (http://blogs.nasa.gov/cm/newui/blog/blogs.jsp): through the multimedia area, users may select on the right column 'Blogs,' which will lead to several spaces where one can read recent postings by NASA scientists with updates about NASA missions. Users can post comments and rate news articles and other contents [114]. There are moderators that analyze all postings in order to filter improper content. There are also blogs for educators. Even the NASA Administrator has a blog there!

Chats (http://www.nasa.gov/connect/chat/index.html): students, educators, and the general public learn more about NASA's missions, current discoveries, astronomical events such as meteor showers, and other interesting topics by sending questions in real time to NASA experts. Each chat has a broad topic such as "earthquakes," and represents opportunities of having several participants (50-100 in average) interacting with NASA professionals, thus allowing more direct knowledge to be shared. All that a person needs to participate is an internet connection and creative questions. Each chat has a Web site containing materials regarding the topic to be discussed. People willing to participate should prepare themselves by reading about the subject with anticipation. There is a moderator who takes questions from the audience, and 'links' the interviewed specialist to the public. From about 300 questions submitted per event, an average of 40-60 questions get answered, because of time constraints.... A chat event usually lasts 60 minutes. Chats are made available on their respective Web sites within two business days after the event; video chats will be in video format, and text chats will generate PDF transcripts. These materials provide long-term value for the chat, as they are available for people who could not participate [115]. For a list of past, current, and future chat opportunities, visit http://www.nasa.gov/connect/chat/index.html. An example of video chat http://www.nasa.gov/offices/education/programs/national/nes2/home/live-chat-Ortega.html, and text chat http://www.nasa.gov/connect/chat/geminids2011.html. During video chats the e-mail for which questions may be sent is usually shown at the beginning of the transmission. All questions sent go into a queue. Individuals from all around the world can participate, and it is not necessary to subscribe.

*NASA Student Ambassadors Virtual Community (*http://intern.nasa.gov/intern/*):* links current and former higher education NASA interns and fellows.

d) NASA Learning Environments and 3-D resources

Interactive features (http://www.nasa.gov/multimedia/mmgallery/): environments that provide information in format of texts, links, videos, images in which users can click and read related content, 360 degrees view of spacecraft, animations, video tutorials, and other interactive features, such as maps where you click on icons to learn more about specific topics (http://www.nasa.gov/multimedia/mmgallery/index.html). Examples:

The Future of American Human Spaceflight: have fun while you learn about NASA plans for future human exploration, http://www.nasa.gov/externalflash/human_space/

Eyes on the Solar System: 3-D environment with plenty of real NASA mission data that you can explore from your computer; you could see the solar system moving in real time, and control space and time parameters, http://solarsystem.nasa.gov/eyes/

Desert RATS 2011: 3-D environment combining NASA models with topography and satellite imagery, http://www.nasa.gov/externalflash/dratsvirtual/.

Goddard Virtual Tour: learn the entire cycle of a space mission, from concept to spacecraft operations and data analyses, http://www.nasa.gov/externalflash/goddardVT/.

Experience 3D Virtuality With Robonaut: use augmented reality and learn about the first NASA humanoid robot, http://www.nasa.gov/externalflash/myexploration/index2.html. This feature is very cool: you will be asked to print a NASA logo, enable your webcam, and when you direct the printed logo towards your webcam, you will have a surprise… Try that!

Send Holiday Greetings to the Space Station: send a little of warmth and love to the crew at the ISS! You can send your own message to current ISS astronauts, and sign your name, http://www.nasa.gov/externalflash/postcard/. You can also follow NASA astronauts on Twitter: https://twitter.com/#!/NASA_Astronauts. They share their quotidian activities, thoughts, and poems… Visit also the NASA Astronauts' home page: http://www.nasa.gov/astronauts/.

3-D Resources (http://www.nasa.gov/multimedia/3d_resources/index.html): you can see several satellites in 3-D (http://www.nasa.gov/multimedia/3d_resources/index.html). There are also games available. Check this incredible 3-D game called 'NetworKing' (http://www.nasa.gov/multimedia/3d_resources/scan.html) with real technological challenges.

You will need to learn a lot in order to build and upgrade a complex communications network in order to support scientific missions. As the game is complex, it is interesting to read the manual http://www.nasa.gov/multimedia/3d_resources/scan_manual.html before playing... Otherwise you will feel frustrated! Another interesting 3-D game is the 'Station Spacewalk Game.' You will feel like a real astronaut as you play! You could execute simple tasks such as exploring the ISS, and more complex ones such as repairing solar arrays. There is also the game 'Eyes on the Earth,' in which you can explore the planet with data from several satellites, studying aspects such as global temperature, carbon dioxide, ozone, and sea level (http://climate.nasa.gov/Eyes/). You can play these games from your browser, or download that to your computer (available for Windows and Mac).

Virtual Learning Environments: the agency's experience of using simulations to train astronauts facilitates the development of technology-rich learning products such as simulations, immersive environments, game playing, and so forth (http://www.nasa.gov /offices/education/programs/national/ltp/home/index.html). It helps motivate and help students learn in innovative manners. NASA pursues environments ready to use, and others under development. Many of these virtual worlds are accessible to the blind and visually impaired students, such as the MathTrax, a graphing tool for middle school and high school students to graph equations, physics simulations or plot data files. The graphs have descriptions and sound so that one can hear and read about the graph. Users can access visual math data and graph or experiment with equations and datasets (http:// prime.jsc.nasa.gov /mathtrax/index.html). *NASA eEducation Island in Second Life*: a space to facilitate NASA Education projects' entry into virtual worlds (http://slurl.com /secondlife /NASA% 20eEducation/125/156/43). There is also a virtual microscope with 90 samples available (http://virtual.itg.uiuc.edu/). Another example is the Virtual Field Trip, an immersive application for exploration of areas on Earth that have been identified as analog sites to regions on Mars. The environment consists of various linked 360-degree spherical stitched virtual reality environments, 3-D based layered environments, global fly downs, along with video files within each area of research (http://quest.nasa.gov/vft/).

e) Access to Mission Data

Science mission data: provision of universal access to archived science data products (i.e., information, images and videos) from NASA science missions. The agency archives all science mission data products, organized by discipline or theme. The purpose is to assure long-term usability of this data by scientists, educators, and the general public, thus contributing to enhanced scientific discovery and public engagement [115]. For instance, the Heliophysics Data Environment[20] provides access to data and physics-based models that facilitate a systems' level of understanding of the Sun and its impact in our solar system. NASA has also education and outreach efforts using science data archives to engage students and the general public to use the data in citizen science endeavors. For example, the My NASA Data Web site (https://mynasadata.larc.nasa.gov/) mentors teachers and amateurs using atmospheric and Earth Science data [114]. Students learn how to select and analyze

[20] Extracted from the NASA's Science Heliophysics Web site, http://science.nasa.gov/heliophysics/heliophysics-data-centers/heliophysics-data-environment-hpde/.

data on areas such as climatology and atmospheric radiation, compare data through the time, and compare different parts of the world. There are instructions on how to find and analyze data (such as PowerPoint material and YouTube videos), lesson plans, and more. NASA also provides science data sets and fosters public participation with software applications on the Web, such as desktop geographical browsers i.e. Google Earth (including three-dimensional Mars, Moon and sky); the World Wide Telescope (WWT), http://www.worldwidetelescope.org/ [114] which works as a virtual telescope, allowing one to use professional astronomical data to explore and understand the universe, and Web applications such as the Galaxy Zoo, http://www.galaxyzoo.org/ [114], where users can play active roles, such as classifying galaxies according to their shapes (tutorials available). These platforms serve millions of users worldwide, and make it possible a broad participation of the general public in science as not viable previously [114], and the most important aspect is that the public can produce and share their own content. Currently, NASA science data can be also accessed on smart phones and tablets, through diverse applications.

*Scientific Visualization Studio – SVS (*http://svs.gsfc.nasa.gov/index.html*):* project that facilitates scientific inquiry and outreach within NASA programs through visualization, animation, and imagery that explain current NASA missions and scientific results. A team of scientists, writers, and animators combine efforts to generate visualization aimed at promoting a better understanding of Earth and space science research. These products serve TV, print media, educators, museums, and the general public. The project's Web site contains more than 4,200 visualizations available in different formats and definitions. Users may choose to visit their gallery http://svs.gsfc.nasa.gov/Gallery/index.html containing a wealth of resources such as visualizations, conference videos, and educational videos by theme. Further, there is a link to the Conceptual Image Lab http://svs.gsfc.nasa.gov/index_cilab.html; podcasts http://svs.gsfc.nasa.gov/vis/iTunes/index.html; and application for iPad http://svs.gsfc.nasa.gov/nasaviz/index.html. There are links close to visualizations allowing easy content sharing via Twitter, Facebook or e-mail. In the classroom, educators may use visualizations about space weather or Earth environment, to enrich classroom and provoke students' questions, for instance. Students may also explore autonomously these resources in order to learn more about specific curricular topics and/or enrich their school assignments.

f) NASA Distance Learning and Teacher Development

*Digital Learning Network – DLN (*http://dln.nasa.gov/dln/content/centers/*):* the DLN offers videoconferencing, webinars and webcasting at no charge for the United States, and sometimes internationally. For instance, in 2009, a Brazilian school had the opportunity to participate in a NASA's special DLN event on occasion of the Apollo 11 40 Years. Hundreds of students could watch in real-time an interaction between NASA experts and other four U.S. school students participating on the screen.[21] The NASA personnel can see the students, and vice-versa. Some students were able to ask questions to the NASA specialists. This network enhances NASA's capability to deliver content to allow students and educators at all levels, to interact in real time with NASA experts from their workplaces at the several NASA field

[21] More information on the Correio Braziliense Web site, "Cientistas da NASA falam com estudantes sobre viagem à Lua," http://www.correioweb.com.br/euestudante/noticias.php?id=7047

centers, and acquire a new appreciation of the importance of STEM education. Two-way audio/video videoconferencing systems either H.320 or H.323 standards based are compatible with NASA DLN, and five schools at different spots are connected at the same time in one DLN event, which lasts around 50 minutes. Students get involved more intensely with things they experience, construct, and understand – in other words, with meaningful learning. Videoconference strongly contributes to this students' emotional engagement with contents. The use of videoconference as didactic resource in precollege education also helps teachers to use concrete situations explored during a DLN event to present, contextualize, identify new connections and re-signify curricular contents. As for webcast, one computer with internet connectivity and a projection device or a computer lab and internet connectivity are necessary; student interaction is possible via chat forum or e-mail [107]. More information about DLN, webcasts, and webinars (virtual classes) can be found on the DLN Web site.

*Endeavor Science Teacher Certificate Project (*http://www.us-satellite.net/endeavor/index.cfm*):* collaborative learning environment with online courses and research. Courses comprise subjects such as life and marine science, astronomy, space science, physics, astrobiology, engineering, and so forth.

Distance education: the NASA Electronic Professional Development Network – ePDN (http://www.nasaepdn.gatech.edu/nasa_about_epdn.php): program that develops electronic professional development courses, bringing STEM content to teachers through modular-based ePD courses, and through shorter self-paced courses and events. Courses emphasize best practices, inquiry learning, case-based scenarios, and data analysis. They are also designed to promote teachers' use of new communication tools such as video sharing, podcasting, visualizations, social networking, and other online platforms. The ePDN is one of three activities that make up NASA's Learning Environments & Research Network (LE&RN), a project within the K-12 program of NASA's Office of Education.

g) NASA Innovative Initiatives

*NASA KLASS (*http://www.nasa-klass.com/*)*: students simulate the situation of being responsible for a real space shuttle launch, as if they were sitting at launch control room at Kennedy Space Center. They use real-life launches scenario, and needs to integrate knowledge learned in class to inform if the spacecraft is "go" or "no go" for launch. On this Web site you will find files to get started with the experience. See a video about the program: http://mfile.akamai.com/18566/wmv/etouchsyst2.download.akamai.com/18355/wm.nasa-global/ksc/ksc_100308_klass_vid.asx

NASA INSPIRE project (http://theinspireproject.org/): provides creative hands-on opportunities for students in the U.S. and other countries to observe very low frequency (VLF) waves, such as lightning and other atmospheric sounds. Students use radio receiver kits to make field observations of VLF signals. There is an INSPIRE VLF-3 receiver designed to receive radio waves in the 0-10 kHz frequency range, which is audible to the human ear when sound waves are converted by either speakers or headphones. Students receive hands-on, real world experience and a higher degree of learning in the fields of physical sciences. It also seeks to enhance educators in their knowledge in the field of space science, providing insightful exploration of the world around us.

NASA Engineering Design Challenge (http://www.nasa.gov/audience/foreducators/plantgrowth/home/index.html): this program poses actual spaceflight related challenges to students; challenges that engineers face as they design habitats, vehicles and technology. In one program's activity, "Seeds in Space" students compare the development of plants from seeds that have been flown in shuttle and/or ISS missions, and seeds from control groups that have remained on Earth. As space faring nations plan to establish colonies on other celestial bodies, it is fundamental to learn about plants and animals' development, behavior and health in space.

h) NASA Collaboration

Collaboration (http://www.nasa.gov/connect/index.html): there are programs that allow students and the general public to collaborate with NASA in several levels. In some cases, students can suggest new photograph targets on Mars (http://www.uahirise.org/hiwish/), participate of an art contest (http://www6.cet.edu/copper/contest_index.php) with prizes in cash, propose cutting-edge solutions for NASA challenges such as the NASA Tournament Lab initiative (http://www.nasa.gov/directorates/heo/ntl/), and the general public learns how to possibly establish partnership with NASA (http://octpartneringtool.nasa.gov/oct/).

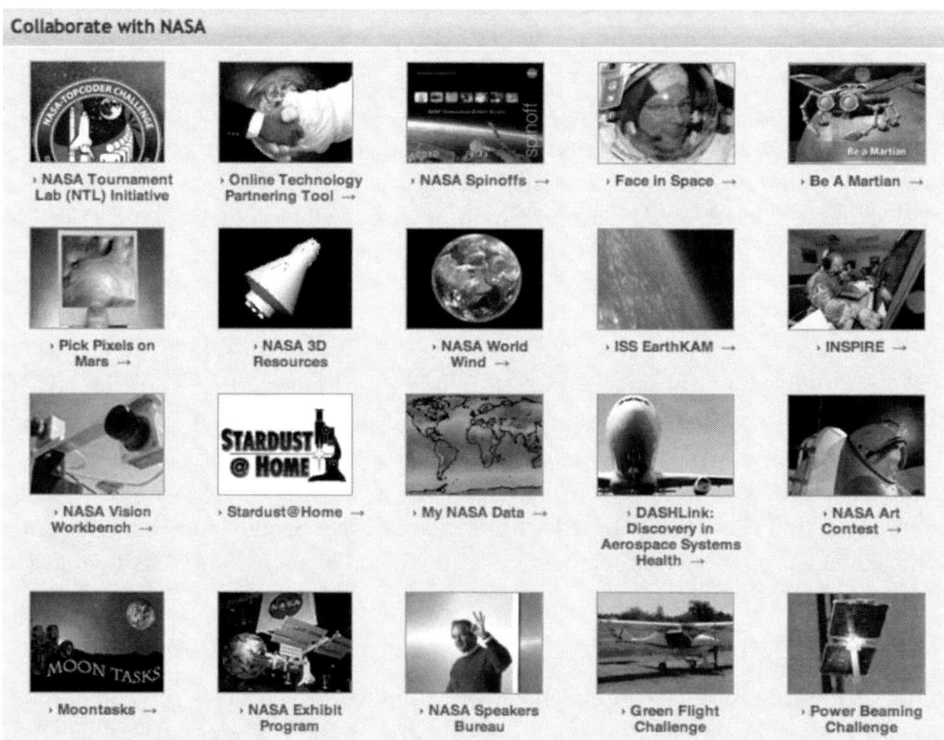

Figure 22. Collaborating with NASA online. There are several efforts that allow one not only to learn about NASA science, but also to develop that. Opportunities range from sending a picture of one's face to space, to programs in which students develop technologies required in NASA's missions. Source: NASA. Credit: Norma T. O. Reis.

Figure 23. Plants in space. There is an educational effort in which students sow seeds that have been to space, to make scientific comparisons. Source: NASA. Credit: Norma T. O. Reis.

The examples presented here open possibilities for enhanced STEM teaching with the use of innovative resources, and for students' autonomous exploration and significant engagement in topics of interest. Sometimes an educator knows very well how to teach, though does not know how to capture students' hearts. If you capture your students' heart, they will not only learn for the exams; they will learn for life. And this learning will transform themselves and the way they deal with the world. Using some of the tools presented here will help you engage students, and make they feel that their work is important. And this contact with NASA content, resources, and personnel will certainly inspire many of them to dream with STEM careers, and believe in a promising professional future for themselves. And if these children follow their own stars, they could make their dreams become reality, as shown in many NASA astronauts' biographies! The personal impression that I have from NASA Education personnel is that they put much *love* in their activities, and this is certainly one of the most important ingredients to mission success at these initiatives that have been involving and transforming the lives and perspectives of countless children and teenagers, regardless of their socioeconomic background, geographical location, physical disabilities, gender, ethnical background, or any other factor. In the sequence, I present selected state-of-the art NASA Education efforts that I analyzed during my ISU internship at NASA Goddard Space Flight Center.

Chapter 4

State-of-the-art Practices in Space Science Education at NASA – Summary and Excerpts of Interviews with Specialists

There are uncountable and different state-of-the-art initiatives in space science education ongoing at NASA, and for reasons of scope it has been necessary to be selective with regard to the ones to present here. The majority of these efforts are carried out at NASA Goddard. Some full interviews and comments of NASA Education specialists are here also featured. The purpose of this section is to provide readers with in-depth analyses based not only on information, but also on personal impressions from those who are ahead these successful enterprises. Many of these people are more than professionals; they are 'grown-up space kids' who once have had their interest captured towards STEM careers, and today encourage many other people to also follow their inspiration. It is recommended that educators or researchers reading this chapter also access the Web sites of each one of these programs to obtain further

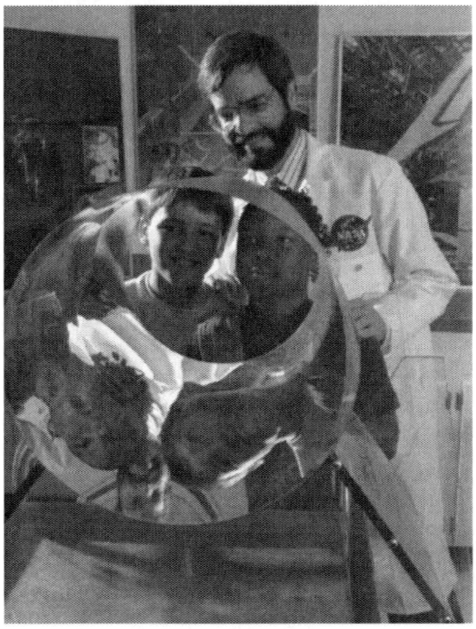

Figure 24. Students interacting with NASA professionals. The direct contact of students with cutting-edge researches and 'hot topics' of space research foster students' interest in STEM disciplines, and motivate students to become autonomous learners. Credit: NASA.

information. Further, it is recommendable that they contact NASA centers by e-mail or letter in order to receive materials related to specific topics of interest. For this purpose, a list of education contacts for NASA Centers is available in Appendixes.

4.1. NASA's Science Mission Directorate Education Projects

In this section, I explore mostly initiatives belonging to the NASA's Science Mission Directorate (SMD), which conducts research on Earth, heliophysics, planets and astrophysics. NASA's Space Science Enterprise is responsible for carrying out the agency's goal of "*explore the Solar System and the Universe beyond, understand the origin and evolution of life, and search evidence of life elsewehere.*" [116] The NASA's Space Science Enterprise is implemented by means of four forums:

- Sun-Earth Connection;
- Solar System Exploration;
- Astronomical Search for Origins and Planetary Systems;
- Structure and Evolution of the Universe [116].

Figure 25. Panel for the Sciences and Exploration Directorate at NASA Goddard. At NASA, areas are divided in codes. For example, the code 600 was the one for that directorate, in which the heliphysics division is located. I was in front of the panel in this picture. Credit: Norma T. O. Reis.

It has the largest educational enterprise at NASA, and also the oldest, with activities since 1993. Indeed, all space science missions at NASA conduct education and outreach (E/PO) activities. As stated in the NASA Space Science Education and Public Outreach Report 2003, "*The NASA space science E/PO program emphasizes high-leverage approaches and is carried out through extensive partnerships undertaken in a decentralized way.*" There is also great commitment with diversity and inclusiveness: "*The (...) effort placed emphasis on increasing the participation of minorities in space science, with the goal of diversifying the space science workforce. It forged relationships with minority institutions and minority professional organizations, and targeted urban schoolteachers working with multi-ethnic populations, providing them with resources and PD opportunities.* [117]" One example is the Exceptional Needs Workshop (ENWS), a weeklong summer event that bring together individuals from the NASA space science community with formal and informal education and communities of persons with disabilities to share expertise and develop ways to make space science resources accessible to persons with special needs. Participants learn about persons with special needs through hands-on experiences and about space science content [118].

I explore here mostly space science education projects of the Sun-Earth Day team at NASA Goddard Space Flight Center, my internship host. This team, in the scope of its educational component, aims at engaging citizens in how the Sun produces space weather effects, interacting with Earth's magnetosphere, and derived effects on our planet. This is accomplished by a synergy of programs and projects that motivate students at all scholarly levels to learn STEM and space sciences motivated by a multitude of Sun-Earth themes. NASA Goddard is the largest NASA center, counting with the largest number of unmanned scientific missions, many devoted to study the Sun and the Earth. From those targeted to study the Sun, we can mention the Solar TErrestrial RElations Observatory (STEREO), and the Solar Dynamics Observatory (SDO). All these missions are helping scientists better understand the inner structure of the Sun, its surface features, sunspots, the solar cycle, space weather effects such as solar flares and coronal mass ejections (CMEs), and so forth.

As the Sun is Earth's principal source of energy, it is easy to motivate students to study the Sun, such as during total or partial solar eclipses, with inexpensive equipment or methods such as projection, and also with digital contents that they can download for free from the NASA Web site as applications to their smartphones, tablets or computers. Students would be curious to learn about sunspots, why they vary, what they indicate, what impact these changes have in space weather and correlated effects down on Earth; they get also excited for learning about violent solar explosions such as solar flares and coronal mass ejections; why we can see in certain parts of the Earth beautiful plasma curtains called 'auroras' (mostly in poles), what exactly cause them, and so forth. Our planet is protected by a magnetic field called magnetosphere. But it is not an Earth's exclusivity to have a magnetosphere. Jupiter, for example, pursues the largest and strongest magnetosphere in the entire Solar System. These are some elements that the NASA's Science Mission Directorate's heliophysics area can bring inside classrooms through its portfolio of educational efforts.

4.2. Sun-Earth Day Team

The Sun-Earth Day (SED) team is responsible for a set of educational efforts intended to bring to students the excitement of the Sun-Earth connection by engaging them in themes such as space weather, topics such as coronal mass ejections (CMEs) and flares, auroras, solar plasmas, the changing Sun, solar wind, as well as their interaction with the Earth magnetosphere, and the impacts of this interaction – sometimes represented by solar storms – on human activities on Earth: disturbances in telecommunications and positioning satellites, and risks posed to extra-vehicular activities carried out by astronauts in outer space and so forth. Strategies carried out by the SED team include events, online resources, hands-on and interactive activities for students. Some main activities comprised by SED:

- The Sun-Earth Day Web site;
- Solar Week;
- Tracking a solar storm;
- Earth to sky;
- Space Weather Action Center – SWAC.

Let us discuss here the example of the Sun-Earth Day Web site, available at sunearthday.nasa.gov/. It is aimed at engaging the educational community and the general public in themes related to the Sun-Earth connection of interest to the population. Several actions are carried out, such as the so-called "Sun-Earth Day," an international celebration that features yearly a different event associated to the Sun. Events are held throughout the year, and culminates with a celebration on/near the Spring Equinox. Themes can be a total solar eclipse, the transit of Venus, auroras, and so forth. For example, the theme for 2008 was space weather around the world (http://sunearthday.gsfc.nasa.gov/2008/index.php). In 2012, the theme was transit of Venus (http://venustransit.nasa.gov/transitofvenus/). There is a map available on the projects' Web site (http://venustransit.nasa.gov/2012/about/eventlocations.php), where event organizers from all around the world can indicate where related activities will occur. It enables those interested in participating in a Sun-Earth Day event to locate their nearest events, according to address, and also to inform locations for their own events. Applications for smartphones and tablets are made available on the Sun Earth Day's Web site, along with podcasts, YouTube videos, and a special live webcast highlighting discussions about the theme and recent results from NASA heliophysics missions. The SED team partners with NASA EDGE to produce webcasts and videos that are shared with audiences in the U.S., and internationally [119]. The SED team also provides professional development to educators, amateur astronomers, museum personnel, Girl Scout leaders and so forth, in order to support them in the implementation of their own actions that use SED resources [119]. To strengthen interaction with the educational community and the general public, social media networking groups are formed on Facebook and on Twitter [82]. Everyone can register to receive information about SED efforts, at http://sunearthday.nasa.gov/registration/login.php.

In the scope of the SED celebration of 2008, I carried out a research to produce a paper about "Eclipses through the Centuries," both in English and Portuguese, along with two corresponding podcasts in the mentioned languages, and an educational sheet on the projection method for Sun's safe observation. A summary of the paper is featured on Appendix 4. The research was conducted as part of NASA's coverage of the total eclipse of the Sun observed in China, on August 1, 2008. The paper illustrates the researcher's immersion in a NASA Education ongoing program, and was also aimed at being converted on publication in scientific and educational journal of physics teaching both the United States and in Brazil,[1] as a contribution for precollege and college Physics and science educators. The paper presents several stories of solar and lunar eclipses, contains several illustrations and references. An adapted version of the mentioned paper is available in Appendix 4.

In order to provide a deeper understanding of the Sun-Earth Day team activities in education, I include an interview with my ISU internship mentor Dr. Timothy E. Eastman.

Interview with Sun-Earth Day team expert (Summer 2008), with permission

Interviewed: Dr. Timothy E. Eastman
Employer: Wyle at NASA Goddard Space Flight Center

What is your position at NASA Goddard?

I am a deputy group manager for science support. Our support functions, among others, include data systems, science support, computer support, and many additional functions, primarily in support of the Science Directorate.

In addition to that, do you have some work with education?

Most of what I do, day by day, is to support the science group or help people prepare proposals, and also support people in scientific research. However, we have closely coordinated science education activities, and other groups providing space science education needs. I do education as my own personal interest, not as a work obligation. Research and education are very complementary. Research into the nature of physical systems has important synergies with communicating that information to others, especially the next generation. Further, we need to keep adapting ourselves to new technologies and scientific results.

And one of the education initiatives would be the Sun-Earth Day?

Yes, close to the spring equinox for each year, we coordinate a special Sun-Earth Day (http://sunearthday.nasa.gov). The theme changes each year.

And where did the idea of creating this event come?

Sun-Earth Day is inspired by the highly successful Earth Day (see http://www.earthday.org). The idea is to leverage and augment Earth-Day activities with Sun-Earth Day associated concepts and problems. Earth Day focuses on Earth system sciences, and Sun-Earth Day focuses on the geospace environment. Here, "geospace" refers to systems coupled with a global environment, which is ultimately connected to the sun because the sun

[1] It is published in Portuguese the paper "Métodos de projeção para observação segura de eclipses solares" ("Projection methods for safe eclipse observation"), in the journal: Caderno Brasileiro de Ensino de Fisica. http://www.periodicos.ufsc.br/index.php/fisica/article/view/2175-7941.2012v29n1p81

provides all energy for Earth systems. In addition to the sun's radiative powers, which permeate the entire solar system, we have particles, fields, and plasmas that are constantly flowing away from the sun. This outflow interacts with all planets, moons, and other objects, creating space environments around everything within the solar system. With our Sun-Earth education programs, we are exploring not only the sun and how that affects our Earth system, but also exploring space environments more generally for all planetary systems, including ionospheres, magnetospheres, and the coupling of these regions with the interplanetary medium, especially the solar wind.

So you want children and teenagers to better understand these connections?

We all benefit as more people gain a deeper awareness of environments — environments that they have for themselves in their own backyard; local environments where a person lives (village, region, nation) with attention to issues of sustainability, clean water and air, and other needs that people often take for granted; environments for an entire city; environments for an entire region and continent; how they couple with rivers and oceans, and the whole atmosphere; how increased pollution in one country can affect another country. This theme is relevant to all scales, including how our Earth system is imbedded in (spatially larger) space environments, and then our entire galaxy within a group of galaxies. Everything in the universe is in some way linked in terms of being within its multiple environments. Thus, the theme of environments is relevant to all different scales, from your immediate backyard to the universe at large. To me this is an exceptionally important theme because we cannot make good decisions with the assumption that we can radically separate what we do in our backyard from what is happening in someone else's backyard. We are affecting one another. It is important to recognize this concept of interconnection between systems and between people: — how one person, or one city, or one country, affects others — the ultimate type of "globalization," frequently used only for the economic component, which is sometimes employed to basically deny such interdependence. The extent to which certain systems can be treated as independent is always an approximation, often a very good approximation, but never an absolute "truth." Much science is focused on how to best characterize and apply our approximations (in turn, imbedded within various physical relationships, or "laws").

In this way, through good education for Earth system science and space science, I think there is opportunity and resources to help people rethink fundamentals about the nature of the world and humankind in the environment, as well as the implications of that broader reality for how we should think about human transactions such as economics. Surely, just because we are humans, we do not have some God-given right to exploit the environment; instead, we have responsibility for that, and we are constantly making decisions on how to act on that responsibility; but we are often not doing that in a positive and productive way. However, good education in Earth system science and space science, combined with this fundamental theme of environments at all levels, can help us all to rethink the nature of the world that we are in, and how best to act on that new understanding. Of course, such analyses have not only science and education implications, but also important social and political implications, including considerations about reflecting critically about our context (within all these environments) and issues of values and meaning.

And what new understanding about space science will our educational community find at the NASA Sun-Earth Connection forum?

The Sun Earth education forum is very active in taking advantage of the many spacecraft and datasets for observations of nearby space environments, including the geospace environment from the sun to the Earth. In studies of space weather, for example, we focus on impacts near Earth of space plasma environments on astronauts and technological systems such as communications satellites. Plasmas are the interactive mix of ions, electrons, neutrons and magnetic fields, electric fields, and combinations of those components that exhibit collective effects and can carry electrical current - sometimes called the "fourth state of matter." The reason we need to make reference to plasmas in a distinctive way is that normal liquids and gases are both fluids as described mathematically to understand their dynamics (viz. the Navier-Stokes equations). But plasmas are described using different physical relationships, including those for electrodynamics (the Maxwell equations), and those for the dynamics of charged particle distributions and fields (the Boltzmann equations). This is different from the fluid formalism used for much more familiar collisional, neutral gases and liquids (the Navier-Stokes equations). Thus, electrodynamics and plasma systems really are distinct systems of matter (see http://www.plasmas.org/basics.htm).

Except at very high temperatures, space plasmas are generally not visible because hydrogen plasma (dominant in the interplanetary and interstellar medium) does not have resonances in visible wavelengths. To help us "visualize" plasmas, we combine direct satellite observations with various representations of such plasma data including, when possible, direct video information like those of the sun, solar flares and coronal mass ejections (CMEs). Recently, such video data has helped us to appreciate how dramatically variable the sun can be, especially when viewing higher energy radiation (UV, X-ray) in contrast to its very steady and mundane appearance in the visible (see daily updates on current views of the sun at the Web site for the Solar Dynamics Observatory (SDO) at http://sdo.gsfc.nasa.gov/data/.

Essentially all sources of the visible universe are made up of plasma. Stars, major portions of nebulae, interplanetary and interstellar regions, are also made of plasma. The fact that only very energetic plasmas, actively emitting photons, are easy to see leads us to emphasize such plasma domains; e.g., SDO videos of CMEs near the sun. However, as such enhancements of ever-present solar wind plasma propagate farther outward, the plasma cools and becomes no longer visible – then we have to piece together other types of data, sometimes direct sampling from a particular spacecraft, and then develop a model and visualization to show it as if we could "see" remnants of that CME as it propagates outward.

Many educators and scientists speak only of "gas" even when the referent is clearly plasma, presuming that "ionized gas" is only a type of gas (i.e., a Navier-Stokes "fluid"). This simplification is occasionally useful for pedagogical purposes, but is almost always misleading (or simply wrong) in scientific applications (e.g., the misleading use of "gas" for "plasmas" in most astrophysics publications).

Once again, **we are just beginning to see, with new "eyes", plasmas in all their glory, both visible and invisible.**

4.3. Space Weather Action Center (SWAC)

The Space Weather Action Center (SWAC) is an extraordinarily interactive, content-rich, low-cost, and innovative space science education program that encourages teamwork, debate, inquiry-based learning, and scientific investigation/method. It is a heliophysics education program aimed at engaging students into the understanding of the Earth-Sun system. Students explore concepts such as *heliophysics* - the study of the Sun's activity and its effects on the interplanetary environment, the planets, other solar system bodies, and the interstellar medium. Another concept is that of space weather, which results of the behavior of the Sun and the plasmas it emits via solar wind, the nature of Earth's magnetic field and space environment, and our location in the solar system. During periods of intense solar activity (solar storms) the Sun releases lots of plasma energy. Several human activities in space can be affected by space weather. For example, severe geomagnetic storms can interfere with communications and navigation systems, and reduce the useful lifetime of telecommunications and observation satellites. It can also disturb spacecraft orbits due to increased drag, and provoke power blackouts [30]. Just as we have meteorologists predicting weather patterns on Earth, it is critical to have specialists forecasting patterns of this interaction between plasma emissions from the Sun with our magnetosphere, as it impacts important technological systems.

The SWAC program is aimed primarily at middle and high school students, though its data resources have also been largely used at university and elementary level. As for scope, SWAC resources have been downloaded and used by a variety of countries i.e., Brazil, Germany, Canada, and India. For example, the Brazilian Education Ministry's Teacher Portal contains a Portuguese version of SWAC (http://webeduc.mec.gov.br/nasa/). The major objective of this program is to enable students to produce space weather forecasts by accessing nearly real-time data from NASA satellites. Students access current conditions of geospace, monitoring how a solar storm progresses since its eruption on the surface of the Sun until it sweeps past our planet, causing substantial changes on Earth's magnetic field [120]. Students produce space weather reports, and share it with peers around the world via the Web. They produce their own weather journal data by creating these reports, and broadcasts. This program complements and enriches theoretical discussions carried out by middle and high school educators in disciplines such as Sciences, Physics and Geography.

Basically with a computer and internet connection, students can build a Space Weather Action Center, a platform from which they periodically access the unique imagery and data from NASA satellites and observatories. Though it is necessary to have an internet connection to participate, students not having computer and internet connection can take advantage of the supporting activities: how to build and/or use a sunspotter, understanding magnetism, and so forth. Schools with limited internet connectivity can still participate. If technology is limited and dedicated access to a computer is only via a teacher's computer, educators can access and share specific sets of data with the class on a daily basis. Students can then simply copy the data to their notebooks or maintain a classroom data folder [30].

Figure 26. An example of SWAC display board. In this picture, there is a display board and an apparatus that represent a laptop in which students carry out systematic research on space weather NASA data. This image was obtained during a course that NASA Goddard offered to teachers. Credit: Norma T. O. Reis.

SWAC at a Glance

- *Teaches space weather*
- *Promotes meaningful usage of ICT in education*
- *NASA award-winning educational program*
- *Requires computer + broadband internet access*
- *Downloadable instructional material on NASA SWAC Web site*
- *Media Viewer – students access NASA satellite data in near real-time*
- *Media Viewer – now available for smartphone and tablet*

In order to facilitate the execution of the tasks, NASA makes available on the Space Weather Action Center Web site (http://sunearthday.nasa.gov/swac) an online Instructional Guide (http://sunearthday.nasa.gov/swac/instructional_guide_V2.pdf) that will help class in accessing, analyzing and recording data, along with a 'step-by-step' Educator's Setup Guide (http://sunearthday.nasa.gov/swac/setup_guide.pdf) containing recommendations and diagrams on how to construct a fully functional SWAC in the classroom while keeping possible limitations in classroom space and technology in mind. It is also available evaluation guidance, and a sample script that students will use to fill in with their forecast information. There is also a so-called 'Additional Data' section, where students can further their studies with data from other links and resources [30]. The program is composed of four stations (learning centers) related to different aspects of the space weather monitoring, namely:

1) Sunspot Region;
2) Storm Signals;
3) Magnetosphere;
4) Auroras.

The first step to construct a SWAC station is to build a display board. Have students working in teams and provide a surface in which they could draw, paint, and glue space weather-related images. For example, they can put a picture of the Sun; they can paint the magnetosphere, and include other elements that would make the board more visually attractive. When students finish working on the display board, the teacher could instruct each group to put the display board behind the station – composed by the computer or notebook. The display board plus the computer will constitute the learning center in which students will carry out their scientific research. The construction of the display board by students themselves provides a feeling of belonging with regard to the station, which motivates students to participate more actively in the research tasks. A list of material required for the display board assembly is available on SWAC Instructional Guide.

We have seen that the SWAC program divides its content in four main areas (sunspots, magnetosphere, solar storms, and auroras). Thus, it would be recommendable to have four stations (one for each main theme), so that groups of five students could rotate weekly, working on the four different stations. Students should move through these stations and follow directions that will enable them to collect and analyze data relevant to each topic. It is recommended that a secure area of a classroom or a dedicated room be established for the SWACs. They should be located in easily accessible spots, offering students a separate space to visit topics of interest and/or to reinforce selected skills. Having a specific place for them assures that it will not interfere in quotidian classroom activities [30]. In one month, each group would have researched all themes. Teams could have four students working as data miners, while one student from each group would act as data analyst. Data miners would collect and record all information needed on the data collection sheet specific to their assigned subject area. The data analyst in each group would monitor student data sheets, summarize information and fill in the required information on the space weather script (available on the program's Web site). In each group, students can share or rotate data collection responsibilities for their specific subject area [30].

After the SWACs are ready to use, the next phase is to have students accessing, analyzing, and recording NASA satellite and observatory data almost in real time in order to produce professional-looking space weather journal data, reports, and multimedia broadcasts. The program is based on well-defined protocols. The principle is to enable everyone to achieve roughly similar results by following the same procedures. For this work of producing space weather reports, the main resources that students will use is the Space Weather Media Viewer, A Web page in which they will find plenty of near real-time space weather satellite data, videos by NASA professionals, visualizations, and tutorials with detailed instructions for data interpretation. It contains several images of the Sun that you can zoom in and out, in order to study features of its surface, as viewed with different filters. Once opened, you may keep the Viewer in a separate browser tab or window for faster data access. The data collecting should be carried out on a weekly basis. Daily data collection would be recommended if the curriculum time allows a minimum of 2-3 times a week. If journaling is not done on a daily basis, students need to review the space weather data from previous days to stay up to date. Teams use SWAC clipboards or notebooks to document data at their action centers [30].

The final phase of the work occurs when students use video resources to share space weather forecast forecasts with their peers, just like weather professionals that we see on TV. The difference is that students would inform weather forecast… of the Sun-Earth environment! Basically, students will produce videos inside a professional or improvised studio with space weather scenarios. As the scenario is presented to the audience on the video, students read a script that is shown on a computer screen. Videos have been used in education as a powerful resource that promote engagement, as students feel that their work is important, and that other people will see that. In SWAC, students feel as if they were working on a TV program. In order to carry out these activities, teachers should acquire inexpensive or free video editing software such as SONY Vegas Movie Studio, Adobe's Visual Communicator 3 (VC3), CCTV, iMovie, Audacity, Garage Band, and/or available school-based broadcast studios. It would be necessary an additional computer, a desk, a webcam or camcorder, and a wall area for use of a 5x6 feet green screen or backdrop [30]. The SWAC Web site contains sample scripts, video clips, sample reports, downloadable graphics, teacher guides, and so forth. These additions are needed to use video broadcasting or movie editing software. A sample space weather video report can be viewed on the SWAC Web site. More information on how to construct and use your action center can be found at http://sunearthday.nasa.gov/swac/.

Figure 27. Another example of SWAC display board. Students can several resources and all their creativity to assemble their display boards. Credit: Norma T. O. Reis.

The Space Weather Action Center (SWAC) is a program that teachers can use to contextualize several space sciences relate concepts. Students are able to experiment all phases of the scientific method, from problem raising – i.e., how should the space weather be like in the coming weeks? – until communication of results through broadcasting.

As for using the scientific method, for example, students can relate the number of sunspots to periods with more intense solar storms, characterized by the occurrence of a larger number of coronal mass ejections and solar flares. Teachers can surely conceive other manners of using SWAC. For example, as students communicate results using videos/broadcasting, teachers could promote a competition in which the best well-produced videos would be published on the schools' YouTube channel. Finally, the SWAC structure, which is one or more learning centers (computer + display board) from which students access databases with up-to-date information regarding a specific topic (in the NASA case, space weather), analyze data, produce a report, and communicate results using video, this structure can be adapted to allow educators to explore topics other than space weather. For example, teachers could create a climate change action center, or a remote sensing action center. In other words, teachers can identify other online databases on topics of interest online, and have students using the same 'action center' architecture and implementation design.

4.4. Radio Jove

It should be amazing to "listen" to sounds of the Universe! It is not everyone who knows that the Universe is a noisy place… And that you can listen to sounds emitted by the Sun, the planets, the Milky Way… Some sounds are ghostly, others sound like a spaceship perhaps … It is just like some special sort of cosmic ID… You can check that simply by exploring some YouTube videos that bring sounds of the Sun, sounds of Jupiter, and so forth. We are accustomed to observing the Cosmos with our eyes and the telescope, but listening to 'sounds of space' can produce a special effect, as we can "feel" the Universe as a 'living thing' perhaps, and not only as a beautiful picture in a portrait. Motivating young students to listen to the Universe could raise interest for an important scientific field named radio astronomy. Further, listening to sounds of space has the advantage of engaging more people, namely visually impaired students. It is important to encourage these students to learn about the Universe in the way that is possible for them. This can encourage these students to follow radio astronomy or other STEM careers, in which they could contribute with their unique skills.

Let us learn about a very interesting NASA Education project named Radio Jove (http://radiojove.gsfc.nasa.gov/). It is one of the few projects I am aware of in which students not only learn Science, Technology, Engineering and Mathematics (STEM) through the motivation of the space thematic, but also use technologies, explore digital resources, have electronic experience by assembling their own radio telescope, collect, analyze, and contribute data on an online global network of observers maintained by the NASA Goddard Space Flight Center, where the entire Radio JOVE community share and archive their data [121]. In other words, it is a very complete and enriching educational experience, in consonance with innovative educational practices required in today's educational scenario. According to J. Thieman [122], Radio Jove is an education and outreach project involving primarily high school and college students in solar and planetary radio astronomy. This project began in 1998, conceived by a group of (mostly) University of Florida graduates working at NASA, who had the idea of building an inexpensive radio telescope kit suitable for detecting radio emissions from Jupiter and the Sun [121, 123]. The project is basically a hands-on inquiry-based activities that allow students, teachers and the general public to learn about radio astronomy. They build and operate their own radio telescope from an inexpensive kit and/or use remote radio telescopes through the internet, without expenses. Participants also collaborate with each other by means of interactions and sharing data on an online global network of participants. Until presently, more than 1100 teams of students and interested individuals have already purchased non-profit radio telescope kits. The project also involves middle school students, science clubs and ham radio operators. The latter can get involved, perhaps with local schools, helping them to get on the air and make observations [123], as in some cases students and teachers will find difficulties in assembling the kit, which is part of the scientific process anyway – finding difficulties and working to identify possible solutions. Radio Jove Kits cost less than 200 dollars and are available from http://radiojove.gsfc.nasa.gov/office/order_form.htm.

Figure 28. Radio JOVE dipole antennas and project coordinator James Thieman. Equipment located in front of Building 26, at NASA/GSFC. Credit: Norma T. O. Reis.

Radio Jove is a successful NASA educational project that counts with a group of volunteers all around the world. It is a hands-on educational project that uses Information and Communication Technologies (ICT) and educational technologies to facilitate teaching-learning process for a wide range of STEM curricular contents related to space sciences through the motivation of radio astronomy. In this project students gain electronic experience and scientific method usage, as they construct and operate their own radio telescope. Students work as if they were real radio astronomers, observing and analyzing planetary and solar radio emissions, and forecasting noise storms. In order to accomplish that, students should integrate knowledge from different areas and follow the scientific method.

Radio Jove is a simple radio telescope aimed at introducing students to radio astronomy. It is composed of a dipole antenna and receiver. According to the project director James Thieman, dipole antennas with 20.1 MHz frequency capture radio emissions from Jupiter (the strongest planetary radio emission source in the Solar System; actually, Jupiter's magnetosphere is the largest object in the solar system!) from the Sun, the Milky Way, and the Earth. Signals are formed by the combination of radio emissions captured in both dipoles, which have a surface sensitive to radio signals. Antenna support can be made of materials such as aluminum or PVC pipes. Aluminum and PVC support poles for the JOVE antenna each last very long times, but the PVC poles are flexible and tend to bend too much if we have the antenna raised to height of 20 feet. Still according to James Thieman, For that height, the metal poles are better. Further, PVC pipes are less expensive. It is also recommended that antennas be places in a 30m x 30m area where there is little radio noise that could interfere with observations. At NASA Goddard, there are two dipole antennas about 02 meters high and 03 meters distant to each other, in front of Building 26, where I had the opportunity to carry out my internship.

Figure 29. Radio JOVE receiver, located in a laboratory at area 202, at NASA Goddard Space Flight Center. Credit: Norma T. O. Reis.

The receiver is where radio signals coming from both antenna dipoles converge to form the final signal. The receiver looks like a small box with several small pieces inside. Teachers and students participating in the Radio JOVE project assemble their own antennas and receiver. I fill not go further on the project, as there is a chapter in this book dedicated exclusively to this project. In the sequence, I provide a summary of the project, along with two important interviews – the first with the project manager James Thieman, and the second with Wanda Diaz, an education specialist who visits NASA Goddard from time to time, and works with Radio Jove and other NASA projects. Wanda is a scientist who lost her vision due to diabetes, but never lost her career objectives and ideals. When she found out about Radio Jove, she said to herself that it was the way to go, and today she inspires and helps visually impaired youth to engage in radio astronomy and STEM overall. Wanda has been a special friend to me during my internship. She is very enthusiastic, professional, hard working, and enjoys a lot what she does. Indeed, an example of dedicated educator fueled by passion and vocation.

Radio Jove at a Glance

- *Contents*: planetary and solar radio astronomy, space physics, electronics, and the scientific method
- *Audience*: middle, high school and college science teachers and students, and radio amateurs.
- *Experience:* hands-on radio astronomy, electronics construction and testing, interaction with other Radio Jove participants to exchange data, ideas, and experiences (NASA, 2008).
- *Inexpensive radio astronomy in precollege schools*
- *Promotes contextualized usage of ICT in education*
- *190-dollar kit + computer with sound card + broadband internet required*
- *Optional online access to radio observatories on the Web*
- *Visually impaired students benefited*
- *Observes Jupiter, the Sun, the Milky Way, and the Earth*
- *International network of observers online*
- *More than 77 countries participating*
- *Self-sustainable NASA educational project*
- *More than 12,000 kits distributed all around the world*
- *Engages students to learn scientific contents*
- *Rather inexpensive in comparison to buying an optical telescope*
- *Electronics experience in building antenna and receiver*
- *Learn about the Universe observing waves other than the visible spectrum*

Interview with James Thieman, project director

Radio JOVE is currently mostly independent of NASA funding, but still receives some support from NASA. The Radio JOVE idea came from a group of us at the University of Florida about 10 years ago who wanted to do something to promote radio astronomy in the educational community. One of them created the receiver that we needed. Then we got together and created the kit, which allowed people to do radio astronomy in an inexpensive way. The kit was about 100 dollars. We got funding from Goddard for the first 100 kits. After building some test versions of the kit we started selling the kits for 100 dollars and used that money to buy the next kits and so on.

What is more fascinating about Radio JOVE? *One of the interesting aspects is that it is rather inexpensive in comparison to buying an optical telescope, which usually costs many hundreds of dollars or more. Here with only about 190 dollars we have a useful telescope! The kids also learn how you can study the universe other than through optical wavelengths. So it opens their eyes to the fact that there are other parts of the spectrum that tells about the universe. This is why space observation is so important because for most other parts of the spectrum, other than radio and visible, you have to go above the Earth's atmosphere to study X-rays, gamma rays, infrared, and ultraviolet. But from Earth you are limited to just visible and radio wavelengths and these are just different wavelengths that we study.*

What did you realize – challenges in working with students? *I think the main difficulty for students is that it is not visible, you do not take pictures, and you do not see images. Instead you work with a chart recorder, and variations in sound basically, so it is a little different than typical astronomy. There is a commitment involved to build this kit, and the kit takes 15-20 hours to build the receiver, and perhaps 10 more hours to build the antenna. We do sell the kit completely built, but we recommend that students build them themselves to get a little more feeling into the nature of science when you build and test your own equipment, often correcting mistakes in the equipment and then make observations, then analyze the data and so forth. This illustrates the whole process of science when building the kit from scratch. I think that it is good but it also makes it a little harder, and some students never make it out of the building stage. So it is a process, but lots of people really like the aspect of building the equipment. Some do not use the system that much but just enjoy building the equipment. We know a number of students that so much liked building the kit itself that we were told of people who have skipped classes and go to build the kit rather than doing their regular education. A lot of people are like that, and one can learn electronics in the course of doing that. You have to learn about capacitors and resistors and how they go together to create a radio. So it is not only radio astronomy - it is also electronics.*

How is the interface with schools, does it happen nationally, regionally… *we advertise the program nationally as much as possible and internationally as well in talks, posters and so forth in a number of places, and we advertise through our Web site. Also the core team that we have is made of people from Florida, Tennessee, Hawaii, and elsewhere. We have one gentleman who regularly comes from Japan, so we have a fairly widespread team and we reach out to those areas. Even at the very beginning we had a pretty widespread project. Kits went out to places all over the United States. So it is not too confined to any particular area. We have somewhere on the order of 77 countries that have the kits so it is well spread out these days.*

What is your greatest satisfaction with this project so far? *I think the fact that we do have many kits going out and being distributed, and that we know a number of success stories where students who have taken the kits and used them in science fair projects and won prizes.*

In addition, we have some people who are using them to do research. There is even one group that took a couple of Radio JOVE radio telescopes and made an interferometer out of them! With time it keeps growing, and it is satisfying to see that.

How Radio JOVE can be also helpful for impaired students? *A few years ago we were asked by the director of Goddard at that time to put an emphasis on the disabled. In particular he was interested to see whether this program applied to the visually disabled and, yes, that makes sense, because with this particular thing it is the sound that matters, it is not what you see. So as, for example, Wanda is completely visually impaired, she could hardly become an (optical) astronomer. However, she can be a radio astronomer (and now an astronomy graduate student) and she does not have to see things to be able to discover things about the universe. So that's good and we put in a proposal to develop this aspect a little better to make it more applicable to the visually impaired. So we had visually impaired students who helped us out and build one of the Radio JOVE kits, who were not completely visually impaired, but impaired to some extent. So we think that it can also be used for other groups as well. We took these results to some conferences and advertised what could be done. It is very interesting to see that development happen.*

How do you see this commitment with diversity, with assisting impaired students, and inclusiveness at NASA? *We try to put some emphasis on the aspect of diversity, in working with the impaired, because that is a national requirement. There are many talented individuals who are disabled so there is no reason why they should not be able to do some of these things, and NASA is always looking for good people. It does not matter if they have some impairment. We just try to make sure that they are capable, competing with the others for other things that we do and hope to encourage a diverse working group that can contribute to the future of science.*

If I am a teacher, a principal etc., what do I need to do to engage and what do I need to have in terms of equipment, technology, etc.? *The main information about how to get involved with Radio JOVE is on the Web site. There you can join the Radio JOVE group and, if you wish, immediately order a kit. You do not have to build the kit if you want to just monitor some of these things. Instead, you can download free software, and then monitor some of these radio telescopes that are sending their signals over the Web. In that way, people can get involved without any money at all. However, if you want to build the kit and have the complete experience, then you order the kit, which is presently 190 dollars plus shipping, and then build the kit and so forth. Most of the information that you need in order to build the kit is on the Web site. But when you receive the kit we also try to make sure that you understand that there are some minor components that do not come with the kit. Thus, you need to separately get the poles and ropes that support the poles, and support structure for the provided antenna for which we recommend PVC pipe to erect the antenna, and if you do not have parts and tools like that you need to purchase them. Also, we do not supply any computers; so you have to have a computer to put the software on it. Overall, other than a computer, there is probably a need for at most 100 dollars for material besides what is in the kit. Further, you need to have room to locate the antenna. Generally speaking, it is still inexpensive.*

What makes this program unique in comparison to other NASA program? *Although there is a number of radio astronomy education and outreach programs, not many have the relatively small expense of startup like Radio JOVE. There are some where you can build a relatively small dish antennas to do radio astronomy, but it is not a NASA offered. You have to do that more as an amateur, and it is expensive. So I think what makes the Radio JOVE kit unique is that it represents an inexpensive radio astronomy project. It is also unique in providing an organized network of observers whom you can interact with via the Web using the software that we have available. These users can meet each other around the globe through that means, and we do have an organized group of people who are involved in this, and who help out mostly voluntarily. We also have coordinated observers on a regular basis where we offer the possibility for people to telephone in and discuss the project with that group for free.*

In the sequence, you find more information about the program, by another education specialist who is visually impaired, and carries out a very singular work with radioastronomy.

Interview excerpts – Wanda Dias, education specialist at NASA Goddard

How can you describe Radio Jove?
Radio Jove has teleconferences and before each teleconference you have the students get information about their hometowns. They have to gather and share the information.

How does that happen?
They get information about the different activities in their hometowns and go to the telephone... the good thing is they are listening to what the kids are saying. A little bit of cultural contacts to the teleconference... after that then they do drawings... on different stories, they have to create stories, local, cultural legends and they have to work to see if it is true, as me as a scientist, what would I ask, what would I ponder so they make drawings and they have to come and convince me that the history they are telling is true. They have to be ready to the teleconference, submit their questions to me before; students are not selected; it is not for the smartest kids, it is for whoever wants to do it, but the questions submitted have to be pertinent... I have a lot of fun...

Do you consider that current schools offer equal conditions to impaired students?
They can, the schools are willing to do that; they do not have the knowledge and the resources. I've met the most capable teachers but sometimes they do not know how they should work with those students, who have another way of learning. We should facilitate their way.

How would you summarize what NASA has been doing to inclusiveness in terms of programs, projects, and so forth...
For NASA it is a priority to make the field inclusive to every kind of learner. From the programs that I have seen I understand there is a big amount of people but they tend to be for a tiny elite. NASA needs to have a way of selecting. Radio Jove is a program open to prepare the teachers and to be able to continue the program in the school but these teachers also need a mentor so the way NASA may improve is by developing a way to mentor, a mentorship to

keep going for many years, not for a couple of months and then disappear. For example, at home, I grab my telephone and ask my teachers how they are doing today? Do you need anything? So mentorship is very important. Radio Jove with the teleconference offers you the opportunity of person-to-person contact and mentorship. The visits to schools are important, but it has to be on a long range. This mentorship has to be constant so that students do not feel abandoned. From my understanding NASA could improve long-term mentorship. They want to contact us, can I call you, have my mobile, my Skype... of course! We need to close the gap.

For students who have some kind of disabilities who have dreams what would you say, you are here, doing beautiful work at NASA... what would you say for those people?
Never give up. Never ever give up, no matter what. There are always means to make your dreams come true. Do not forget your dreams. Do not get your eyes away from the goal. Never give up.

4.5. Educator Astronaut Project

The project consisted basically in the selection of outstanding educators to become permanent members of the NASA Astronaut Corps. The idea is that the astronaut uses his or her visibility and pedagogical communicative skills to inspire the scholar community in space exploration, to foster careers in STEM and to elevate the esteem of the teaching profession. The program introduced new content into education, advanced technological tools, and several educational services, such as direct participation in space research along with interaction with NASA scientists, engineers, and astronauts.

Currently, the project is over, because its purpose has been achieved. According to C. L. McArthur [124], the purpose has been to enable teachers with technical (scientific, engineer) background to be eligible to become part of the NASA Astronaut Corps without a special mechanism for that. It means that currently any teacher in this condition may candidate directly, without the necessity for a specific project.

Another unique contribution of the project is the fascination of sharing space travel experience with students and teachers in a language that is peculiar to teachers and also the importance of communicating the achievement to other members of the scholar community. According to C. L. McArthur [124], despite of the fact that all astronauts are good communicators, and they all have a formal obligation to dedicate part of their agenda for communicating their experience to the public, it is a fact that it makes a difference when it is an educator who is interacting with another teacher, due to the fact that they share the same concerns, commitment, and experiences.

The Educator Astronaut Project has been carried out in the scope of a larger initiative named Teaching from Space - TFS[2]. This exciting program, based at NASA Johnson Space Center, is responsible for facilitating the flight of educational activities on the Space Shuttle and the International Space Station. Activities focus on having ISS crewmembers to perform curriculum-based activities in space to demonstrate basic principles of STEM and geography in a microgravity environment and answer questions from students on Earth, so that

[2] The project's Web site, http://www.nasa.gov/audience/foreducators/teachingfromspace/home/index.html

promoting real-time interaction. Most of them involve video recording of the demonstrations and/or photographic documentation of astronauts onboard ISS performing experiments [125].

Figure 30. NASA's first educator astronaut and former school teacher Barbara Morgan meets students during "Meet an Astronaut Day" at Johnson Space Center, Houston. Credit: NASA.

Figure 31. Cake in format of shuttle. This cake was part of a celebration of the first U.S. woman in space, Sally Ride. The party took place at the NASA Goddard Visitor Center, in 2008. Credit: Norma T. O. Reis.

4.6. NASA EXPLORER SCHOOLS

This project has suffered changes since my investigation in 2008. Today, its activities are carried out online. At the time of internship research, the project used to establish a three-year partnership between NASA and up to 50 school teams yearly, consisting of teachers and education administrators from diverse communities across the country. Focusing on underserved communities, NES joined educators, students, and families in sustained involvement with NASA's research, discoveries, and missions. The project was designed for education communities at the 4-9 grade levels to help middle schools improve teaching and learning in STEM through significant structural techniques such as professional development, stipends, grants and curricular supports based on NASA's resources. An integral part of the NES is availability of the NASA Digital Learning Network that provided NASA personnel, technology, facilities, programs, and resources, delivering interactive learning opportunities via videoconferences to teachers and students [107].

The initiative served communities that otherwise would not have access to the benefits of space education approach. Encourages students who have never thought about pursuing careers in STEM to do so and even consider the possibility of working at NASA.

Figure 32. NASA's Explorer School Web site. The project is now carried out with a virtual campus. Source: NASA. Credit: Norma T. O. Reis.

In order to provide a better panorama of this effort, I share an interview with a NASA education specialist working with that program.

Interview with a NASA Explorer School expert (Summer 2008), with permission

Interviewed: Trena Ferrel
Position: NASA Explorer School Program Coordinator
Employer: NASA Goddard Space Flight Center

How long does this program exist?
Since 2003-2004.

Where has the idea of creating this program come?
The program has been created to have kids interested in science careers.

Is it targeted to low-income communities?
Yes, and minorities.

What is NASA's mission with these students? What makes this program unique?[3]
I guess it is opportunities that are offered to teachers and students. Over a 3-year period, each school receives 7,500 dollars, and they can use that for technologies, so a lot of schools purchase videoconferencing equipment, and then use that to reach NASA scientists, to do DLN, also for videoconference with other people, and this is the big plus. The program also offers professional development for educators.

Is this a national program?
Yes. We have schools in the entire U.S., Virgin Island, and Puerto Rico. The program is managed in Glenn, Ohio. And there is a coordinator in every NASA Center. We serve about 250 schools right now [2008] including active and former partnerships.

How do schools get involved?
There is an application online; schools answer a bunch of questions. There are criteria for selection. We have to make sure that the application is good, that the school really needs us, etc.

What happens in the schools involved, how are activities implemented?
We have teachers in aerospace education, NASA educators, and certified teachers who go to the school to provide support. When a school gets involved, these teachers do a kick-off for the school and community; they also organize events called Family Night, when families can come and interact with the students. NASA personnel also support these events.

Do you involve other professionals such as astronauts?
Yes, we try to get astronauts to schools; in many cases they have gone to schools.

What about the NASA's Digital Learning Network - DLN?
There are DLN modules delivered throughout the year, if the school is doing microgravity, for example, DLN can support that. That is why we want all our schools to

[3] This type of on-spot partnership is no longer active; today, activities are carried out online.

have videoconferencing equipment; that is why we want them to purchase this 17,500 dollars equipment. DLN is used very frequently; our schools do that a lot. It is very interactive; it is real time, students can talk to, ask questions to scientists in the NASA centers from their hometown, which gives a personal feeling.

What are the main difficulties you have been facing so far?

Budget. We want to pick up more schools.

What are your criteria for selecting these low-income schools?

We pick up schools that have free or reduced lunch for students; it is low-income schools where probably there are minorities. So it is a high percentage.

Do you have some partnership or is it a NASA program only?

It is a NASA program, but once the school becomes a NASA Explorer School, they form partnership with communities, so this forms sustainability, especially to assure continuity of efforts, after the three-year partnership with NASA is over.

What have been the main accomplishments so far?

We got kids interested in science, and now a lot of kids can actually see themselves working at NASA, and before they never thought about that.

The program is targeted to…

Elementary and middle schools.

And why did you choose elementary and middle school, and not high school?

We wanted to target people early. At middle school, for example, kids are thinking about what they want to be when growing up. At high school, many have already chosen their professions.

What do you consider one of the main ingredients for the success of this program?

Communication. We are always in touch with the schools, and families are involved, through events such as Family Nights. It is strong communication between NASA employees and the school.

How is this communication?

The coordinator is responsible, for example, I am responsible for all the schools in NASA Goddard region, there is one like me in every NASA center, so this leader is in constant contact with the school, to know what their needs are, and work towards fulfilling them; we also visit schools, to see kids' faces, walk through the building - this is important.

What was the moment you felt more fulfilled?

When you go to schools, and the kids know you and ask you questions (...) Kids love space, so we bring many space things to them (...) we see their eyes shine.

Some lessons you learned…

We need to have lots of professional development for teachers, such as workshops where they can see other teachers and discover what is new and hot, and go back to school. Another thing is the DLN program, because it is a major way in which we can reach a large population of students.

And do you give materials to schools?

Yes, lithographs, posters, things like that, two or three times throughout the year, and also the school can call us and request if they need something. For example, a teacher needs an educator guide about rockets; then we send the specific material.

What about the money you grant, can the school use the way they want?

No, we monitor. It has to be for technology, and similar purposes. Schools receive a list of what they can use the money to, and what they cannot.

How do you monitor that?

Schools have to fill out a report every year, and we evaluate it. See if it is approved or not. They need to specify what they invested in. We have system with a coordinator, like myself, and we have a tracking system. We visit schools, and we know if they purchased the videoconferencing equipment because they are doing DLN. For example, we see if they purchased a robotic kit because they are carrying out robotic activities, after having attended a NASA workshop about robotics. We also have a system called e-Folio where the schools report what they do. For example, when they have a NASA speaker coming in, it goes to the e-Folio; there is a tracking system, they can take pictures, tell how many students and community members were there, things like that, so we can keep track of all that with the e-Folio. This e-Folio is also useful for schools, because they can see what they have done.

Do you communicate with the Department of Education?

Yes, we do. All NASA materials follow national educational standards, so we base our lessons on that.

When you were a teacher, did you already work in low-income school?

Yes.

What would you recommend, as an educator, for teachers in the U.S. when working with low-income students?

I would just say, get the kids interested, and let them know that they can do it, he/she can finish college; he/she can finish high school, and go to college.

4.7. NASA Academy

NASA's premier leadership program, it consists of an immersive and integrated multidisciplinary experience, an intensive summer of higher learning for undergraduate and graduate students with various backgrounds interested in pursuing careers in space-related fields. Consists of a 03-month internship in a NASA Center, during which students work in specific fields of space research and exploration. The idea is also identify talents that may become part of NASA workforce after the completion of the experience [125].

The curriculum provides opportunities for direct contact with advanced science and engineering R&D and promotes awareness of the complex managerial, political, financial, social, and human issues faced by the current and future aerospace programs [126]. During the summer, students carry out some concomitant activities. It is required that students know how to conduct research, because they need to develop a group project and an individual project. In addition, they should attend seminars and colloquia, and present posters [125].

The program is open to both U.S. and non-U.S. citizens. A non U.S. institution may be able to have students at the NASA Academy provided that the institution becomes part of the International Space Education Board[4] (IESB), of the International Astronautical Federation (IAF). Another advantage is that, different from the majority of similar programs, the NASA Academy does not charge academic fees from their students. The program also makes it possible that the student visit several NASA facilities in special trips, in which they have the singular opportunity to learn more about what NASA does and interact with its personnel.

Figure 33. NASA Academy students. The curriculum contains several visits to facilities tnat enrich learning. Credit: NASA.

Interview with a NASA Academy expert (Summer 2008), with permission

Position: Operational Manager for the NASA Academy at NASA Goddard
Employer: NASA Goddard Space Flight Center

What is the NASA Academy?

It is a 10-week program, carried out at four NASA Academies at different NASA centers: Goddard, Ames, Marshall, and Glenn. The one here at Goddard begins in early June; students

[4] The board's Web site, http://www.iafastro.com/index.html?title=International_Space_Education_Board

come from very different backgrounds, most likely aerospace engineering, or some other science and engineering background, and interest in space. The idea of the Academy is to prepare the students in the program for space careers. It is a very interesting summer. Students are usually rising junior to graduate students. Most have past research experience and past internship experience at NASA or outside NASA; some of them have professional experience. The point being that most come already knowing how to do research. During the summer students have an individual research project, and by the end of the summer they do a final presentation. They also get together during the course of the summer, and decide upon a group project that they also need to carry out. The group project is not limited; it could be an outreach project - whatever the group decides. the deliverable for the group project is a paper by summer's end. The ISU summer program is somewhat similar to our program. During the course of the summer, students come to work every day in their individual research project, but they usually are on site visits Thursday to Friday. We may travel to NASA Wallops, NASA Langley, other NASA centers, and students have the chance to meet the NASA center directors, top researchers, visit facilities, thus having experiences that regular interns are not exposed to. Students also visit local space industries like Lockheed Martin or Applied Physics Laboratory, attend events (...) a lot of different activities. Ultimately, the summer is really intense. Students not only produce a group project, but also conduct individual research in their laps while traveling to meet various important industry executives.

And how long does it take?

10 weeks for everything, it is pretty intensive.

What is the purpose, the mission of the program?

Gerald Soffen was inspired with the ISU summer program. He therefore founded the NASA Academy in 1993, which mimiced the ISU program but he added some other aspects to that ISU experience. Its intended purpose is to provide a unique experience for the best and brightest to have access to NASA scientists and engineers, even the NASA Administrator, and have a casual conversation with them – a truly unique experience. The Academy looks for students who are very serious about aerospace careers and connect them with industry executives where they are able to sit down and talk casually. The purpose of the Academy is also the alumni association. Usually each Academy builds relationships that are long-lasting. The Alumni Association is just one way for them to remain in touch. Most keep in touch following the Academy for 10 years or 15 years down the road, (…) opportunities that may lead to future work opportunities.

And who is the public of NASA Academy?

Most of them are undergraduate students, but it depends. We have a few graduate students. Ages vary. Students are from rising junior to graduate students.

And… can some of these students be hired by NASA, by the end of the summer?

We had a student this summer asking about co-op position; she was interviewed, and got the co-op position. The co-op is one of the most common ways of becoming a NASA employee, so we do have a student this summer that got the co-op position. We definitely would like for our Academy students to become NASA employees; if they go into industry it is OK, but we really want them to become NASA employees.

And how do you evaluate the program?

Each Academy has a director, thus the way we evaluate our Academy here at Goddard is to solicit students to provide feedback regarding which visits, trip, speakers during the summer they had enjoyed. We also ask whether or not the summer has changed their career goals. They usually realize something they really would like to do, they are passionate about, and that sometimes makes them change their career plans or even change their college major. This is the manner we measure the program; to have students saying that they had a great summer, and met people they would not have gotten the chance to speak to (...)

And which are some difficulties faced by the program? Lessons learned from the past...

One of the difficulties this year was the fact that the Office of Higher Education (...) they are unifying all the education offices, there would not have a separate office of higher education; it will be incorporated to the office of education. And these changes in management implies financial issues so there are things this summer that we could not do, that was easily available last summer. And the other difficulty as a manager, most of the time, in order to be able to manage, operational manager like I am for the NASA Academy, they are mostly students, they are usually students coming back so I was in the NASA Academy last summer 2007, and became a manager, coming back this year. Sometimes it is hard to be an authority for your peers, because the ages are similar, graduate, undergraduate students, people who have more research experience than you, maybe more experience in your field; it is kind of hard to manage people coming in the Academy like that. I guess this is one of the difficulties most managers experience. But for me (...) I just try to make sure that I manage the Academy, and not try to be in the Academy again.

And who composes the staff? Technical people, educators...

The NASA Academy is mostly a mixture. The director has an engineering background, but he has been in the office of education, so he is an educator; he has worked in the office of higher education for a while now. The academic dean is a Physics professor in the University of Maryland. He helps students with their research projects and posters; he helps them decide their careers; helps them do their research during the summer. The two staffers include me as operational manager; I have engineering background: aerospace engineering; and the logistics manager has engineering background: computer engineering. Managers will be technically based, and some of them educators.

So they are an interdisciplinary group...

Yes.

And how many students do you usually have?

18 to 20 students in the Academy.

Not so many...

No, not so many at all.

Why?

Because the way the program is designed, we want a limited number of students, for that they get the change to know each other, because if you have more than 20 it is kind of hard to organize the group project, as well as it is hard because it is lot of stuff, there is no way within 10 weeks to do all that they need to do as well as to get to know each other together, thus the idea is to limit to 18-20 students, no more than 20.

And do they attend classes?

No. A typical week in the Academy is... On Monday they will get up about 7 am, catch a bus from where they live, and go to individual places where they do research. They are working with various people here at NASA Goddard in their individual research project, and the main part of their day consists of that. After work, they will catch the bus back. Once back at the housing, there is usually a short time to relax before the evening activities. For the evening, on Mondays they will probably have a big group meeting with me, and the other manager when we give them an overview of the week, let them know of any changes, what they need to be prepared for the week, what business to do, any research they need to get in, this type of thing. Then we will have a dinner speaker. So all get together for dinner speaker; we invite them to speak to us about their jobs, about what they are doing at NASA or within the aerospace industry, they talk about their career, where they got their education, and then students can ask them questions about their interests, ask career advices (...) and it is really informal. Some speakers we invite want to bring a PowerPoint presentation with them; however, we prefer they really engage with the Academy when sitting face to face. Following dinner, the Academy has their group project meetings. Group project time is outside of work at Goddard, and it is outside meetings, so about 9 pm they are free of work and they do group project meetings, which goes until midnight or one o clock (...). On Tuesday and Wednesday they get up at 7 am, come to Goddard, work here and go back, and maybe have another speaker, or sometimes when they come to work during the day they will have a speaker here at Goddard, maybe talking to one of the engineers working here is easier to speak here as a group (...) On typical Thursdays and Fridays, maybe they leave on Wednesday night for visits; this is not every Thursday and Friday, but on a typical Thursday or Friday they will take a trip to Johnson Space Center or maybe to learn about what they do there, visit facilities, and return for work on Monday. And the other weekend when we do not have trip we will go to, maybe, a local place such as downtown, such as NASA HQ, the Applied Physics Laboratory (...) the Federal Aviation Administration.

And who can apply?

It is open to anyone, from any country in the world; we always have at least one international student in the program, such as students coming from space agencies such as ESA. The only restrictions we have are academic; we only accept students with a 3.0 GPA or above. We usually do not bring students just entering college; they are usually very young and still did not have all the core classes that they would need. Two big requirements, we look for students who have a strong demonstrated interest in space; we also look for students with previous research experience and previous internship experience; if they never had internship before, it becomes overwhelming for them. Those are the major things.

And how can individuals apply?

There is a website, if you do Google search and type 'NASA Academy' you see the NASA Academy webpage coming, then you see the different NASA Academies in different

NASA centers, and you apply for ours, at Goddard, for example; or you could Google 'NASA Goddard + Internship application,' and the page will come up.

And which benefits do you see for students who attend NASA Academy?

I see plenty of benefits. There are more, more people who have been to NASA Academy and now... I do not know if you know about the company Space Adventures, the founder is Academy alum. We have also had students to go to the International Space University. Usually employers are eager to hire Academy alumni immediately (...). The Academy helps you define your career, and if you really take advantage of the experience and talk to scientists and engineers, it brings lots of networking opportunities, you can call them later and say, "Do you remember me out of the 2008 Academy? I am really interested now in what you have done, and I remember you talking to me. Can I fly down and meet you, at work, or can we have a coffee, to talk," and the contact can help one that way".

Do you pay tuition fees for the NASA Academy, or do you have support for the course you are attending?

Tuition fees are not paid for the NASA Academy alumni; and international students usually have support from their country (space agency, for instance), for living expenses for the summer.

4.8. Solar System Radio Explorer Kiosk – SSREK

The Solar System Radio Explorer Kiosk (SSREK) will teach visitors about radio waves from the planets and the Sun and what they may be telling us about these worlds. The SSREK will also show that the senses of hearing and touch are viable ways to learn about these emissions. Through the innovative rendering of radio astronomy data, the SSREK will impart to the public the excitement of scientific discovery and inspire and motivate a new generation to pursue careers in science, technology, engineering and mathematics, especially students that are currently underrepresented in these fields, according to L. Garcia [127].

The Solar System Radio Explorer Kiosk <http://radiojove.gsfc.nasa.gov/ssrek/> is a NASA outreach project designed for informal education institutions such as museums and science centers. This educational technology teaches visitors about solar and planetary radio astronomy. SSREK also demonstrates that the senses of hearing and touching are viable channels to teach and learn more about radio emissions.

Shaped like a video arcade game console, but designed to be wheelchair-accessible, the interactive software is controlled via a simple interface developed to be accessible to a wide audience. All key pressed provide visual and auditory response. The surfaces of the kiosk present colorful graphics over which are transparent, durable sheets embedded with tactile graphics and Braille characters. One panel provides a representation of Jupiter as seen across the electromagnetic spectrum. The other provides a scale representation of the planets of the Solar System. The unit renders the radio data as sound, and as visual display on the screen. The project is currently presenting visitors with samples of radio emissions from Earth, Jupiter and the Sun. Through animations tied to vocal descriptions and captions, the SSREK allow visitors to learn more on what these bursts inform scientists about conditions at Jupiter or the Sun [125].

The Kiosk motivates scientific discovery and benefits impaired students to learn more about the Solar System and related subjects.

Visitors can use that Kiosk system to hear sounds from the Sun and planet Jupiter. The project's goal is to benefit from audio and video resources to support different learning styles, facilitating teaching-learning process of scientific content through planetary and solar radio astronomy. Through animations tied to vocal descriptions and captions, SSREK enables visitors to learn more on what radio emissions tell scientists about conditions at Jupiter or the Sun. SSREK is targeted to high school students and the general public, notably visually and hearing impaired students. All sounds played by the kiosk are simultaneously sent to three speakers. Two of the speakers are small while the third is 10-inch diameter. This speaker is attached to the back of a drumhead. The kiosk has a circular hole of the exact same size as this drumhead, which is mounted to the front immediately behind the large image of Jupiter viewed from the front. The purpose of this is to provide deaf visitors with multiple means of learning about radio bursts. They see the sound pattern moving across the screen while being able to feel these same sounds in form of vibrations. In particular, Jupiter S-bursts will look and feel much different from Jupiter L-bursts just like we can hear a great difference between them. These are real data and are in no way artificial.

The Educational Kiosk is an interactive technology roughly like a video arcade game console but designed to be wheelchair-accessible. The software is controlled via a simple interface developed to be accessible to a wide audience. All keys provide a visual and auditory response when pressed. The software provides a menu-driven, simple, high contrast interface. The surfaces of the kiosk have colorful graphics over which there are transparent, durable sheets embedded with Braille characters and tactile graphics. One of the panels provides a representation of Jupiter as seen across the electromagnetic spectrum. The other panel provides a scale representation of the planets.

The unit renders radio data as sound, as vibrations from a membrane mounted to the front of the unit, and as visual display on the screen.

The Kiosk works in partnership with the Radio JOVE project, using its network of telescopes that offer real time data for radioastronomy in Europe and in the United States.

The Kiosk also accesses Radio JOVE files, which contain hundreds of sounds fromJupiter and the Sun. The project is executed through partnerships such as the U.S.

Federation for the Blind – NBF <http://www.nfb.org/nfb/Default.asp> and the Maryland Science Center – MSC <http://www.mdsci.org/>. These institutions have been collaborating to design, develop, test, and execute the project. The estimated cost to replicate the Kiosk is approximately 5,000 dollars plus shipping. The manufacture of the panels is done by Touch Graphics Inc. <http://www.touchgraphics.com/>. The Kiosk structure is wood, and it is necessary to purchase additionally a laptop and buttons.

The project aims to provide audio and visual capability to support different learning modes, thus making learning easier for the physically or developmentally challenged, by working along with organizations devoted to reaching out to underserved/underutilized groups in science and technology, such as the National Federation of the Blind (NBF) Jernigan Institute [125].

Interesting to everyone, the Kiosk in an inclusive initiative that provides blind and deaf students with opportunities to learn more about the solar system by using other senses. Further information can be found at: http://radiojove.gsfc.nasa.gov/ssrek/

Kiosk at a Glance

Radioastronomy outreach project for impaired students
Visually and hearing impaired students assisted
Designed for science centers and museums
Requires the Kiosk (5,000 dollars) + shipping + laptop
Project uses Radio JOVE resources
Learn about radio emissions from Jupiter and the Sun

Interview with Sun-Earth Day forum expert (Summer 2008), with permission

Interviewed: Leonard Garcia
Employer: Perot Systems at NASA Goddard Space Flight Center

What is your position at NASA Goddard?

I have many different positions; one of them is technical editor at the Science Proposal Support Office (…) and I have educational duties, one of them being the Kiosk. This initiative takes radio wave data from Jupiter, the earth, and the sun (…) using those specific examples as a way of teaching in informal, museum environment, teaching about these planets and the sun; we teach what these radio emissions might mean, and what we can learn about the planets from them; we use the sounds, we take the radio emissions and turn them into sounds. And we use these sounds as a manner of first engaging the audience and then hopefully we make them want to learn more about the background (…) the idea of studying radio emissions in educational context.

Can you talk a little bit more about the Kiosk? What is that and how that works?

The Kiosk was developed as a means of presenting this data in (…) a museum setting rather than a formal educational program (…), one of the objectives was to use sound (…) as a way of engaging the audience, but also audiences that may not normally be receptive to exhibits in a museum, especially those that are blind. In a museum setting there often are not exhibits that are specific given to this audience, and there is very little educational materials accessible to them, so we had this in mind as we developed this. And since we started with that idea of presenting the sounds in this fashion we incorporated Braille into the surfaces of the Kiosk so those who are blind can feel the text that we present, and we did not stop there. We also looked at other ways this could become accessible, we made the units wide enough that a wheelchair could be moved up to the unit, and we also made that in a manner to allow those who are deaf to be able to still participate because we have a captioning for all the local descriptions. Whenever the voice comes on, there is some written text that those who are deaf can read, and we also pick the sounds, and render them as vibrations out of a membrane. There is a drumhead that is mounted to the front of the unit, so that those who are deaf can put their hand on the drumhead and (…) they can sense the vibrations, and there is also visual representation of the sound that is playing as a little animation sequence on the screen so those who are deaf can see the display of the sound, feel the sound and think of the video presentation. There are many audiences we are trying to benefit here. As well as those who, you know, are not blind or who are not deaf (…). Hopefully, all audiences will be able to participate.

And how many Kiosks do you have so far?

We have just one [2008]. We started with a prototype, an earlier version of this (...). We presented that to some educators who are deaf, and we went to a conference of educators, and incorporated very good ideas, and one of the ideas was to shrink the size, because originally it was 6 feet 4 inches tall, and we shortened it to about 5 feet tall (...) for the wheelchair accessibility, and other changes (...), the software to make it more accessible, so we are now second generation, but currently there is only one unit.

Do you intend to multiply this?

The idea is that we have the software and descriptions of the hardware, and hopefully the graphics, and the software that could be adapted or used by others and have all of this information posted on a website. We do not anticipate at this point unless we obtain new funding to create one more of its units, and we hope that by the end of our funding period we could have other thing available on the web so that someone else can come along and duplicate this effort.

Is this expensive? How much does it cost in average?

Well, much of the development that we have done hopefully will not have to be replicated by other people; we have something that they can start with. The development of the graphics panels, and the Braille and tactile graphics that cover the sides of the unit cost a few thousand dollars to develop in terms of design and manufacture. That design works if you have to replicate (...) I would estimate an order of, perhaps, including a laptop, the order of 2-3 thousand dollars to do this once again, with the audio, speakers, computer equipment, and the wood, because this is a wood unit that we painted, and it is on wells. So, that is on the order, I have not really estimated exactly how much a replica version of this would cost.

Can you 'sell' this idea to schools, and other organizations because it is not so difficult to afford, right?

We certainly try to keep in mind the affordability of this. We have not really done anything too sophisticated. The panels (...) are the only things that are really special made. The wooden box that is within is crafted. We had someone build that, of course, a volunteer, really, so that was quite inexpensive, maybe 500 dollars for that, but, yes, I expected, this really would not cost very much. And I think it can be adaptable because the software could potentially stand alone from the hardware, you know, if you were just to run the software application on computer with speakers. That would be sufficient to get some other message across. Thus, I think there are some means of scaling down this project to an even more affordable level. But, again, I have not really priced that out, but I would imagine that to be cut at least by two thirds, in that case just using computers and speakers.

Maybe the idea would be perhaps to introduce the Kiosk, to sell the idea to other organizations, in order to let more people know about this product?

We are going to be posting on a website all this material so they can download it and try for themselves, and that is something that we are in process of doing right now. We have really just completed the software to the point that I feel comfortable presenting to the public. We are in a stage now that I have something that I would not mind posting on the web and sharing. There is certainly room for improvement; we want to add more modules, right now we are only talking about two objects, Jupiter, the sun and the earth, there are many more planets in the solar system, there are more aspects to study, there is more in-depth science that can be done, there is discussion on the history, and interviews with scientists…

Can you talk a little bit about the excitement that students have when they go to the Kiosk, for example the blind students, how do they experience that?

We have presented this to groups of blind students twice; there was an event last summer in Baltimore, Maryland, where there were 700 students from all around the country by high school age. And there was a tremendous amount of enthusiasm, they really enjoyed; they liked hearing the sounds, or they were interested in what was going on, they were asking us questions, so I thought that was a great example of the potential of this project. We went back again in January of last year, it was a smaller group this time, but again, we had blind individuals, older people (…) perhaps a broader audience, or smaller audience, and again, we had a lot of good response. And we have not clearly defined in which things there might be room for improvement, we are still uncertain about what could be improved, changed, redesigned, but I think we are at a point where people can really appreciate this. Again, I am sure that there are always things that can be changed, but in general the response from blind individuals has been very positive, they really enjoyed that. Thus, I think that stage has already been done and has been good. There has not been a careful study about the amount of educational advancement, for example (…) I would like to see also how this unit would behave in a completely stand-alone environment, which is where it is designed to be. Whenever we presented this Kiosk it has been attended, I was there or someone else was there with me. It would be interesting to see how someone who is blind left alone would be able to appreciate this. I have a strong sense that they would be able to progress through the menus and they would understand how to access the units, I think that we put a good job in enabling someone who is blind to take the first steps into navigating and using the device.

There is a NASA official commitment towards inclusiveness, right?

Yes, and it consists of many other directives that NASA has been promoting, such as Web site design, other efforts, and other projects, other fields. NASA serves many different scientific communities, and each one has its own interest, earth sciences, astrophysics and so on, in order to make sure that communities that may not be normally served, those with blindness or deafness, will also be able to appreciate the science that NASA is doing. Thus, to a great degree that has been promoted by NASA.

4.9. GLOBE Program

Hands-on, school and community-based science and education program that unites students, teachers and scientists in study and research about the dynamics of the Earth's environment. Over a million GLOBE students in more than 24 thousand schools located in 112 countries [75] have taken important environmental measurements. Their data is used in their own research activities as well as by scientists around the world. More than 44,000 teachers have attended GLOBE professional development activities. It is important to point out that although GLOBE is funded by NASA and the National Science Foundation, it is an international initiative.

Students interact with colleagues and scientists worldwide, sharing data, following exactly the same protocols and making real science. As students realize meaning and usefulness in the activities they are carrying out, they tend to become more motivated and engaged. Furthermore, this initiative promotes teamwork not only in the scope of the school but also with international colleagues. Further information can be found at the Web site: http://www.globe.org/

Interview with a GLOBE expert (Summer 2008), with permission

Interviewed: Todd Toth
Position: Education Specialist at NASA Goddard and GLOBE teacher
Employer: NASA Goddard Space Flight Center

What is your position at NASA Goddard?

I am an education specialist, and have been teaching for 31 years. I have been doing GLOBE since 1994, and I have been one of the first 25 teachers trained for the program. We designed an entire science course around the GLOBE program. I like GLOBE because students have to use what they should have learned in all the different science areas.

Which subjects do you teach?

My background is on earth and space sciences, but in 1994 our school principal asked some of us to create this new class, GLOBE, and I thought I would try. I also have environmental sciences background; the earth and environmental sciences are the perfect set for GLOBE.

And basically you teach high school, elementary school...

I started with junior high school, 7th, 8th, 9th grades, for about 10 years, and then I went to a middle school. But most of the time, I taught to 9th, 11th, and 12th grades.

Tell me a little about your work as a GLOBE teacher.

The GLOBE program I taught was also sponsored by the Californian University of Pennsylvania and if the student got an A or B, and did a research based on the GLOBE program they would get three college credits for 25 dollars, which is a real bargain (...) About supplies, we had a professor of the University of California who saw our supplies, and how I taught the class. He joked that I was better supplied than him. By being one of the original GLOBE teachers, every time they came up with a new piece of equipment we got a daily test (...) I said, we are high school and we are better stuffed than the college! You have everything that GLOBE develops.

How do you think the GLOBE program can contribute to make science teaching more engaging, considering that fewer and fewer students are choosing science careers?

What I like about the GLOBE program is that you take all sciences, the biology, the chemistry, physics, and students have to use that to do a certain protocol. They need to have a certain background of understanding, read and understand what they have to do, solve problems, go out, get data, then analyze and synthesize it, make some sort of conclusion, and many times it is not right. I think that the problem in high school is that you have some sort of what I call 'can lab,' you know what the answer is going to be, it will always going to work, or teachers will not do it; but it is not real science. With the GLOBE program, sometimes things do not work, and then students have to find out why it does not work; you have kids thinking.

When kids participate in 'can labs' they do not feel they are doing real science. When kids come to my class, I send a note to their parents, telling that we are going outside on the rain, or snow, and students are going to get muddy, they need clothes, the necessary stuff to go to school, and parents do not like the idea of kids going outside and getting dirty, but science requires that. Kids got surprised one time we went out, and it was raining, "are we going to go out?" they asked. That is real science. Kids go outside, use equipment, and apply their thinking, reasoning skills that scientists are going to be using. We have a situation where a lot of data is collected for specific sciences (...) kids do it all; we get soil temperature (...) capture data for the entire month such as temperature, the pH. Then they download that to a computer and send to the professor. A lot of our atmospheric stuff dealing with ozone and things like that we send directly to university. Students run everything, equipment, battery, if something goes wrong, they do it all.

So your students have lots of hands-on activities...

Yes, and when I teach two or three kids how to use an equipment, when they are done they have to teach somebody else how to do that; and they have to do it right, because they have to follow protocols.

Can you talk a little more about these protocols?

The students log in; this is the high school where I taught in [Todd logs in his computer]; so any data is sent through this school website [he shows]. This is the general page; anybody logs on the GLOBE program and can look through the stuff, but to put data you have to log in. There is a place for registration for workshop on how to teach the protocols, there is also a teacher's guide for download, educators find background on each of the protocols and suggestions on how to teach it, along with answer sheets. I never used GLOBE assessment tools because I always used my own tests. We have been here for a long time, so we have lots of data stored, we can see all data we put, and the last time we entered that; and they (seniors) should be able to use graphs, and they can take their own data and plot graphs (...) we have math teachers working on this.

The protocols are something to have the entire world following the same methodology to reach similar results?

I do not think it was originally set up that way, but it is now, everybody is following the same steps to get the same data, so a research scientist here could use the data from somebody in Asia or South America, and they follow the same steps at his location, which is perfect for science. But the data entry, and the protocols in atmospheric and climate change a lot, ozone, clouds, aerosols, surface temperatures (...) And something else I have to teach students beforehand is that they need to know solar noon. All data should be submitted between 11 am and 1 pm. In most cases, scientists use only data sent during this period in order to meet the parameters (...) There are gadgets for measuring the atmospheric optical thickness, you measure how much sunlight is able to come through, how much pollution. We put latitude, longitude of the school and you send that information, get students informing the temperature, how much solar radiation (...) There are protocols for converting your school's elevation, what the pressure is right here. Some time ago, these measurements were done by hand, now students use computers. Rain fall, we keep track of that for the entire year, and every month we keep track of the weather, every month I make a report for the national weather service and they keep track of our weather (...) Everyday I include the high temperature, low temperature, solar noon, the cloud cover and type of cloud, precipitation, pressure, and send it to GLOBE. I also double check to make sure students did not get something incorrect, so I can go back and change that. So they are all involved in that. It is so neat, because it is so hands on (...) In the beginning it was tough for administrators and even teachers, because our kids were on the room for 10-15 minutes, we gave them instructions, they got the equipment and went outside. We had a difficult time seeing kids going in and out. Parents were complaining (...) For me this program is perfect... now you are applying the science, you see there is a reason why you learn this math, algebra, why you use graphs, because people do. This is real world; you do use that. So GLOBE program brings all this, kids see why they learn the math, why they learn the science. We measure the snow, we keep track of it (...) and the precipitation and then send it in, normally we have to put in the current temperature, the maximum, minimum (...) that is the type of thing you do for each of the protocols (...) And this is our school site... Meteorology and atmospheric measurements, we usually do that everyday, or as much as we can; the water we try to do maybe two or three times a week, some of the climate stuff we do maybe once a month, for soil temperature stuff we do twice a month. Some of the stuff we do everyday, some once a week, some once a month (...) Thus, we do not have to do everything everyday; you pretty much do whatever works for you, whatever fits in your curricula. To me it is so easy to implement the GLOBE program because it covers what you already teach, what kids already use, which is neat.

How can people get engaged in different countries, those that are not doing GLOBE yet...

If you are going to do that in large scale, say an entire country in a large number of schools, you should contact the GLOBE office, and they will send trainers there, they will help you train. If it is just one school in Brazil, for example, probably the easiest way we have on the website is the called partners...

So you can engage as a school?

Yes. About 8 years ago, we did a study with the University of China, Hong Kong, and they had one student that understood English, and we did not understand anything of Chinese, but we did a lot of study with them (...) they did a scientific experiment. It was kind of hilarious because we did a telecon with the student, and he spoke English, but it was difficult to understand, it was funny (...) Kids have a great time, I had a class with Argentina, and it was so neat to see, because my class said, ‘hey, they are just like us'... we had teenagers on the camera here, and nobody said anything, until finally two girls from Argentina started to talk, and eventually, the teacher there had to shut them up because they were talking informally, it was so neat, that is a neat part of the GLOBE program, not only the data, but also 'hey, what do you watch, how do you do this', the culture exchange, which is worthy (...) I like this part, you can work with students and teachers in other countries. You can work with one school at a time, or a bunch of schools.

If I am a school in Brazil, for example, how can I get involved? Should I go to the website...

OK, I would go to the website, and send questions to a GLOBE help desk; they answer everything, they are the nicest people there…

Basically, GLOBE is for elementary and secondary level schools, right?

Yes.

And what would you say is the mission of this program?

To make students much more fluent in sciences and math, and to make them see that there is an end product, that there are practical uses for sciences and math, and they can lead the job. It is to have students getting a practical, hands-on use of math and sciences.

I am a teacher, and I want to get engaged with GLOBE. What are the resources needed?

It is very basic. If you just want to do some of the atmospheric stuff, you need water, a thermometer. It depends on how much you want to get in to it, and your financial resources. If you want to start from scratch and do all that we do here it would be very expensive.

But basically, they will need computers...

If you want to record your data and send that as we did, you need at least one computer; you need some thermometers, and they are not expensive. With those, you can measure both water and soil temperatures; some handheld things (...) like barometer to do the barometric pressure (...) I think the basics would be a computer, some thermometers (…) it really depends on which protocols you want to do, in most schools such as elementary in our area, we do the atmospheric, measuring the air temperature, a can or something to collect the rain, or snowing, and we can measure that (...) The big expense would be the computer if you want to send your data. You could just collect that, you do not have to send that in, but it would be important to have kids working on the computer, the use of technology involved, it would be nice to have at least one computer connected to the internet. That would be the big expense, and you could come up with other ways of getting relatively cheap, inexpensive equipment to do the protocols (...) You can use supplies that schools have (...) You can use the same thermometer to measure the air temperature and the water temperature. It does not have to be super special. (...) I think you can do it fairly inexpensively, and do it well. And if you have near some college or university that would be great because then kids could see the next steps of what they are trying to do (…) to see professors using the data to do research (...) to keep connection with professors who are using the data.

If I am a teacher and have difficulty with a protocol or whatever, how can I have help?

The help desk; you can call or email how many times you need, they answer everything, since little questions. If you telephone you have an immediate response, and if you email you have a response on the next day. They have been very good, they answer since simple questions, until complex questions about why equipment is not working.

You should have many people, many demands, many questions, since it is an international program...

The helpdesk we have here is not the same helpdesk they have in Europe. I am assuming that as they help all these people they should work 24 hours a day...

While you are working with the program, which have been the main difficulties you faced since the beginning, implementation...

Following the protocols is much easier now; when they first started this, the protocols were written by the scientists themselves... In the early days, I had a hard time to understand them, and that is why I learned to use the help desk, I did not know what they were asking for, so I did not know what I should have the kids do. But I think the protocols today are very much improved. Other than that, once in a while, there would be a problem with entering the data. They have re-written the curricula at least three times since the original, and it got much better. And one of the things they did, were teacher links, some pages on how to do, background on the protocol, possible examples and worksheets. On the student side, there is a blank worksheet with directions on what to do, how to do, and place inform answers. So they really made it much better to answer the data than before. When we send great amounts of data we can email this; if you do this once a month or two months, it is a problem. Further, many scientists have their own format of inserting data so they want you to change yours. You can put that on an Excel spreadsheet and put that in columns (...) many scientists do not use that, they have their own version, so you have to convert that, which takes time. Most of the protocols you can do in class period. The conversion time takes 2 or 3 days. At times, the teacher does not have the time to stay some 2-3 days with the entire class only to convert data. I found a way to get around it; I had some students who learned how to do that conversion, in the following year they just happened to have a study hall kind of a kitchen, so they came in and talked to students on how to do it, but that is usually a problem. But they are just a few issues, most of the time students go to the website put the data, if they realize the data is not right, they take care of that (...) there are not big problems anymore.

At this point at the GLOBE program, which do you consider to be some challenges?

Gathering data, I guess it depends on the level you want to get. If you are using just thermometers and the basics everything is pretty improved, but if you go to other levels there are more expensive equipment that you cannot buy; it was not given to us. It is more difficult because if children break, you cannot replace (...) We are doing a nice job, we designed the website, we are doing the actual protocols, because kids have the worksheets and I make sure that before leaving the class kids know exactly what to do and how to do before we go outside. That probably is going to be a problem for some teachers if they are not comfortable with kids out of the classroom. You have got to be very organized, cover everything before you go outside, because explaining stuff during the data collection is difficult. Students are in the water, with the equipment, they are not going to listen to you. Our program is really streamlined, it flows, but once you have a plan and the kids know what to do then they could be outside. As for supervision, at least my kids are not always supervised. I have 25-30 kids, they are spread the entire way around our building campus, I cannot see them all (...) You have to make kids get aware that they cannot get in trouble, otherwise they could not leave the classroom again, I keep warning them of that point. In so many years, I had one kid doing something wrong that he should not have done (...) Teachers have a problem in the beginning ruling the kids, teachers are used to having kids in their seats, quiet, on tasks (...) when they are outside, there are five or six different things going on, different protocols at one time, and teacher has to make sure students are doing what they are supposed to do, and there are any problems, but many teachers ask me 'how can you do that?' 'Don't students try to get away?' In the beginning I had to set up a camera in one spot, and had some teachers back on the room looking down. I could go across the hall and see them all, and they knew I was seeing them all without being with them, so I planned that early on, they knew I saw them, there was a point I was not seeing all groups, but they did not know that. So that was the way to get around that. I was very fortunate; I just had very few problems with the kids outside. I told them we were taking field trips in local dams, in reservoirs (...) they wanted to do that.

You have been with GLOBE program for more than 10 years...

And enjoying that, it is fun class to teach if you like being outside, and you like doing the science; it is a perfect class to teach and the kids, they have fun enjoying stuff, and they are actually learning all the time, you are not going to have many problems. They like what they are doing, the field trips... Something else you should know about GLOBE, it has students having fun while learning sciences, they go into the water, get dirty, they love all that, record the data, work with a piece of equipment, there is always something for everybody, every level. The GLOBE program reaches all academic levels, since kindergarten, which is great. The teacher's guide instructs on how to teach the protocols; they really come up with a lot of ways to help teachers come through any problems.

4.10. INSPIRE

The Interdisciplinary National Science Project Incorporating Research and Education Experience – INSPIRE maximize student participation and involvement in STEM, enhancing the STEM pipeline from high school (grades 9–12) into the undergraduate level. A solicitation for this concept was initiated in FY 2007 and will be awarded in FY 2008. Students work along with NASA professionals, collaborating in their ongoing researches and acquitting unique know-how. Please note that there is another practice at NASA Goddard named INSPIRE, which refers to the Interactive NASA Space Physics Ionosphere Radio Experiments, existing since 1989. Further information can be found at: http://image.gsfc.nasa.gov/poetry/inspire/.

Figure 34. High school interns at NASA. Summer inside NASA are a unique opportunity to students. Credit: NASA.

The initiative allow high school students to have the opportunity to work during the summer with real NASA scientists and engineers in their diverse fields of expertise and to develop high level capabilities that will be helpful for their future academic and professional life. The idea is also identify potentialities to constitute part of NASA workforce.

4.11. NASA TV EDUCATION

As we have seen, NASA TV Education Channel features International Space Station and Space Shuttle mission coverage; live special events; interactive educational live shows; electronic field trips; aviation and space news; and historical NASA footage. Programming has a 3-hour block and includes the Video (News) File, NASA Gallery, and Education File – beginning at noon eastern time and repeated four times throughout the day. This channel offers programs appropriate to each grade-level. Further information can be found at http://www.nasa.gov/multimedia/nasatv/ [128].

Figure 35. NASA TV Education studios at NASA Goddard. Credit: Norma T. O. Reis.

Interview with NASA TV Education expert (Summer 2008), with permission

Interviewed: Ryan P. Fitzgibbons
Position: Education Producer, NASA TV
Employer: NASA Goddard Space Flight Center

What has motivated you to work for NASA TV?

I was looking for a position that would fit my background, because I have always loved sciences, and my background is in micro-life sciences. I have always enjoyed learning about any kind of sciences, astronomy, medical sciences; anything that is new and upcoming. So I heard there was an opening for a producer here, then I applied... My big motivation was to find a position where I could use my science background, passion and skills in TV production. And they heard me!

It must be really exciting to work at NASA TV.

It is amazing to work in a place where we have the brightest minds in the country (...) Everybody wants to get messages of what they are working on, and we have to figure out the best ways to communicate the research with whomever we are talking to - kids, casual TV watchers, people surfing the web.

How do you work with technical people in order to have their messages understood by the general public?

It is a lot of talking to the scientist and the team. I assume that if I can understand, if I can rely that to a kid, a parent, another producer, and someone like that understand, then it is a start. The first thing, I need to find out what I need to understand about the subject, the research, in order to explain that to people who may not have any science background. And I believe the biggest challenge is to find out how we can get something that is very technical into the understanding of everyone, how to make it engaging and understandable to a very wide array of people who have different educational and cultural background; it is try to find a common ground, stripping away all the technical specificities of it, and try to get to the big message, the implications of that, and how it can relate to the ordinary people. It is a little bit of magic (...) a lot of magic... Some scientists are very good at talking to the average person, the next door neighbor, teachers etc., while others have all that up in their heads only; they understand, but we need more people to understand and get excited about that, so that is magic...

Which techniques do you use to communicate in order to get the content engaging?

I always use myself as a reference; if there is a way I find something interesting I think others will also do, and I always ask: How will this thing that happened two days ago or two thousand years ago, how will this tie to me, my life, and my world today? I think that finding these connections is a broad philosophy of mind that a lot of why we search for things using science is to find out how we as humans fit into this world, into this universe. It is kind of a selfish thing to think about, I think people want to find out how they fit into a world, into a little community, into a solar system - it is all about relationships. So my technique is just to figure out what about this, no matter what it is, may it be plasma physics, or ocean currents, some fact that occurred long ago, why would I find this interesting, why would I find it relevant in my day-to-day life? And by finding this link I can figure out how to make the connection between scientific content and people in the living room, who have lots of channels to choose, and if they are surfing on the web, they have even more.

There is something unique about NASA, and the way it communicates with the public. How would you describe this uniqueness?

From my experience, NASA really wants the public to know. Researchers and NASA's Office of Public Affairs want the public to know what is going on because NASA is a governmental institution, so people should know; NASA wants the average people to know why this certain development is important, how it can lead to some cutting edge technology, to make their life easier, whatever impact that could have in their livelihood. The commitment is to be open, to engage the public. There are just so many amazing things going on at NASA, technologies emerging, 3-D images of the Sun, and we want the public to get excited about it.

How did NASA TV start?

I think it was with the Apollo; they wanted to record what was going on with the astronauts.

How would you describe NASA TV Education mission?

I think it is to reach not only the general public, but also other scientists around the world; it is just to create communication among different people; the ones who are casually interested in space exploration, and those who are doing similar things in other countries in space exploration. The mission is to communicate in a way that engages people, and that is exciting. Even if the actual research is not coming totally new, there are always greater possibilities coming out of research, and we want people to get excited about it, such as new methods of waste disposal, heating and cooling etc. The main point is to create bridges of communication among very different pockets of people in our country and worldwide.

Who is the general public of NASA TV Education?

That's a big question; some say it is the general public, I say 'general public' is such a broad term, it is really hard to define, it could be your next door neighbor, a mechanic, a fifth grader, another researcher, so to say the 'general public' mixes a lot of people. In the education area, where I work, it is school age kids, from kindergarten, first grade, all the way up to high school, and undergraduate. We try to reach people without technical, science background. Our standard is that if you have a 6th grade education at least you can understand what NASA TV is trying to convey. It is popular science, so it has to appear to the populace. If it were too technical just a few people would watch. The way NASA TV education is designed, we want to inspire people to search for more information. For example, they see something about eclipses and space weather, find that cool, but decide to find out more about that, to go beyond that sixth grade level of comprehension. And maybe some of them will end up working here, will want to devote their careers to something that they saw on NASA TV.

I suppose you intend to contribute to teachers, and get feedback about the interaction with the general public and teachers. Do you have an instrument to receive feedback on what people would like to see when they go to NASA TV?

The Education Department is responsible for finding out what the teacher wants to teach and what they want from NASA, how they can integrate it in their lessons, and they have to make sure that we meet some educational standards set by states. These people get the feedback; I think they get feedback directly from teachers, from informal education institutions like museums, parks and so forth. I think if they had seen something that NASA TV has done, they can say that was great, that really fitted in well in our lesson about measurements or something, or we would like to have something to explain about eclipses, or other themes. I think the actual education people who are in the TV production do work figuring out what educators need. Then the education people at NASA come to me and say, hey, we need something to fit in a lesson about measurements. We also have the Digital Learning Network – DLN and that is like a teleconference but you have a camera with the presenter here and they can see the students and teachers at school, and they can see the presenter, and they can hear each other. We just did one (...) for young kids, we have integrated the character WALL-E and presented some stuff (...) They explained some missions coming for Goddard mostly, so they had WALL-E, and showed things about missions to the sun, to the moon, and there was little tricks and so forth, but that is integrated to a live presenter, it is funny because the animation is all done beforehand, the people at NASA TV personalize that, so they make a WALL-E showing pictures and things related to NASA Goddard. It is all done beforehand and we have the live person, and you have to rehearse because, it is like doing a play, and you have to interact with people that are not there, and then as it is broadcasted it is all put together as one. So the live presenter is talking about something that is going on at NASA, such as satellites we are sending out, but he also interacts with the pre recorded, animated character. When it comes up it looks great when ready, when children see that, kindergarten up to third grades, they love WALL-E. Children sometimes interact with the astronauts, ask questions, such as how do you go to the bathroom in space, and for the most part these DLN modules are designed to be to grade-specific, about what they are learning in their science or math classes, and there is a lot of feedback on that.

Is DLN part of NASA TV?

It is a hard question; there is NASA Education, and there is NASA TV education. The DLN is where the two come together, but it is more TV and production serving the purposes of education, so it is a big teamwork between TV, media people, and education.

Which educational levels does DLN serve?

It serves elementary and secondary education. It is like a guest speaker coming in your class, our speaker stays here, and the classroom is wherever they have the technology.

Is DLN serving all US states, municipalities? How do they contact you, how is the interface?

Not all schools has access to the equipment, and I think NASA has helped set up many hubs for schools around the country. About schools infrastructure in the U.S., there is no standard, we have schools with lots of resources, and schools with a couple of rooms and some books; but in this case they can interact with DLN in other spaces where they find the actual facilities, and at the very least, there are DLN lessons on the web, and you can download that, it is not interactive, you do not have that live person and students asking questions. But that is an alternative for reaching those NASA contents.

And who contacts you - the school, state, municipality...

It is the school.

Which do you consider the main challenges of NASA TV education?

Defining the audience, to be more specific. General public is something nebulous. For example, when doing a program about eclipses, we need to be specify - is it for everybody or are we making something that teachers can use? Another challenge – finding out an engaging way to communicate the research to the audience.

Which do you consider qualities of a good communicator?

They must know well their field, and be able to communicate in simpler terms, avoiding technical jargons. Also, they should use metaphors and analogies, visual elements, putting the content in a tangible, everyday situation, so that people can understand. It is difficult, because these scientists spend all the time in their research, they are not thinking about explaining that for the neighbor or for a school child. Goddard has a media-training course lasting over a few days to these researchers, for this purpose of public dissemination. For example, they should know that explanations on TV have to be short. And the researcher should be excited about what he or she is talking about, otherwise how could them engage people?

4.12. The AEB School Program – A Brazilian Initiative for the Enhancement of STEM

A pilot project named Project AEB School has been carried out in a Brazilian low-income city nearby the Brazilian capital Brasilia. The purpose has been to access its performance and verify possibilities of replicating the practice in other parts of the country, disseminating the idea. As it presented notable success in engaging students and educators in the excitement of discovery and exploration, the Brazilian Space Agency decided to adopted strategies to reproduce the initiative in other cities of the Federal District and subsequently to other parts of the country, turning the project into a Program and promoting a) in-service training for teachers, some carried out in partnership with the Brazilian Ministry of Education; b) attendance and organization of scientific and educational events; b) production and distribution of didactic materials to schools, by means of strategic partnerships such as the one with the Ministry of Education, which has been a strong partner to the Program; c) joint organization of the so-called Astronomy and Astronautics Olympics – OBA.

According to I.M.S. Rodrigues [129], manager of the program at the time of this research, the AEB School Program started in 2003, firstly by her personal desire of contributing to national education. And secondly due to the fact that the Brazilian President, Luis Inacio Lula da Silva, stated in his first mandate that he would like to see elementary and secondary level schools visiting the Space Agency and having more contact with space activities. Then she started to work in the conception of the program. The program currently attends middle and high school levels nationwide and has several important partnerships.

AEB School offers activities for both students and educators. Students attend lectures, participate in interactive expositions, astronomical observation, contests for the construction of telescopes with low-cost materials, construction and launching of rockets, amongst other hands-on, interactive activities.

The strategy of in-service training for teachers aims to develop competencies to enable those professionals to effectively work with space-related contents, which are presented in a contextualized and interdisciplinary approach, so that allowing its utilization in the diverse scholar disciplines.

One of the main strategies of the Program is to relate, by means of interactive activities, contents of the several disciplines taught in the classroom with themes of the space area and its applications in the quotidian of the student. Hence, it is expected that the Program can contribute for that students "learn how to learn" and that teachers seek for innovative and creative teaching approaches.

The actions of the program are characterized by the integration of activities of public and private entities dealing with STEM teaching and space thematic, and/or interested in supporting this initiative. Actions in cooperation increase the effectiveness of processes, reduce costs, and extend benefits for a larger quantity of individuals (AEB, Diario de Bordo).

All interviews here presented have been recorded and are available in full version, under direct request to the author, on the e-mail: norma.reis@gmail.com.

Chapter 5

PEDAGOGICAL ANALYSES AND RECOMMENDATIONS

Space science education has become a pedagogical practice more common and widespread both in developed and developing countries in the last decades. It comprises a set of pedagogical practices, including methodologies, approaches, techniques and contents utilizing the unique motivation, curiosity and fascination exerted by space and its exploration in order to capture students' interest towards STEM and to promote public awareness of space sciences, exploration and its benefits. Last but not least, these practices also have the purpose of encouraging students to consider career paths in STEM and in the space sector. The nature and scope of these practices in each country is a function of the level of investment in the corresponding airspace sector and of the nature of each given space program. For instance, in countries such as the United States and Russia, in which space programs comprise a large and complex set of initiatives with several purposes, including exploration, the generation of high technology products/processes/services, and the expansion of the human presence in the cosmos, practices in space education are equally ample in scope, purposes and resources. As for developing countries, in which space programs are targeted to solve specific societal demands, such as Earth observation, for instance, such practices are usually targeted to more specific purposes, as well as to promote awareness of the benefits of space exploration for societal needs, as their corresponding space programs have limited budget and usually need to obtain further social and political support. As space educational practices are commonly carried out by governmental organizations, majorly space agencies, they are also concerned with inspiring and preparing their future employees. Another reason why space agencies carry out space education and outreach programs is that such practices are capable of gathering and strengthening sustainability for their programs, as they reach teachers, students, general public, decision-makers and other key societal actors. As a matter of fact, the more people understand the space enterprise and are aware of the benefits space exploration brings to society, the more they will be willing to support the maintenance and amplification of these activities.

As an emblematic example of practices in developed countries, NASA Education programs have been serving educators and students of all educational levels, as well as the general public, both in the United States as well as internationally, as long as a wide range of resources are made available online so that they can be downloaded from the NASA

Education Web site. Additionally, the Agency has established international partnerships so that specific NASA programs/projects/activities are also carried out in other countries, such as the Globe Program and the NASA Explorer Schools. On the other hand, in developing countries, with some exceptions, space educational practices do exist, but are considerably restricted in scope and resources. Fortunately, this is not the case in Brazil where an impressive program is carried out with sustainability thanks to the organization and strategic partnerships that the AEB School Program has been establishing with several public and private, national and international entities. Partnerships are vital for developing countries; they amplify possibilities of actuation, make possible the injection of additional resources for the initiatives carried out, and in some cases are also capable of bringing the advantage of sharing unique expertise from public and private entities. Another essential difference between the NASA and Brazilian programs are structure and management. NASA initiatives are to a great extent decentralized, due to the large number of activities and to a structure in which each center is responsible for providing support to a specific region of the country. In Brazil, initiatives of the AEB School Program are more centralized in the Brazilian Space Agency, but as the program tends to expand, it is possible that it may become more decentralized as a result of the large geographical dimensions and diversity of the country, as well as of the appropriation of such practices by the education systems and the resulting achievement of autonomy in the conduction of their own independent space educational practices.

In the case of NASA, a model of excellence in terms of resources, facilities and personnel has been presented. Structure, implementation and a sample of state-of-the-art programs have been discussed. These initiatives have pedagogical possibilities and constraints, which will be hereby analyzed.

One of the furthermost advantages of NASA Education is its policy of promoting awareness of the Vision, Mission, activities and accomplishments of the Agency. NASA is unquestionably open doors to society. Moreover, NASA unique resources, facilities, and personnel, makes it possible that its space educational practices capture the interest of students of all educational levels in a singular way. As NASA works with four levels of engagement – inspiration, engagement, education and employment, the Agency is committed with the student since the early stages of his/her human development until this he/she is fully prepared for a job occupation, possibly at NASA – actually, one of the purposes of NASA Education is to prepare the next generation of explorers – part of this generation will certainly work for NASA. In this context, the Agency is also preparing its own future employees.

NASA resources for teachers – online and printed – constitute valuable tools for teachers of all levels: Hubble images of the universe, hands-on experiments, didactic books, and so forth are helpful for educators as they contribute to the enhancement of pedagogical practices in STEM and other scholar disciplines, such as geography and history, helping teachers prepare more exciting classes. In this sense, for instance, it makes a great difference whether a mathematics teacher presents a trigonometry problem out of context, or if he/she presents it in the scope of a real space exploration problem. It also makes a difference whether the geography or science teacher presents environmental issues such as deforestation and climate change only by discussing the contents with students or if the teacher, in addition to that, also brings to the classroom satellite images and instructs students on how to "read" those imagery, comparing data on environmental changes in different countries and in different historical periods, for example. It also makes a difference if the Physics teacher discusses space weather with students and if this teacher also presents movies, Podcasts, and Webcasts to students so that they can visualize the interaction between solar wind plasmas and Earth's magnetosphere, and how this interaction can affect technological systems whose operation are vital for telecommunications, navigation and astronaut, for instance.

In addition to resources of quality, interactions with NASA Education specialists and other NASA professionals such as astronauts, *in situ* or by means of other communication strategies such as the Digital Learning Network, are of fundamental value to help teachers consider other possibilities for the teaching/learning process, making classes more dynamic and interactive. NASA professionals, through short-term courses, workshops, lectures, events, expositions, visits and other activities act as a bridge between the physical/electronic resources and teachers. The success of this interaction has positive impact in the subsequent interaction of educators with knowledge, students and the scholarly community. Another strong point of NASA Education is the concern with evaluation of their materials and activities. In fact, many NASA materials accompany a questionnaire that educators are asked to answer and send back to the Agency, about the perception and effectiveness of that specific resource, so that they can be adapted and enhanced based on users' feedback.

The NASA Offices of Education also dialogues with the Department of Education in order to find out what are the system goals and principal needs in education for each level and how the agencies can act with synergy to offer activities in response to those demands.

The Agency has also a strong commitment to inclusiveness and diversity, by means of programs such as the NASA Explorer Schools, which aims to strengthen STEM among underserved communities, and projects such as the Kiosk, along with the development of special didactic materials that aims to facilitate the understanding of space sciences by impaired students – blind and death students, so that contributing to promote effective inclusiveness of these special students in the educative process, in recognition to their legitimate rights of access to education with quality.

NASA also takes great advantage of hands-on activities and experimentation, and promotes interactive activities using new education technologies, such as the Space Weather Action Center. Another especially interesting hands-on NASA initiative is the Globe Program, because students have the singular opportunity to do real science and interact with scientists and international students. The Agency also brings students inside its walls by means of strategies such as the INSPIRE program, through which high-school students carry out internships at a specific NASA Centers, contributing to ongoing research and activities developed by NASA scientists and engineers. Part of those students, as they demonstrate excellence in their activities, could be hired by NASA and follow up with their research at college and graduate levels. One current constraint of this program, though, is that it is still contemplating mostly students coming from elites. It is necessary, thus, to identify mechanisms by which excellent students coming from underserved communities could also be identified and included in this program, in order to assure also in the scope of this program the NASA principle of promoting inclusiveness and also because it is a fact that talents can be found in all societal segments, although not at the same percentage. One potential idea would be to use national exams carried out by the Department of Education to identify and select those students, so that making the process even more transparent.

NASA Education is large and complex structure that requires effective communication strategies both between NASA Education specialists and between these specialists and the scholarly community. In the first case, communication plays an essential role as it is necessary to effectively coordinate initiatives and interactions among the Assistant Administrator for Education at HQ, the EOs, EPOs, program/project managers and other NASA Education actors. In this perspective, in the internal NASA Web site there is a place in which NASA Education personnel can post their experiences – the section is named Lessons Learned and it is of fundamental relevance in order to replicate successes and avoid the reproduction of failures. In the second case, that is, the communication between scientific and scholarly communities, it is essential to make activities as clear as possible, otherwise the entire enterprise would be devalued. It should be strengthened and enhanced by the establishment of multidisciplinary teams to discuss, implement and evaluate altogether the efforts carried out before, during and after the development of space education initiatives. In addition to that, special courses should be organized for scientists and engineers who are supposed to "teach the teachers." These courses should focus on improving the scientist's communication skills so that he or she would be able to have his or her message understood by the scholarly community and by the general public. These courses should desirably be taught by specialists in education or by scientists with a graduate background in education. Another idea would be to include scientific journalists as instructors, due to the fact that they pursue a unique and singular know-how in communicating and engaging the public that, associated with the pedagogical skills of the educator and the knowledge of the specialist, could result in very high quality communication.

Another aspect to be highlighted is the valuing of continuous mentorship. Although NASA courses/workshops for educators are usually effective in accomplishing their mission of providing in-service preparation of teachers in specific space contents, it is recommended that NASA conceives strategies to extend the period of interaction between their specialists and the educators by means of maintaining channels of mentorship after those activities are concluded. Actually, it is very common in education that teachers attend courses/seminars and become fascinated by the possibilities of using new resources in the classroom, but many times when he/she returns to school and faces the same difficulties imposed by their specific scholarly reality, they end up giving up trying to implement the new strategies. The maintenance of mentorship after the course period makes it possible for the educator to share his/her successes, difficulties and concerns with the NASA specialist, and will receive feedback for at least part of them. At the same time, the maintenance of this interaction makes it possible for NASA to access, after the end of the pedagogical intervention, how successful the practice and the communicative process has been in terms of leading teachers to understand the topics presented, and exploration of pedagogical strategies. This interaction will therefore contribute to future courses in the same topic that will have enhanced maturity, generating better responses for new educational demands, in a cycle of improvement and mutual collaboration. The simple application of questionnaires in order to access teachers' satisfaction and understanding is insufficient to access how those professionals will use such knowledge and methodologies in classroom.

Another recommendation is related to the consideration of audiences' previous knowledge and experiences when designing courses for teachers. As individuals with different backgrounds usually learn in different ways, it is desirable to avoid putting together, in the same audience, for instance, both elementary and secondary level teachers. These professionals deal with very distinctive educational realities. They have been prepared to work with a specific type of student, at very dissimilar levels of human development, and their filters of reality to understand technical content are also different. Also, courses for elementary level teachers need to be presented at a different language level than those offered to secondary school teachers. As usual in education, specialists too often do not consider the audience to which their speech is being targeted, so that although they teach, the teachers do not effectively understand and are then unable to replicate the knowledge in their schools. Thus, putting together different audiences that learn differently can compromise the effective understanding of one of these audiences.

When teaching teachers, it is also fundamental to consider the audience that these professionals work with in a daily basis. It makes a substantial difference whether they work with underserved communities, for instance, or if they work in a school that does not access sophisticated technological resources, if there are high indicators of criminality in the school area and so forth. In this case, NASA Education specialists should adjust their pedagogical proposal for each specific audience and present it in accordance to their needs, interests and in a manner that the ideas presented can be effectively translated into practices after the end of the intervention. It is also fundamental to find out if the school pursues the basic resources necessary to implement the practices and, on the contrary, verify if NASA has possibilities to fill in these gaps by providing financial or physical resources for the school in order to make it possible that it participates in the effort proposed.

Space sciences and exploration should be also strengthened in the scientific curricula. I recommend here that NASA also discuss with the Department of Education about updating national guidelines for contents and methodologies for STEM teaching in schools. This initiative would assure a national impact and facilitate even more NASA Education initiatives overall. Although NASA pursues autonomy to do education and establish partnerships with schools and education systems, it is also important to synchronize efforts, discussions and articulations with the Department of Education as a manner of promoting better STEM education for the interest and benefit of the entire country. This initiative would be closely related to the strengthening of pre-service courses for teachers. As a matter of fact, several teachers' practices are gathered at this initial stage and, if one has the purpose of changing curriculum, a national effort to change the preparation of teachers at college level should also be carried out, and NASA in this case would be the strongest partner to promote this substantial historical educational change. This is especially relevant now, as NASA has a purpose of expanding human presence in the universe by means of its Vision – if we want to explore more, we also need to promote more awareness and engagement to this enterprise.

As for the manner in which contents are presented on the NASA Education Web site, I recommend the conception, pilot-project and further implementation of a unified Web-based interface for NASA space science education projects/programs, which could be established by consolidating a new database, somehow simpler than the present general database. Through such a project, teachers could find the category of resources and information they are looking for more easily and effectively. NASA Education is extremely ample and complex and finding information and material sometimes can be difficult because the user in many cases does not know exactly how to encounter what he or she is looking for. For instance, one could organize space science education programs/projects by specific categories, such as: a) education levels; b) type of material: fact sheet, Podcast, video, book, and so forth; c) theme: e.g. astronomy; geology; space weather and so forth. The idea would be to centralize all that NASA does in space science education in one single Web-based interface. The link for this database should be visible in a home page for NASA space science education, as educators and students might find it difficult to find what they are looking for whether they are not familiar with NASA's Web page structure.

As an alternative to assure a deeper understanding of NASA Education pedagogical practices, I recommend development of medium-term research on NASA programs/projects as one alternative to effectively access the possibilities and constraints of activities, courses, and workshops for teachers. This initiative could be implemented by granting scholarships for college and graduate students belonging to the education field so that they would develop their masters or doctoral theses by investigating in a more comprehensive and systematic manner those programs by interacting with NASA Education professionals, school teachers and administrators, as well as students, by applying questionnaires, conducting interviews, and other research strategies in order to better access that reality. Participative research is also recommended due to the fact that it allows the immersion of the researcher in specific scholarly realities, allowing the easier identification of difficulties and potential channels of solution.

Another practical recommendation would be the amplification of the Agency budgets targeted to underserved communities by means of projects such as the NASA Explorer Schools, so that more children would become excited about STEM in those segments. Partnerships not only with the Department of Education but also with private institutions/organizations that have foundations committed to the advancement of education in such communities would be recommendable. In Brazil, for instance, in the scope of specific activities carried out by the Brazilian Ministry of Education, partnerships with the private sector have been especially helpful not only in terms of promoting synergy of efforts but also of taking advantage of the singular ideas and expertise they have developed. Another recommendation would be to spread ideas such as the Kiosk and other projects/products targeted to impaired students so that teachers can replicate them in schools, strengthening the commitment of NASA Education with inclusiveness and diversity, by amplifying the engagement of that special public in the excitement by STEM and encouraging careers in this sector amongst those students.

Another suggestion would be to amplify the dissemination of NASA Education activities by means of strengthening marketing strategies, such as announcements in open television, as well as to strengthen NASA Education TV presence in open television, to distribute magazines about NASA for elementary school children throughout the entire country, in coordination with the Department of Education, in order to reach all schools in the country more easily and less costly.

Finally, it is an unquestionable fact that NASA has reached a stage in which its brand is also international. Actually, its insignia is the best known among all space insignias worldwide. NASA represents much of the human effort to explore and understand space. In this panorama, NASA can be also somehow considered as a human exemplar, a universal symbol. By leveraging this unique branding, NASA could helpfully amplify its international activities through enhanced partnerships around the world. The amplification of international partnerships and the promotion of international events would be helpful to access what other countries are doing in space science education and education overall, both developed and in development, in order to share knowledge and expertise and also to take advantage of the beneficial ideas and initiatives that could substantially leverage the spirit of NASA's reality and purposes.

CONCLUSION

Education is a complex and multifaceted process that has been facing several challenges presently, intrinsic or extrinsic to the educative process. It is necessary to promote inclusiveness in schools, to value and integrate diversity, to reform the curriculum, to review methodologies, techniques and education technologies, as well as to transfer the focus from disciplines to themes due to the transversal nature of contents and the increasing level of societal complexity. As a matter of fact, students currently need to develop the competency of permanent learning due to the fact that knowledge changes incredibly quickly and, with Internet, individuals also need to be selective with regard to what is effectively relevant to learn for their professional, academic, and personal development.

With the current scenario of globalization of economy and competitiveness, the work market has become increasingly selective and demanding, requiring more complex skills in all professional fields, even the more simple ones. Scholar institutions and academia are making a special effort to pay attention to these demands, although the educator is not always prepared for the new scenario, due to the fact that he or she has been educated in a different context and his/her academic preparation did not cover these new demands. In-service short-duration trainings, by its turn, although extremely important to keep educators updated with advancements in their respective fields and to present knowledge and skills are not always enough to promote the necessary qualitative changes in the teachers' *habitus*.

Another difficulty arising from this is the fact that education has not been successful in preparing the integral citizen – not only for the work market, but also for life overall. This includes preparing students in knowledge, competencies, values, and attitudes for life. It also includes promoting an education committed to the betterment of society, by helping students become actors of positive changes in the societal contexts in which they act and interact. In this perspective, subjects are supposed to be critical with regard to their realities, not accepting situations passively as if they were unchangeable, but by identifying potential failures of the system, unfair contexts, then identify and take effective actions that can transform that reality. This education clearly comprises the development of confident attitude and the promotion of intellectual autonomy.

Hence, education needs to radically change in order to fulfill the needs of preparation of integral citizens (indeed, global citizens). It is clear that this type of education is much more challenging and demanding than education for work market only. Investments in this kind of education cannot weaken the teaching of scientific contents, though. It is necessary to find the ideal balance between teaching contents and promoting integral education. The main reason for this concern is the fact that knowledge is essential for that students coming from underserved and underrepresented societal segments can empower themselves and become capable of breaking the cycle of economic subjugation in which their social class usually fall. Knowledge is a highly important element, although not the only one, essential to break the cycle of reproduction of societal inequalities.

In this panorama, space science education is a powerful pedagogical practice capable of contributing to that fundamental goal of fostering intellectual autonomy amongst students of all educational levels, by strengthening science education overall, due to the fact that, as this practice captures the interest of students for space, they also improve students' understanding of STEM and, the more easily they can appropriate of STEM knowledge, the more easily they can become interested in learning more or even following career paths in those fields. These practices, as a result, contribute to raise the level of scientific and technological literacy of both scholar community and general public.

Activities in space science education promote space awareness and are especially relevant in the current panorama in which it is necessary to make populations understand the urgency of taking effective actions to preserve environment – from infinite universe to local environment – someone's garden, neighborhood, village, and city. It is necessary to promote this awareness which, whether starting from immediate environments, will end up including wider ecosystems; that is, ultimately Earth and space environments, at all temporal and spatial scales, so that such awareness would generate concrete attitudinal changes such as national and international actions towards the mitigation of problems as space debris, for instance.

Since the beginning of the space exploration enterprise, NASA has been working towards engaging and exciting students, teachers and general public about its Vision and Mission, although the latter are subject to change with shifting political administrations (e.g., the Vision for Exploration focused on the Moon and Mars could be revised under a new President).

NASA practices in space science education present the uniqueness of counting with a large variety of high quality online and printed didactic materials, courses, events, personnel and infrastructure in order to make that happen with the highest level of quality. It does make a significant difference the fact that students know their country has space shuttles, the ISS, an Astronaut Corps, realistic plans to go back to the Moon and on to Mars and beyond, and a considerable amount of job opportunities in the sector. Whether one is aware he or she can work in the space sector someday, citizens could invest more on that and follow careers in STEM or other space-related fields.

Currently, NASA has been investing in the development, introduction, training and implementation of new technologies in education to be used as powerful resources to capture students' interest in STEM. Actually, the integrated use of resources such as NASA TV Education, the Digital Learning Network, Podcasts, Webcasts, Vodcasts, animations and so forth has been contributing to fulfill one of the educational missions in the current panorama of globalization, that is, to promote scientific and technological literacy and inclusiveness. However, there are constraints in this panorama, including the preparation of teachers to effectively use those tools. There is also the constraint related to the fact that many schools do not count with necessary equipments in order to implement such activities.

Some other aspects were identified during this research. Even currently in a highly technologically developed country such as the United States, as it can be verified on the news, a considerable number of citizens believe that the budget invested in space exploration would generate more benefits by being applied in other societal demands. Other individuals believe that the Vision of going back to the Moon and Mars does not bring sufficient contributions to society. It means that it is necessary to strengthen not only activities that highlight the excitement of exploration and discovery, but also of actions and products that help the general public understand that space exploration brings singular socioeconomic benefits, and also assures a privileged position in the worldwide scenario, due to the fact that it is cutting-edge science. Actually, it generates uncountable direct and indirect benefits to almost all other sciences and technologies overall. Top-level scientific and technological development, allied to education of quality, strongly contributes to the assurance of autonomy and sovereignty for countries, as well as strengthens security.

The introduction, in every single course, training or other pedagogical strategies promoted by NASA, of a module or debate on the benefits of space exploration – in the past, currently and the predicted benefits derived from the implementation of the Vision would be helpful in order to make the public even more aware and supportive to the medium and long-term NASA plans, strengthening societal support.

Another constraint identified in the scope of some NASA Education initiatives, also very common in educational research, has been the difficulty of accessing results of short-term activities in space science education, such as courses and workshops, in which the teacher is left with some new knowledge and printed material, but there is no certainty whether he or she effectively understood the content and, most importantly, the pedagogical proposal to use it effectively in the classroom.

Furthermore, I believe that some budget decisions should be revised in order to fulfill the special needs of initiatives targeted to underserved communities, due to the fact that these actions are vital to the development of those students and teachers. As a matter of fact, these initiatives are capable sometimes of changing some of these students' realities, allowing them to consider other fascinating possibilities for their lives and careers in STEM, which they would never conceive otherwise. Programs such as the Explorer Schools should receive more resources in order to amplify the number of schools assisted and bring more young kids from those segments engaged in the excitement of STEM and potentially also in careers in the area. Likewise, initiatives such as the Space Action Weather Center and the Kiosk should develop mechanisms to reach more schools in the country and eventually internationally, due to the fact that they encourage students' autonomy and, as for the latter, it also help impaired students to better understand STEM and potentially become professionals in the area.

The case of the AEB School Program, by its turn, consists of an example of the manner it can be carried out space education with quality and at the same time with limited budget. The administration of the program invests largely in partnerships with public and private sectors, national and international institutions. Currently, the Program is centralized in the Brazilian Space Agency, and its foremost activities comprises in-service training for teachers, activities and events for students, promotion of events and national contests for students, production and distribution of material for educators and students nationwide, with the cooperation of the Brazilian Ministry of Education, and other actors. The activities conceived in the scope of that Program takes into consideration the reality of the Brazilian education systems, permeated by considerable disparity in terms of financial and technological resources available in different regions of the country. In fact, Brazil has significant internal differences in terms of distribution of wealthy, that is, some regions are much more developed than others. In this panorama, the Agency's priority is to develop low-cost experiments that could be replicated in any school of the country, regardless of socioeconomic development indicators. The main purpose is to benefit students belonging to underserved and underrepresented societal segments.

Finally, it can be summarized that space science educational practices at NASA, primary object of this project, has been achieving growing success nationally and internationally, notably at present with the increment of the adoption of new technologies in education. There is still considerable space for improvement, in terms of evaluation, communication and feedback collection of teachers/students/communities, and amplification of national and international partnerships. Actually, when it is about education, it is always possible and desirable to make things better, as the more education develops, new difficulties arise, challenging educators to conceive, design and implement innovative and creative solutions towards an education with quality and for all. It is the constant flux of pedagogical practice, which is never complete. As society changes, education also needs to change. And as society is in permanent process of transformation, it is necessary that education overall and space science education specialists and decision-makers be attentive to what is new and what is necessary to change or adapt in order to provide support to new necessities for the motivation, updating and preparation of teachers and students in space sciences, STEM and beyond.

How it All Started – A Childhood's Dream

In the pathway to the stars there are angels of light.
There are poems and songs on beautiful things.
There are warm and friendly embraces.
The pathway to the stars is made up of love,
Since it has been drawn by God in one instant, for all Eternity.

How are dreams born in someone's mind, in someone's soul? It is an intriguing mystery. Would that be God's inspiration? Just imagination? Why do some people go through everything and accomplish their dreams, and others let then die while facing difficulties? Possible answers would be financial resources, opportunities, and persistence (also called 'resilience'). Another possible answer would be: education. Yes, education can transform someone's life. Here I share some of my history, a history that brought me to NASA for the dream of a wonderful summer internship, which inspired me to prepare this book.

I was born in South Brazil, in a nice city called Curitiba. It was known as the Ecological City, because of its beautiful parks and concern with environmental issues. The city was also called University City because of its concern with education. It was also known as the Smile City, perhaps due to the friendliness of inhabitants in spite of the cold weather… Curitiba is one of the coldest Brazilian cities.

My space dream was born inside my soul and encouraged by my parents. I am only child, and my mother Mrs. Antonia Reis was an elementary school teacher; she passed away more than ten years ago, and keeps on inspiring me with her unique example of courage, hard work, love, and determination. My father, Mr. Antonio Oliveira Reis is a retired teacher, who used to teach visually impaired students. He lost vision in an accident, and he is my great master and partner in my pathway. He always encouraged my space dream, and is my eternal hero and friend. My father provided me lots of encouragement as I was preparing this book.

Brazil used to be marked by substantial social inequalities, though this scenario has been changing dramatically in last years, thanks primarily to public policies committed to better income distribution and sustainable socioeconomic development. We are not a "rich" family. It was mostly thanks to my parents' hard work and vision that I was able to attend very good public schools. My parents were both retired when I was studying, and as their income was not sufficient to sustain my studies they decided to work with national lottery business to complement our family income. They always said that my education was their priority. Unquestionably they are heroes, and after God I owe my accomplishments to their continued support through my entire life. For example, my mother taught me how to read and write when I was less than five years old. I loved listening to her stories, and she inspired me to enjoy studying. When I was a child I also used to like drawing, thus my mother and father used to buy me several colored pencils, and other artistic materials for painting for example… I also enjoyed playing with my dolls as if I were a teacher, as many young girls…

Figure 36. Mrs. Antonia Reis, my mother, my eternal inspiration and strength. She has been an exceptional elementary school teacher in South Brazil, who encouraged my career in education and my 'space dream.' Credit: Norma T. O. Reis.

The beginning of my "romance" with *space* started when I was just five years old I think, and I was home. I remember one day very clearly. That was an ordinary morning, but I had an extraordinary vision. I was in my bedroom. In Curitiba, there are many wooden houses because of the cold weather. My house, where my father still lives (that I visit always when I have a chance) is one of those partially wooden-made houses. The window was not opened yet that morning, so it was dark inside. Suddenly, I clearly saw the entire solar system! The Earth, the Moon, the Sun, the other planets… inside my bedroom! And I was not sleeping. That must have been the vision of a passion that would accompany me through my entire life – if God allows me to achieve my *space* dream.

My next "contact" with *space* occurred when I was exploring some boxes with old books that my parents used to keep in our basement. I found by chance a book that my father used during his middle school studies. And… I found a chapter about the solar system! I got delighted to read about the planets, the Moon… I do not know why, but I felt ashamed to tell my parents how I enjoyed space! Sometimes children are so strange and silly… Sometimes I wonder why we like some things. It seems that love is something that is born in our soul without a clear explanation… But I wonder there must be a reason for most of our feelings. We live in a Universe governed by action-reaction laws! There must be a reason why we love specific themes, things and people! In other words, love must be logical; it must make sense! It is just that we do not understand that.

Third contact – it was 1986, and I was six years old. I had just entered preschool. We used to enroll at the age of six in preschool at that time. Currently Brazil' enrollment age is six for primary school. Anyway. The Halley comet would be visible that year, and at night I attempted to see that. I guess I was able to see the comet only once, and I always found a manner to go to the backyard in the evening with some excuse just to stare at the sky, look for the comets, and date the stars… Sometimes I talked to the Universe, literally. I talked with the stars, with the Moon… Again, I felt embarrassed to let my parents know that I was amazed by the Universe. I do not know why, but it seems that I wanted to keep that secret…

But my *space* "secret" did not survive for so long! When I was seven years old, I was enrolled in primary school. Every year students would have to present some specific project in science fairs, and I used to prepare in teams a presentation about the Solar System. My colleagues and I used some Styrofoam balls, figures, drawings, and assemble mockups to explain to the public characteristics of the planets – size, composition, and so forth. At that time, my parents were aware about my "space passion," and they brought me a small telescope that I still keep.

I do not remember exactly when my parents found out about my passion for space, I do not even remember if we had some first conversation about that. I do remember though that at the age of five years old I used to glue several pieces of paper I "stole" from my parents, which I used to construct "wings." Then I wore those wings. There are stairs at our house in Curitiba. I used to go upstairs, and jump from a reasonable height. The jump was not enough to cause me any harm or concern to parents, but enough to nurture my dream of flying in space someday. When I first saw a NASA space shuttle launch on television I got absolutely crazy, ecstatic. I was definitely, irremediably in love with space, and with NASA. That was love at first glance! Indeed, I had never felt like that before, and anything else in the material world captured my love, inspiration, efforts, and loyalty the most. I knew what I wanted to become when I grow up: an ASTRONAUT! I was (and I am still) a 'child of the sky,' because I cultivate my dream inside my soul and daily efforts. When I feel tired, sad, or when some of my plans do not go right, I look inside my heart until I find my dream. And when I cannot find that, it is easy to remember. I just need to look up the sky, and there is my inspiration. Then I can find again the strength to proceed.

At school, I used to get high scores because I studied a lot, and enjoyed doing so. I used to think that was my merit. Currently, I understand that it was much more my parents' love, advices, support, and material conditions than my own merit! Indeed, all of us owe lots of our success to our parents in special, spouse (for those who are married), our mentors, and friends. I know that my parents used to talk with me before I was born, and they encouraged me ever since to become a great person. I hope I did not disappoint them, at least not so much! Last but not least, *space* was my inspiration to study. I wanted to become a NASA astronaut, and astronauts must work hard. I do not know if I work that hard, but at least I try to! And I believe that one's willingness to do something counts a lot.

One of the things I enjoyed doing when I was a child was reading. I used to be a loyal "client" of our Parana state library. My father used to bring me there almost every day after school, where I spent hours exploring games, drawing, but mainly reading. When I was seven years old, I started reading about Astronomy and Astronautics. For instance, I used to explore newspapers in which I found out very interesting things about the Apollo manned missions, regarding the first astronauts and cosmonauts in space, and about the astronaut selection, training, career, and curiosities. I also enjoyed it a lot reading a Carl Sagan book called

Cosmos, and watching Sagan's videos. Learning about the Cosmos is something fascinating. I remember reading about the Voyagers – NASA's interstellar mission – I found that very interesting that humanity sent so many messages into space, just like sending message in a bottle for an eventual extraterrestrial civilization, and I wonder if someday those messages would be read, understood… and if someone will respond! Another 'hot topic' for me was the astronaut career. That was exciting to learn about the hard astronaut training in early times of human space exploration, and the work in space.

Another theme I was curious about when I was 7 years old was extraterrestrial life. I started reading books and magazines regarding possible UFO abductions and things like that. One thing I enjoyed doing was telling those stories to school colleagues. I stopped doing that when they complained that they were having nightmares because of that! They were getting scared… but I was just excited about the topic. I wondered what I would do and say if I met an extraterrestrial face to face. And I wondered how that would be to be invited by an extraterrestrial person to travel to another planet aboard an alien spaceship. I also dreamed about that, and thought about such a meeting! Yes, I enjoyed Star Wars, Star Trek, and others… That interest for extraterrestrial life matters did not last for so long, though.

We did not have a computer at home when I was a child, even because it was not so usual as it is today among youth. However, in my hometown Curitiba there are institutions called "Knowledge Lighthouses" where any citizen can connect to the internet, and access several medias for free. There was one of them near home, and I became a loyal user. My primary focus on those visits was to explore the NASA Web site, and multimedia related to space exploration and Astronomy. I would spend hours every day exploring these amazing topics.

When I was 12 years old some neighbors and I decided to establish a club about Astronomy and Astronautics. I became the funding president, and the club was named "Space in Evidence." We were around ten members, and some of our main activities included reading and discussing Astronomy and Astronautics magazine and newspaper articles, visiting the city's planetarium almost every Sunday, organizing ourselves to carrying out periodical astronomical observations using a small telescope (my parents bought me a larger one when I was a teenager – the 'official' one!), and observing astronomical phenomena such as solar and lunar eclipses. We used to spend several hours identifying constellations and having enjoyable, interesting conversations about the mysteries of the Cosmos!

During one of those planetarium visits, an astronomer shared with me a NASA address, because I mentioned that I wanted to send a letter for NASA, and request some materials to learn more about the agency that was (and is) my great passion. At that time, I did not know how to write in English, and needed to pay an English teacher to translate a letter that I previously wrote in Portuguese. I remember that I paid the equivalent to 15 dollars for the translated letter, which was a lot of money for me at the time! For that purpose, I had to use the allowance that my mother gave me once per month…

I do not remember exactly when, but I did receive a response from NASA! It included several posters, a solar system lithograph, stamps, and other interesting materials and educational resources containing information about the agency. I got very excited about that! As a response, I told my parents that I wanted to learn English in order to explore the NASA materials I received, and learn more about space. The fact was that my parents did not have enough money to pay the total amount for an English course, which used to be expensive. However, a school adviser recommended that I tried to obtain a scholarship from some

English school. I am sure that God helped me obtain a total scholarship for an English course in the largest language center in the city!

After I finished the basic course the school closed. I tried to get a scholarship from another very good English school. They provided me with a partial scholarship. It was a very dynamic school where I met a special teacher who encouraged my *space* dream. He agreed to translate into English some letters I wrote in Portuguese to send for NASA. I used to study English very hard. I played games in English, listened to songs, watched movies, and did all that I could to improve my language skills. It was a great emotion when I started being able to understand what was written in the NASA materials I received, and when I could write my own letters for NASA. I kept this "tradition" of sending letters to NASA throughout all my precollege education. There have been so many NASA materials I received, and I became particularly interested for NASA efforts in the realm of education.

At primary and secondary school, some of the activities I used to carry out related to Astronomy and Astronautics were science fairs, and role-plays. In one occasion, I constructed a "launch pad" with inexpensive materials, and managed to construct with cardboard a space shuttle, and we presented that model in a school exposition. I also remember carrying NASA stamps with me, and telling my colleagues that I loved NASA, and that someday I wanted to work there. Most of them thought that I was crazy, but my father mentions that some teachers thought that I could reach that goal. I dedicated to all subjects except for physical education in which I was intermediate. One of my favorite subjects was Arts. I used to draw and paint very well, and was able to win some contests at school, state, and national levels. However, I did not consider following a careers in Arts. But who knows – a future project, perhaps…

Figure 37. My mother and I at a science fair. I was a high school student, and was presenting at school a project about Astronautics. My mother always by my side in my dream! Credit: Norma T. O. Reis.

When I finished primary education, due to my interest in *space* many colleagues thought that I would enter some kind of technical school, but as I appreciate the education area I decided to become a teacher. In 1997, I received the Certificate of Elementary School

Teacher by a secondary school named Paraná State Institute of Education Professor Erasmo Pilotto (http://www.ieppep.cjb.net/). Perhaps in the future I will consider a career in Physics, for example… However, my focus currently is to use my educational and space background to teach basic space-related concepts in a manner that precollege educators can understand.

Let me share a little of my scholar life. In my school time my priority was studies. It was my family's priority, and also my own. My parents always said that the heritage they wanted to give me was a good education that would allow me to become an independent person. I must say they were successful in giving me this so important heritage, and I am proud of them. However, dedicated students are not very well accepted by their peers sometimes. It is not always easy to be ourselves and follow our dreams in this world, at times even as grown ups. Some people interpret this as "someone trying to be different or better than others." It is good that currently in several parts of the world there is increasing awareness about the importance of peace in schools, and that we are all different, and at the same time all equal as human beings. Until the fifth grade I had some difficulties with colleagues due to getting high scores, but I also had good friends and teachers who were supportive of my dreams. There is one thing teachers should not do in classroom that is to praise one student so much, as this contributes to make the person unpopular to many colleagues' eyes.

After we were all more mature and these problems diminished. In the 8th grade, I became a students' representative. When I started "working," I put my skills to benefit of all. Elected, I helped colleagues in their studies, helped our class get organized in terms of deadlines, exams, tasks, etc. I became so popular that we all cried together in our last school day, and in our graduation I was the student who received more clapping and shouting! Life changes a lot – I started school unpopular and ended up the most popular, and for the very same reason…

Speaking about my education is speaking about my mother. She taught me how to read and write, and she told me several stories. I learned how to write my own stories with her. As a teacher, she was proud to give me a graduation ring when I also became a primary school teacher. That was a very important moment for me. My father was my mentor, advisor, and best friend. He brought me to and from school, and discussed lots of academic and life subjects with me. In a time when parents do not have time for their children, I am eternally grateful for that my parents provided me with full love, attention, and support during my entire life.

Getting back to *space* in my life… After I finished high school, I wanted to enter university. In Brazil, the best universities are the public ones. Mrs. Gioppo managed to obtain a scholarship for me simply in the best preparatory school in the city. At the time each Brazilian university had a specific selective process, thus several preparatory courses were available for that purpose. It was one year of studies, and I was in doubt about the career to choose. That was a very important decision. I switched from Medicine (which was my father's dream for a child, though he never imposed that), to Aeronautical Engineering, then Physics, and finally Pedagogy. Although my father would like to have a medical doctor in the family, he always said that I should choose what I liked, and that he did not want to influence my decision. I chose Pedagogy because I enjoyed learning about education, in special theoretical aspects, curriculum, teaching methods, and so forth. Further, I think that I would be more helpful as an education specialist able to translate space-related scientific concepts to the general public than as a medical doctor or other professional without vocation.

I used to spend all day long studying at that preparatory school, and when back home I studied in the evening as well, along with weekends. By the end of the year I applied for the

Federal University of Parana State - UFPR (http://www.ufpr.br/portal/), the first Brazilian university in history, and I have been approved in first place for the daytime Pedagogy course. It was a great happiness to me to be approved in the best university in my home state, and one of the best in the country. For a not "rich" family, it was not expected. I remember that many people who knew us became surprised with my approval at that federal university, and even more surprised as I was approved in first place.

Figure 38. Mr. Antonio Oliveira Reis, my father. He is a retired teacher and a great leader, my hero, eternal mentor and supporter. He lives in Curitiba, South Brazil. In the picture, I am by his side in our house in South Brazil. Credit: Norma T. O. Reis.

Already at the university, I started the Pedagogy studies and was very excited about that. However, even public studies need some investment, and I should find a way to raise funds to help my parents with my studies' expenses. The university opened selection for research scholarships in several areas. Basically, students would spend 20 hours a week working in a given project and receive a monthly amount for that. I applied for a scholarship in the Physics department, in the subject Astronomy. For my great contentment I was selected, and I enjoyed learning about space. I remember that I used to spend several hours in my English school library exploring resources about space, such as books and tapes. Further, I used to bring university texts and Astronomy materials to study there. My English teachers all knew about my *space* passion and dream.

When I finished the intermediate English course, my mother started working towards obtaining a scholarship to allow me to study the advanced level in a very recognized English school in Curitiba. Everything was going fine, when in the year 2000 my mother Mrs. Antonia Reis passed away. That was a big sadness and loss for us. My father and I were alone. My parents had been married for 40 years, and they loved each other a lot. Further, my father has total visual deficiency and my mother's death represented a big difficulty for him because they used to work together. It was a shock for the two of us. However, thanks God and my father's support I could keep studying. Things have been very difficult for the two of us at the beginning, but that year I obtained a new scholarship in the Physics department of that university, and found a primary school where I started teaching in the evenings. At the Physics department, I carried out research about Cosmology with a scholarship from our National Council of Scientific and Technological Development – CNPq (http://www.cnpq.br/). The project's title was "Quantization of the Electric Charge and Gauge Extensions of the Gauge Standard Model." In the second year, the research title was "Numbers, 'Clusters' and the Variation of Universal Constants." Yes, both scaring for a Pedagogue… although I did not understand the mathematics behind that, I focused on concepts and their translation to education. During that time at the Department of Physics, I participated in local and national events presenting and publishing research results.

As I studied in the afternoon, I used to have free time in the mornings, and managed to find a job in a private language institute, when I finished my English studies. I also worked autonomously translating materials from English to Portuguese and vice versa for different professionals at the university where I studied. From 2001 to 2003, I worked as primary school (5th to 8th grades) English and Portuguese language teacher in two public schools and in two private English language institutes. During that time, I used to bring NASA materials to my classes, and share the excitement of space with my students.

Afterwards, it was Mrs. Olga A.W. Gioppo – who belongs to the *Soroptimist International* (http://www.soroptimistinternational.org/), an organization that promotes women and children worldwide – the support person (who also became my friend) who opened doors for that scholarship in a school named Inter-American – Cultural Center Brazil-The United Stated of America of Curitiba (http://www.interamericano.com.br/site/), recognized by the Embassy of the United States of America. It was the same school from that my mother was requesting a scholarship to me, before she passed away. I met Mrs. Gioppo in a juvenile forum promoted by her organization, when I was 12 years old, and our trajectories never separated. Who knows – maybe my mother connected to her spiritually to help me with that important accomplishment, as my mother had been already trying to obtain that scholarship for me. As a student, I carried out a bibliographical research about NASA Educations using some materials that I received from NASA, and presented the content to my English teachers in one of that school's pedagogical activities. I wish my mother could have been present to my English course graduation ceremony, but I believe she was there spiritually.

I think that my parents nurtured in me a great passion for studies and career. And I should mention that space and NASA have always been my inspiration to accomplish more. For example, I had a space shuttle big poster that I received from the radio Voice of America. I put that poster on my bedroom's wall and kept it there throughout many years. I would bring that poster to Brasilia capital city many years later, and she stayed with me until I embarked in a mission abroad – the mission that would bring me to NASA for a summer internship. Thanks God any difficulty could stop me from seeking my dreams. Just after finishing advanced English course, I started studying for the Proficiency Examination of the University of Michigan. For that purpose, I used to study individually at home, and undergo simulated tests at home using past exams. By doing that I was able to estimate my progress and fill remaining knowledge gaps. After several months of preparation, I applied for the exam and went through the written, listening, and interviewing phases. To my great happiness, I have been approved! I got really proud when I saw that certificate in my hands... I owe that certificate to God, my parents, and to Mrs. Gioppo's support. She supports me since I was 12 years old, and has been making a tremendous difference through my entire life. In the year 2000, I obtained the Certificate of Proficiency in English as a second language by the English Language Institute of The University of Michigan.

Figure 39. Mrs. Olga Anna W. Gioppo, a distinguished lady, my great supporter, my star. She belongs to the Soroptimist International, and helped tremendously write many, many pages of my professional and personal story. Recently, she has become my Chrism godmother. Credit: Norma T. O. Reis.

These facts are all related to my major *space* dream. During my university years, I was able to carry out research in our university's Physics department, and to teach English in private and public institutions. These actions were all aligned with my vision, as I needed to keep my English skills, and to build knowledge in sciences in order to better understand the space enterprise. At the university, I always shared with colleagues and professors my excitement for space, for NASA, for Astronautics. Many of them did not see a correlation between Pedagogy and *space*, though some professors and students encouraged my efforts and aspirations. In the final year of the Pedagogy course, I carried out investigation in Education, in a research entitled "Conception and practice of evaluation in the Pedagogy course of the Federal University of Paraná State," with scholarship from the university. By the end of the Pedagogy course, the Brazilian Education Ministry applied an examination to access performance of students in their respective courses. For my surprise, I obtained the first position in my home state, Parana!

One of the most remarkable moments in my life was when I majored as licensed in Pedagogy. That was an unforgettable day for my father and I. However, I am sure that my father was even happier than myself. I understand how big was his sacrifice to provide me conditions and guidance to fly higher, and turn my dreams into reality. When I was a primary school student, he used to wait for me every day as I returned from classes; he waited for me close to my school walls, it did not matter if it was sunny or rainy, hot or cold. He used to spend his entire salary with my education. He used to say that we could not have delights such as traveling to the beach, going go fine restaurants, and things like that, because the priority was my education. I am sure that my mother was also present in my graduation, and that she was very happy. After my mother passed away, my father kept on supporting my studies, and being my great advisor, providing me with love, and encouragement. I am a "byproduct" of their efforts and love! And I am very happy to have corresponded in some way.

Figure 40. NASA, my great inspiration. NASA is everywhere in my life, including my birthdays! Credit: Norma T. O. Reis.

In the last year of the Pedagogy course I applied for a specialization course in education and was approved. I started studying and was enjoying that. However, in the middle of the course, a different opportunity arose. It was launched a public national contest for a temporary international contract for an educational project in a Brazil-UNESCO partnership at the Brazilian Education Ministry. I thought it could be interesting, and applied. I traveled far in my country to do the exams. When results were published, for my surprised I was approved in first place for our Federal District (there were positions available in other parts of the country, but the closest one to my home state was the federal capital). I was not expecting to be approved in that contest, and for that reason I had several conversations with my father, Mrs. Gioppo, and a professor called Mr. Nilson M.D. Garcia, and they all encouraged me to embrace the opportunity. I always enjoyed my father's company, and I was sure that I would miss him a lot in the Brazilian capital. However, my father encouraged me and I decided to give that a try. As that was a temporary contract, my initial plan was to go there, earn some money, and return to invest that in a masters' degree back in Curitiba. It was truly difficult to leave home. I missed my father a lot, and I know that it was much harder for him. He had only a person to take care of the house during the day from Monday to Friday, and it was only some years later that my father met his current wife Mrs. Dejanira Reis.

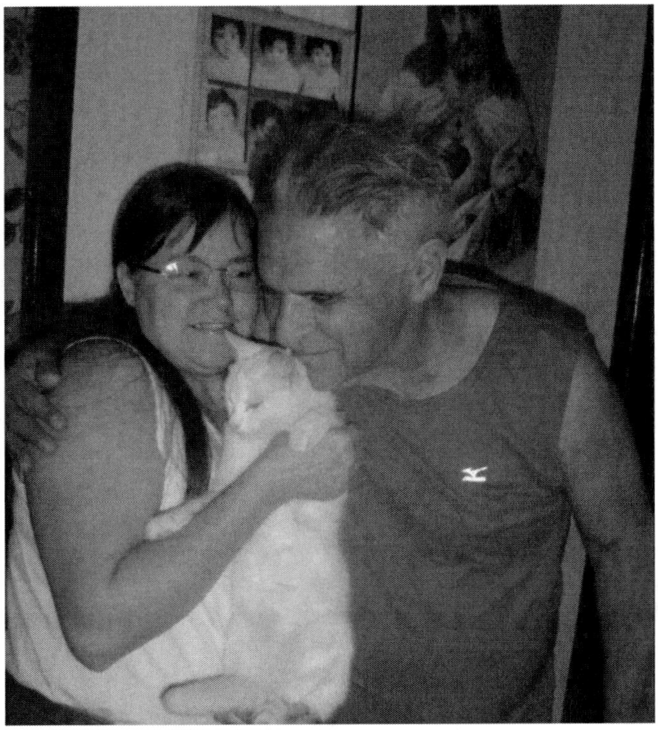

Figure 41. Mrs. Dejanira Reis, my father's wife. After some six years of widowhood, my father met her; they married and live happily in Curitiba city, with our six cats (and more coming soon!). We have always been a family that loves cats. Credit: Norma T. O. Reis.

My *space* dream would have continuation in Brasilia, where I live since January 2004. When I arrived everything was new to me, and I started working in the Office of Studies and Evaluation of Materials of the Brazilian Education Ministry (http://portal.mec.gov.br/). One day, I heard of a cocktail at the Brazilian Space Agency. That was my first contact with official space activities in Brazil. At the event, I met two persons who would make a substantial difference in my career, Mrs. Ivette Ribeiro, former manager for the agency's precollege education program, and Mrs. Meire, who would become a good friend. My initial duties at the Ministry basically comprised accompanying the evaluation process of didactic books. There was an occasion when I talked to Mrs. Lucia H. Lodi, who was the secondary education director at the time, and shared with her my interest and background in *space*, and she included me in a team that she had formed in partnership with the Brazilian Space Agency to prepare high school teachers in the didactic utilization of remote sensing and space sciences. As part of that team, from 2004 to 2006, I had the opportunity to teach courses for teachers in several Brazilian states. We covered topics such as remote sensing, satellites, space exploration, and space science education, including NASA accomplishments in these fields, and indicating for teachers several NASA public websites and resources.

While at that Coordination, another national public contest was opened to the Education Ministry, this time for permanent positions. I knew that public contests to become civil servant are difficult and competitive, but I decided to try anyway. Thanks God I was approved in October 2005. Then I left my temporary contract, and switched to the permanent position, in which I work until currently. Simultaneously, God opened an important door for me – the door that would lead me to my space dream at NASA – I had been invited to work as special adviser to the national secretary of precollege education at the Ministry. After some thoughts and talks to family and friends, I said "yes" to the new opportunity, although I imagined it would be difficult to leave my former colleagues and supervisor, as they were very friendly. However, I managed to make good friends at the new workplace, and in December 2005, I was nominated division chief, working as special adviser to the education secretary. Although that was a very demanding position, I was able to keep on teaching my courses on remote sensing for teachers in Brazilian states, thanks primarily to the support of that mentioned secondary education director, Mrs. Lodi.

Simultaneously to my professional activities at the Brazilian Education Ministry, I carried out independent research in space education with guidance of Dr. Nilson M.D. Garcia, Physics professor at the Technological Federal University of Paraná State (http://www.utfpr.edu.br/). From 2004-2007, we produced scientific papers together, and participated in national and international events presenting research results. We still work together in some academic productions currently, and he has always been an important supporter to my projects.

At the Ministry, my colleagues and supervisors knew about my *space* passion, not only because of my teacher training activities, but also because of several NASA posters I had in my office. There was a day that was decisive for my *space* dream. It was when a friend at the Brazilian Space Agency, Mrs. Meire, shared with me information about the International Space University – ISU (http://www.isunet.edu/), located in Strasbourg, France, and encouraged me to contact that institution. It is an international, intercultural university where students with different professional and academic backgrounds have the opportunity to study several space-related subjects such as medicine, law, policy, engineering, sciences, and so forth. The idea is to prepare professionals to work in different areas of the space enterprise, and dialogue with space professionals with diverse backgrounds. In the beginning of 2006, I applied for the M.Sc of Space Studies at that university. The reason for choosing that university has been the fact that it is international and supported by diverse space institutions, including NASA, and that it allows that master candidates study only space-related topics, so that I would fell more confident to carry out my research project in space science education. Further, the masters' course comprises a three-month internship that can be carried out in diverse space public and private institutions, including space agencies. That could represent an opportunity for that I could attempt to carry out my internship at NASA, and make my childhood dream come true. I could develop research on space science education and have the singular opportunity of contributing to explore and develop space education, which has been having significant impact on the education of million of students.

In October 10th 2006, I received their letter of acceptance for the period 2007-2008. This was just the beginning of my efforts, though. In order to submit a proposal for liberation from my Ministry, it was necessary to collect more "pieces." In other words, I would still need reference letters and financial support. I needed to obtain a scholarship for academic fees, airline tickets, and scholarship for student's expenses. Thanks God, several doors opened for me on the way. InMarsat (http://www.inMarsat.com/) provided me a scholarship for academic fees; the Brazilian company TAM Airlines (http://www.tam.com.br/) provided me with tickets to and from France; the French Embassy in Brazil (http://www.ambafrance-br.org/) provided me a 6-month student scholarship; the Soroptimist International, Brazil Region (http://soroptimistbrasil.com.br/), especially in the persons of Mrs. Olga A. W. Gioppo and Mrs. Norma do Valle, for providing me with the remaining 6-month scholarship; and the Region Alsace (http://www.region-alsace.eu/) provided me partial support for my NASA internship expenses.

Figure 42. Dr. Timothy E. Eastman, plasma physicist and deputy group manager at NASA Goddard. He has been a great friend and remarkable mentor, until current days. Credit: Norma T. O. Reis.

December 2007 was the most special moment in this journey. I found Dr. Timothy E. Eastman's contact information in a NASA catalog by "chance" (?) on the internet, and decided to send him an e-mail mentioning that I had been approved to ISU masters, sending my resume, and verifying possibilities of carrying out an ISU internship at NASA. After that, Dr. Tim (as I call him today), a plasma physicist working at NASA Goddard Space Flight Center, responded to e-mail and mentioned that I would be welcome to call his office. It was a nice, sunny December 2007 morning when I first called him, and consulted about that possibility. It was the best day of my life because he said an initial "yes!" He would certainly need to submit proposals and negotiate internally and so forth, but I felt that he would be my mentor. Since talking to him on the phone someway I saw in him a friend and supporter. In other words, someone I could trust on the way. I am very grateful to the friend Mrs. Paulo de Tarso Milanni, who also works in Brazilian government and studied at ISU, and advised me in advance to contact a person from NASA before starting the masters' studies, because several students (himself included) tried an internship at NASA at ISU, but were unable to obtain approval in time. I owe a lot to his important recommendation.

My *space* dream also belongs to those who believed in me and supported me. I would like to mention here those who prepared recommendation letters to me. It was Mrs. Nilson M.D. Garcia, Physics professor of the Technological Federal University of Parana State (UTFPR); Mrs. Olga A.W. Gioppo, president at the Soroptimist International of the Americas (Curitiba-Batel); Mrs. Lucia H. Lodi, former director for secondary education at the Brazilian Education Ministry; Mrs. Maria do Pilar L.A. e Silva, former secretary of precollege education at the Brazilian Education Ministry; Mr. Francisco C. Fernandes, adjunct secretary for the Minister at the Brazilian Education Ministry; Mr. Ronaldo Mota, former secretary for higher education at the Brazilian Education Ministry; Mr. Timothy E. Eastman, plasma physicist and group manager at NASA Goddard Space Flight Center; Mrs. Ivette M.S. Rodrigues, former manager of the AEB School Program at the Brazilian Space Agency; Mr. Himilcon C. Carvalho, Director of Space Policy and Strategic Investments at the Brazilian Space Agency; and Dr. Sergio M.B. Gaudenzi, President of the Brazilian Space Agency. I owe my *space* achievement to each one of these important persons.

After submitting my liberation request along with the mentioned financial support and recommendation letters, I still needed to wait for the Ministry's official authorization to be published so that I could finally travel to France and start studying. I was able to spend a couple of days with my father in Curitiba before that mission. I remember that it was a rush to prepare all logistics, and several local friends helped me with different logistical items. Friends are fundamental to everyone's success in life. We need to cultivate good friends because they support us on the way, and help us fly higher. I would not have accomplished so much without my friends. Thanks God my official liberation came on the day in which my tickets had been bought. I had never traveled abroad before, so it was a great expectation and I also had some fear. Although my father was happy about my accomplishment, it was difficult for him to see me traveling for another country. And to complicate things even more, I was going to France, but I did not know how to speak French! Then I bought a traveler's guide and was able to learn some basic sentences. In the end, I used some of those sentences, mimics, drawing… but I survived!

Studies at the International Space University were hard. I had a bad time with the engineering contents. Further, it has been very difficult for me to be away from my family and friends. I used to have lots of supporters around, and abroad I needed to learn how to face all challenges alone. Thanks God I won all battles. I must mention some people at ISU who made a tremendous difference in my trajectory. I will start by Mrs. Nadia Repussard, who sent me all information about ISU by post and encouraged me a lot to enroll. Then, Mr. Walter Peeters, ISU Dean, my academic adviser. I had the privilege to work under his guidance as an ISU student, and this brilliant professional that worked training astronauts before in the European Space Agency helped me a lot with recommendations to succeed. Further, he opened doors for that I could carry out an internship at NASA Goddard, as he authorized that from the ISU side. I recognize it should have been difficult for him to work with a person with educational background, because I was not familiar with the engineering subjects. Further, I had a hard time getting adjusted to living alone in a different country, and this impacted my performance to a certain extent. Finally, I am grateful to Dr. Hugh Hill, an ISU professor who believed in me, and provided me fundamental support to my success.

My finest hour was surely my internship at NASA Goddard Space Flight Center under supervision of the distinguished scientists and mentors Dr. Timothy E. Eastman, Dr. Louis A. Mayo, and Dr. James R. Thieman. NASA Goddard located in Greenbelt, MD, and hosts a vigorous program in heliophysics education and public outreach. I worked closely with their heliophysics education group to learn several facets of their efforts to improve postsecondary education in Science, Technology, Engineering, and Mathematics (STEM). In particular, I did a comprehensive survey and analysis of eclipses, resulting in a 62-page paper "Eclipses through the Centuries" as a contribution to their resource materials. In addition, I carried out research on student observations of the Sun and wrote a 8-page concise summary of the "Indirect Observation Method." Finally, my ISU internship work led to a 12-page paper on their Space Weather Action Center program that is published in the 2009 Proceedings of the International Federation for Information Processing. In parallel to that Eclipses research, I completed my thesis in fulfillment of my Master in Space Management Degree, on "Pedagogical Analyses of Current Space Science Educational practices at NASA and the Brazilian Education Agency." A final oral presentation in August 2008 summarized research results obtained from the ISU internship at NASA.

There are not enough adjectives to describe my time at NASA Goddard. NASA Goddard was indeed much better than my dreams. That summer internship was more than an unforgettable, marvelous, and remarkable time. My mentor Dr. Timothy E. Eastman, a plasma physicist and deputy group manager at NASA Goddard, in December 2007 opened doors for that my NASA dream turned into a beautiful reality, and provided me full guidance and support before and during my internship. I still remember when Dr. Tim brought me to the NASA gate for the first time. Crossing that gate and getting my NASA badge was a great emotion that I will always remember. I had waited my entire life for that moment! Working with powerful computers, friends, the cafeteria, the meetings, the library, the volleyball team, pizzas... I remember everything. I remember that I began learning Martial Arts with my mentor Dr. James R Thieman, a NASA director who is truly great master! He also manages Radio Jove (he is the one in the pictures between the Radio Jove antennas), an outstanding program in radio astronomy, and much more. My mentor Mr. Louis A. Mayo taught me a lot on Astronomy, provided guidance, and shared his unique knowledge in management as well. I also met Dr. Fred Espenak, known as Mr. Eclipse, for being the world's greater authority in the field. And I met several other special people who made a tremendous difference at that time and beyond that. I could realize that NASA people have a special energy, it is something of harmony, it is something of love, something unique that I could not find anywhere else. It is all about NASA. There is a smell, there is a special atmosphere, there is a special feeling of being at NASA. I miss those days, and especially those people a lot. Currently, thoughts about that time inspire me to fly higher. I just wish I had been in a better condition when I got there, though. In other words, I was not accustomed to living abroad, and I believe that I could have done better if I had learned earlier how to manage my own life in France. Sometimes we need to make some mistakes, and learn lessons from them to improve future accomplishments. This is what happened to rocket engineers who tested their prototypes. We would not be able to fly into space today if they had decided to give up when their rockets did not fly at first trials. They insisted in spite of all criticism they received, and for this reason now we are able to send objects and humans beyond the atmosphere.

Figure 43. My workspace as an ISU intern at NASA Goddard. Interns work most of the time in cubicules. In that great time, my mentor Dr. Tim Eastman provided even a dedicated printer for my activities! Credit: Norma T. O. Reis.

After concluding my internship at Goddard I returned to the International Space University, and presented research results. The graduation ceremony has been a great emotion, as I got a masters' degree from a space university! It was a dream that became the most beautiful reality. It was a very rich year when I met persons from different nationalities, and backgrounds. In the process, I learned a lot as a professional and as a human being.

After, I returned to the Brazilian federal government (the Brazilian Education Ministry). Thanks God and to supportive supervisors who foster innovation, I have been able to apply a considerable part of my *space* knowledge associated to education and technologies to important projects, including one collaborations with NASA, which is the Space Weather Action Center (SWAC), currently translated and available into Portuguese.

May I also express my eternal gratitude to my mentors Dr. Timothy E Eastman, Dr. James R. Thieman and Dr. Louis A. Mayo, from NASA Goddard, for their continued support, guidance, partnership, and friendship. They have helped me write one of the most beautiful stories of my life, which was the accomplisment of a special mission, a dream that came true. There is no truly important mission in this world that may be accomplished alone, and what I have learned from these three incredible mentors has become a very important part of the person I am today. I regard myself a better person after NASA. Dr. Timothy E. Eastman has had lots of patience with me, as I was a foreign student with difficulties getting adjusted to the country (streets, culture etc.). Sometimes I have gotten literally lost in Greenbelt, and he made his telephone available to get me out of trouble! Also, it should be mentioned that the NASA people work really very hard, and I had learned a lot with that experience. And even when I returned to Brazil, Dr. Tim Eastman kept helping me a lot, with many professional advices, and also because I felt immense 'NASAsickness' for a long time! Our e-mail and phone conversations have helped me in many occasions to survive missing NASA and my so special mentors. I had also a very special moment, when my mentor Dr. James R. Thieman visited South Brazil in 2010, and invited me to coordinate sessions in an international event with him, and in my home state, Parana! On the occasion, I came to know his wife Barbara, another incredible person. We had a very nice, funny time together, and I could re-live some of the so special moments I had with them back at NASA Goddard. Indeed, I learned a lot from all of them, and what I have learned has helped a lot in my career, and in my life. They are part of my story, and also part of this book! I am so lucky and honored to have met these three exceptional mentors.

Much of the inspiration to prepare this book came as I was sitting beside one of our cats in Curitiba city. His name is "Filhote," which in Portuguese means a small, very young cat. He gave me much strength me prepare the Book Idea for Nova, just by being by my side!

Figure 44. My cat "Filhote," at my family home in Curitiba. He has inspired me a lot to write this book for Nova! "Filhote" today is a very big, charmy cat. Credit: Norma T. O. Reis.

May God allow that this history to be continued; may I be able to achieve more in the realms of education and space, and be able to inspire other children and teenagers to believe in their dreams and fly higher to make them come true through creative applications of the 'space education phenomenon.' My biggest dream is to become an educator astronaut, someday, somehow. As a child with paper-made wings, I still believe I can fly. Yes, similarly to Dr. Robert H. Goddard, I believe it is possible. Like him, I choose to believe that the dream of today can be the reality of tomorrow. I also believe that it does not matter how many times we fall in life; it does not matter the difficulties we face and the criticism we suffer; we should always stand up and keep on walking in the direction of our deepest, purest dreams. Life is for those with courage to be themselves, and who protect their dreams. I hope this real history of how space education can inspire a child, could inspire you to turn your dreams, and your students' dreams into the most wonderful realities.

BIBLIOGRAPHY

[1] R. S. Waack, S. Amoroso, *Parcerias Estratégicas: Seminarios Tematicos para a 3a Conferência Nacional de Ciência, Tecnologia e Inovação*. Ministério da Ciência e Tecnologia, Brasília (2005).

[2] R. Amaral, *Ciência e Tecnologia: Desenvolvimento e Inclusão Social,* UNESCO, Ministério da Ciência e Tecnologia, Brasília (2003).

[3] P. Bordieu, *Os usos sociais da ciência: por uma sociologia clínica do campo científico*, UNESP, São Paulo, SP, Brazil (2004).

[4] N. T. O., Reis, N. M. D. Garcia and P. Baldessar, *Revista Brasileira de Ensino de Física*. 30, 1401 (2008). http://www.sbfisica.org.br/rbef/pdf/301401.pdf

[5] OECD, *PISA 2009 Results: What Students Know and Can Do - Student Performance in Reading, Mathematics and Science (Volume I)*. OECD Publishing, Paris (2010).

[6] P. Campbell, *Physics Education*, 41, 514 (2006).

[7] M. I. Monteiro, *Poiésis: Revista Científica em Educação*, 1, 131 (1999).

[8] Brasil, *Orientações Curriculares para o Ensino Médio: Ciências da Natureza, Matemática e suas Tecnologias*. Ministério da Educação, Brasília (2006).

[9] J. Carrascosa, D. G. Pérez, A. Vilches and P. Valdéz, *Caderno Brasileiro de Ensino de Física*, 23, 157 (2006). http://www.fsc.ufsc.br/cbef/port/23-2/artpdf/a1.pdf

[10] E. M. F. Ramos and N. C. Ferreira, *Pesquisas em Ensino de Física – Educação para a Ciência*, Escrituras, São Paulo, SP (2004).

[11] G. Soussan, *Didática e Formação,* UNESCO Brasil/OREALC/MEC-MCT, Brasília (2003).

[12] R. Sutherland et al., *Journal of Computer Assisted Learning,* 20, 413 (2004). http://onlinelibrary.wiley.com/doi/10.1111/j.1365-2729.2004.00104.x/abstract

[13] M. Trucano, *Knowledge maps: ICT in education,* infoDev/World Bank, Washington, DC (2005).

[14] J. Cummins, K. Brown and D. Sayers, *Literacy, Technology, and Diversity: Teaching for Success in Changing Times,* Pearson, Boston, (2007).

[15] J. G. M. Ribeiro, A. C. A. Carvalho, N. T. O. Reis and E. M. V. Rosario, Public policies for information and communication technologies provision and utilization in Brazilian schools, in *Computers in Education*, 1, Nova Publishers, New York (2012).

[16] M. Pontes, *Missão Cumprida – A História Completa da Primeira Missão Espacial Brasileira,* Chris McHilliard, São Paulo, SP, (2011).

[17] United Nations, *Report on the United Nations International Meeting on the Applications of Global Navigation Satellite Systems, held in Vienna from 12 to 16 December 2011*, The United Nations International Meeting on the Applications of Global Navigation Satellite Systems, 12-16, Dec. 2011 (Vienna, Austria). http://www.oosa.unvienna.org/pdf/sap/2011/un-gnss/55COPUOS-1019_E.pdf
[18] P. Freire, *Ação cultural para a liberdade e outros escritos,* Paz e Terra, São Paulo, SP (2001).
[19] D. Saviani, *Pedagogia histórico-crítica: primeiras aproximações,* Autores Associados, Campinas (2005).
[20] D. Gillard, *Education in England: a brief history* (2011). www.educationengland.org.uk/history
[21] D. Bell, *The coming of post-industrial society: a venture in social forecasting,* Perseus Books Group, New York (1999).
[22] United Nations, *Millenium Goals,* UNESCO, Paris (2010). http://www.un.org/millenniumgoals/pdf/MDG_FS_2_EN.pdf
[23] J. C. Libaneo, *Democratização da escola pública: a pedagogia crítico-social dos conteúdos,* Loyola, São Paulo, SP (2006).
[24] L. S. Vygotsky, *Mind in Society: Development of Higher Psychological Processes,* Harvard University Press, Cambridge, MA (1978).
[25] P. Freire, *Pedagogia da Indignação,* UNESP, São Paulo, SP (2000).
[26] OECD, *Education at a Glance 2011: OECD Indicators,* OECD Publishing, Paris (2011). http://www.oecd.org/dataoecd/61/2/48631582.pdf
[27] P. E. Davis-Kean, *Journal of Family Psychology,* 19, 294 (2005). http://www.mikemcmahon.info/ParentEducationIncome.pdf
[28] M. Caspe, M. E. Lopez and C. Wolos, *Family Involvement Makes a Difference: Evidence that Family Involvement Promotes School Success For Every Child of Every Age.* Harvard Family Research Project, Harvard Graduate School of Education (2006/2007). http://www.hfrp.org/publications-resources/browse-our-publications/family-involvement-in-elementary-school-children-s-education
[29] P. J. Palmer, *The courage to teach: Exploring the inner landscape of a teacher's life,* Jossey-Bass, San Francisco (1998).
[30] N. T. O. Reis et al., NASA Education and Educational Technologies Exemplified by the Space Weather Action Center Program, In: A. Tatnall, A. Jones (Eds.) *Education and Technology for a Better World*, IFIP advances in information and communication technology, Springer, Berlin (2009).
[31] W. L. P. Santos, *Revista Brasileira de Educação,* 12, 474 (2007).
[32] OECD, *Assessing Scientific, Reading and Mathematical Literacy: A Framework for PISA 2006,* OECD Publishing, Paris (2006).
[33] J. P. Baptista and L. Ferracioli, *Da Physis à Física: uma história da evolução do pensamento científico,* Edufes, Vitória, ES (2003).
[34] M. Schenberg, *Física e Sociedade,* Landy, São Paulo, SP (2001).
[35] D. Teresi, *Descobertas perdidas: as raízes antigas da ciência moderna, dos babilônios aos maias,* Companhia das Letras, São Paulo, SP (2008).
[36] History for Kids Web site, "Daedalus and Icarus," http://www.historyforkids.org/learn/greeks/religion/myths/daedalus.htm.

[37] Encyclopedia Mythica Web site, "Icarus," http://www.pantheon.org/articles/i/icarus.html.

[38] R. Argentière, *Astronáutica,* Fulgor, São Paulo, SP (1961).

[39] Astronomia no Zênite Web site, "Do sonho à conquista," zenite.nu?astronautica

[40] N. T. O. Reis and N. M. D. Garcia, *Revista Brasileira de Ensino de Física,* 28, 361 (2006). http://www.scielo.br/pdf/rbef/v28n3/a13v28n3.pdf

[41] N. Luca, *A Astronáutica e Seus Grandes Pioneiros*, UFPR, Curitiba, PR (1990).

[42] S. J. Pyne, Seeking Newer Worlds: An Historical Context for Space Exploration, In: S. J. Dick, R. D. Launius, *Critical Issues in the History of Spaceflight,* NASA Office of External Relations, Washington, DC (2006).

[43] G. Haskell, M. J. Rycroft (Eds.), *Proceedings of the Space of service to humanity: preserving Earth and improving life,* International Space University's Symposium Space of Service to Humanity: Preserving Earth and Improving Life, 5-7, Feb. 1996 (Strasbourg, France). Kluwer Academic Publishers, Norwell, MA (1996). http://isulibrary.isunet.edu/opac/index.php?lvl=notice_display&id=4659

[44] P. Diamandis, R. Richards and J. Pelton, The Future of Space, In: J. N. Peltron and A. P. Bukley (Eds.). *The Farthest Shore – A 21st Century Guide to Space,* Apogee Books/International Space University, Burlington (2010).

[45] H. E. McCurdy, *Inside NASA,* Johns Hopkins University Press, Baltimore and London (1993).

[46] The European Space Agency Web site, http://www.esa.int/esaCP/index.html.

[47] R. Chipman, Relevance of space activities to economic and social development. In: R. Chipman (Ed.), *The world in space: a survey of space activities and issues: United Nations Conference on the exploration and peaceful uses of outer space*, Aug. 1982 (Vienna-Austria), Prentice-Hall, Englewood Cliffs, NJ (1982).

[48] NASA, *NASA Earth Observations Serving Society,* NASA Goddard Space Flight Center, Greenbelt MD (2005).

[49] USA, *Grand Challenges for Disaster Reduction,* Executive Office of the President of the United States, National Science and Technology Council, Washington DC (2008).

[50] D. Lockney (Ed.), *Spinoff 2010: NASA Technologies benefit society*, NASA Center for AeroSpace Information, Washington, DC (2010).

[51] UNESCO, *The state of science and technology in the world,* UNESCO Institute for Statistics, Paris (2001).

[52] The Statslink Web site, "The Space Economy at a Glance © 2011 OECD 2011," http://dx.doi.org/10.1787/888932400190.

[53] The Secure World Foundation Web site, "Existing and Emerging Space Powers," http://swfound.org/space-sustainability-101/existing-and-emerging-space-powers.

[54] B. Harvey, H. Smid and T. Pirard, *Emerging Space Powers: The NewSpace Programs of Asia, the Middle East and South-America,* Springer Praxis, Chichester (2010).

[55] N. Bhandari, *The Mysterious Moon and India's Chandrayaan Mission,* Vigyan Prasar, Uttar Pradesh (2008).

[56] The Telegraph Web site, "*India to spend £1.7bn sending man to the moon,*" http://www.telegraph.co.uk/news/worldnews/asia/india/4788143/India-to-spend-1.7bn-sending-man-to-the-moon.html

[57] Agência Espacial Brasileira, *Programa Nacional de Atividades Espaciais 2005-2014*, Agência Espacial Brasileira, Brasília (2005).

[58] M. Bazin, *Educar em* Revista, 14, 27 (1998). http://ojs.c3sl.ufpr.br/ojs2/index.php /educar/article/viewArticle/2027
[59] H. E. Newell, *NASA's space science and applications program,* NASA, Washington, DC (1967).
[60] E. Morin, *Ciência com consciência,* Bertrand Brasil, Rio de Janeiro (2005).
[61] R. Gibson, Space science, In: R. Gibson. *Space*, Clarendon, Oxford (1992).
[62] NASA. *2011 NASA Strategic Plan,* NASA Headquarters, Washington, DC (2011).
[63] J. Rosendhal et al., *Advances in Space Research,* 34, 2127 (2004).
[64] The White House Web site, http://www.whitehouse.gov/
[65] O. Zhdanovich and D. Pieson, Space Education and Space-Based Education: The Russian Experience, In: G. Haskell and M. Rycroft (Eds.) *International Space Station: The Next Space Marketplace*, Kluwer Academic Publishers, Norwell, MA (2000).
[66] R. R. Milliron, *Worldwide Satellite Magazine*, Jan., 28 (2010). http://www.satmagazine.com/2011/SM_January_2011.pdf
[67] A. H. Hansen and M. Ostbo, *Acta Astronáutica,* 51, 617 (2002).
[68] The U.S. CanSat competition Web site, http://www.cansatcompetition.com/.
[69] The CanSat in Europe Web site, http://www.cansat.eu/.
[70] The International CanSat Competition Web site, http://www.cansat.leem.es/.
[71] The JAXA Web site, "Space Education Enriches Children's Heart," http://www.jaxa.jp/article/special/education/nakamura01_e.html.
[72] The SANSA Web site, SANSA – Science Advancement & Public Engagement," http://www.sansa.org.za/science-adv-and-public-engagement.aspx.
[73] K. A. Campbell, S*T*A*R*S-utilizing space and space research for educational initiatives. In: *Space Technology and Applications International Forum – 1999*, AIP Conference Proceedings, Volume 458, pp. 397-399 (1999).
[74] United Nations, *Highlights in Space 2010,* United Nations Office for Outer Space Affairs, New York (2010).
[75] The Globe Program Web site, http://www.globe.gov.
[76] L. D. Wigbels, Acta Astronautica, 55, 701 (2004).
[77] The United Nations Office for Outer Space Affairs Web site, "World Space Week," http://www.unoosa.org/oosa/wsw/index.html.
[78] The GRAIL EarthKAM Web site, https://earthkam.ucsd.edu/.
[79] The GRAIL MooKAM Web site, https://Moonkam.ucsd.edu/.
[80] The GigaPan Web site, http://gigapan.org/.
[81] The NASA Great Moonbuggy Race Web site, http://moonbuggy.msfc.nasa.gov/.
[82] The NASA's Sun-Earth Day Web site, http://sunearthday.nasa.gov/.
[83] The Observe the Moon Night Web site, www.observetheMoonnight.org.
[84] The Global Friendship Through Space Education Web site, http://www.gftse.org/.
[85] The UNESCO Web site, "Natural Sciences," http://www.unesco.org/new/en/natural-sciences/science-technology/space-activities/related-info/about-sep/
[86] Y. Berenger, Space Education and the Decade of Education for Sustainable Development (UN-DESD, 2005-2014). In: *15th UN/IAF Workshop*, 14-15 Oct., Kitakyushu, Japan (2005). PowerPoint presentation.
[87] A. Fraknoi, Space science education in the United States: The good, the bad, and the ugly. In: S. J. Dick, R. D. Launius (Eds), *Societal impact of spaceflight*, NASA Office of External Relations, Washington, DC (2007).

[88] European Science Foundation. *Demography of European Space Sciences. Results from an ESSC-ESF Study*, European Science Foundation, Strasbourg Cedex (2003).

[89] GovDelivery, *NASA achieves "out-of-this-world" results since launching multi-channel communications approach*, GovDelivery, St. Paul (2011). http://www.govdelivery.com/docs/pdfs/NASA-SuccessStory.pdf

[90] NASA. *NASA Spinoff 2007*, Innovative Partnerships Program, NASA Headquarters, Washington, DC (2007).

[91] The NASA/GSFC Solar Dynamics Observatory Web site, http://sdo.gsfc.nasa.gov

[92] United States, *National Space Policy of the United States of America,* The White House, Washington, DC (2010). http://www.whitehouse.gov/sites/default/files/national_space_policy_6-28-10.pdf

[93] D. Cooke, *Plans for Human Exploration Beyond Low Earth Orbit*, NASA, Washington, DC (2011). http://www.nasa.gov/pdf/524774main_COOKE.pdf

[94] S. Y. Park and D. D. Mazanek, *Mission Functionality for Deflecting Earth-Crossing Asteroids/Comets*, NASA Technical Reports Server, Washington, DC (2005). http://ntrs.nasa.gov/archive/nasa/casi.ntrs.nasa.gov/20050186572_2005187880.pdf

[95] European Space Agency Web site, http://www.esa.int/SPECIALS/ Mars500 /SEMGX9U889G_0.html

[96] The "Diario da Russia" Web site, http://www.diariodarussia.com.br/tecnologia/noticias/2012/02/16/marte-500-podera-acontecer-na-estacao-espacial-internacional/

[97] The Mars Society: Human Exploration and Settlement of Mars," http://www.Marssociety.org/home/press/tms-in-the-news/ zubrinhumanscouldbeonMarsby2018

[98] NASA, *Human Space Exploration Framework Summary*, NASA, Washington, DC (2011). http://www.nasa.gov/pdf/525162main_HEFT_Final_Brief_508_20110309.pdf

[99] B. Clay, *ESMD Overview: A Vibrant Future for Human Exploration of Space*, NASA, Washington, DC (2011).

[100] H. R. Quinn, H. A. Schweingruber, M. A. Feder (Eds.), *Committee for the Review and Evaluation of NASA's Precollege Education Program*, National Academies Press, Washington, DC (2008).

[101] NASA, *Native People of NASA Exploring Mother Earth and Father Sky*, NASA, Washington, D.C (2006).

[102] T. E. Eastman, personal communication, 12 August 2008.

[103] NASA, *NASA Office of Education Staff Handbook*, NASA, Washington, DC (2011).

[104] NASA, *Informal Education Program – 2011 NASA Research Announcement*, NASA's Office of Education, Washington, DC (2011).

[105] NASA, *NASA Education – Strategic coordination framework: a portfolio approach*, NASA, Washington, DC (2007).

[106] J. M. Maher, personal communication, 13 August 2008.

[107] NASA, *2007 NASA Education Highlights*, NASA, Washington, DC (2007).

[108] The NASA Education Web site, http://www.nasa.gov/offices/education/ about /index.html

[109] NASA's educational product "The Solar System." http://solarsystem.nasa.gov/multimedia/downloads/21_Solar_System_FC_opt1.pdf

[110] The NASA's Open Government Plan Web site, http://www.nasa.gov/open/plan/

[111] The NASA's Web site, "Education Technology and Products Leveraging Technology http://www.nasa.gov/offices/education/about/tech_prod_e_edu_overview.html

[112] NASA, *Budget Estimates Education 2012*, NASA, Washington, DC (2012). http://www.nasa.gov/pdf/516643main_NASA_FY12_Budget_Estimates-Education.pdf
[113] The NASA's Education Web site, "Learning Technologies, http://www.nasa.gov/offices/education/programs/national/ltp/about/index.html
[114] NASA. *NASA Open Government Plan*, NASA, Washington, DC (2010).
[115] The NASA Team Collaboration, http://wiki.nasa.gov/cm/wiki/NASA%20Chats.wiki/home/home.html
[116] NASA, *NASA Space Science Education and Public Outreach Report 2003*, NASA, Washington, DC (2004).
[117] NASA, *NASA Space Science: Education/Public Outreach - Summative Evaluation Report*, NASA, Washington, DC (2007).
[118] The SERCH Web site, http://serch.cofc.edu/special/workshops.htm
[119] The NASA's Sun-Earth Day Web site, http://sunearthday.nasa.gov/2012/about/about.php
[120] The NASA's Space Weather Action Center Web site, sunearthday.nasa.gov/swac/
[121] M. L. West, N. Frissell, M. Papalos, *Relationship Between the Radio Bursts from the Sun and Ionospheric Propagation*, http://pages.csam.montclair.edu/~west/beacons.pdf
[122] J. Thieman, personal communication, 12 August 2008.
[123] J. Wallace, R. Flagg, *Amateur Radio Astronomy Projects — Radio Signals from Jupiter*, QEX – May/June (2010), http://www.radio-astronomy.org/pdf/qex/radio-jove-proof.pdf
[124] C. L. Mcarthur, personal communication, August 12, 2008.
[125] NASA. *NASA Education Communication Strategy*, Washington, D.C. NASA (2008).
[126] NASA, *NASA Academy Program Descriptions*, Washington, D.C. NASA (2011). http://astrosun2.astro.cornell.edu/specialprograms/spacegrant//misc/AcademyDescriptions.pdf
[127] L. Garcia, personal communication, August 15, 2008.
[128] Educative material: "Native People of NASA Exploring Mother Earth and Father Sky," EW 2006-03-149. NASA Headquarters, Washington, DC.
[129] I.M.S. Rodrigues, personal communication, August 12, 2008.

APPENDICES

APPENDIX 1 – NASA CENTERS

NASA Center/Facility		State	Mission
1	Ames Research Center	CA	Promote innovative partnerships with technological, academic and entrepreneurial leaders.
2	Dryden Flight Research Center	CA	Flight research, technology integration, validation of space exploration concepts, conduction of airborne remote sensing and science observations, and support to operations of the space shuttle and the ISS.
3	Gleen Research Center	OH	Space flight systems and technologies development, aviation propulsion research. Leads the development of the service module and spacecraft adapter for the nation's crew exploration vehicle.
4	Goddard Institute for Space Studies	NY	Global climate change studies. It is a unit of the Earth Sciences Division at Goddard Space Flight Center and a unit of the Columbia University Earth Institute.
5	Goddard Space Flight Center	MD	American largest organization of scientists and engineers dedicated to studying the Earth, the Sun, the solar system and the universe. Established in 1959, NASA's first space flight center.
6	Jet Propulsion Laboratory	CA	Lead for robotic exploration. Managed by the California Institute of Technology.
7	Jonhson Space Center	TX	Home of NASA Mission Control and the Astronaut Corps; center for human space flight and related scientific and medical research efforts. Also manages the development, testing, training, production and delivery of all U.S. human spacecraft, as well as the program office for the ISS.
8	Kennedy Space Center	FL	Launch of missions around Earth and beyond.
9	Langley Research Center	VI	Space exploration, aeronautics and technology transfer.
10	Marshall Space Flight Center	AL	Space transportation and propulsion technologies, including the Ares I crew exploration and Ares V cargo launch vehicles; manages space shuttle propulsion elements and science aboard the ISS and conducts research in space aiming to improve life on Earth.
11	Michoud Assembly Facility	LO	One of the largest manufacturing plants in the world, responsible for the assembly of the space shuttle external fuel tanks.
12	NASA Headquarters	DC	Development of NASA policy.
13	Plum Brook Station	OH	The home of Gleen Research Center's four testing facilities containing space environment simulation facilities to test every element of a space vehicle.
14	Software independent verification and validation (IV&V) Facility	VI	Provides safety and cost-effectiveness for mission-critical software.
15	Stennis Space Center	MI	Home to U.S. largest rocket engine test complex. Testing of space shuttle main engine and where future engines and stages will be tested for returning astronauts to the Moon, Mars and beyond. Stennis also helps partner agencies make more informed decisions through science research results, remote sensing and other capabilities.
16	Wallops Flight Facility	VI	Main NASA facility for the management and implementation of suborbital research programs, managed by Goddard
17	White Sands Test Facility	NM	Test and evaluation of potentially hazardous materials, spaceflight components and rocket propulsion systems.

APPENDIX 2 – MAIN NASA'S EDUCATOR RESOURCE CENTERS

NASA's Ames Research Center
Educator Resource Center Mail Stop 253-2
Moffett Field, CA 94035-1000
Phone: 650-604-3495
FAX: 650-604-3445
http://www.nasa.gov/centers/ames/education/ERC/
Alaska, Northern California (southernmost counties of Inyo, Kings, Monterey, Tulare), Hawaii, Idaho, Montana, Nevada, Oregon, Utah, Washington, Wyoming

NASA's Dryden Flight Research Center
Educator Resource Center
38256 Sierra Highway, Suite A
Palmdale, CA 93550
Phone: 661-276-3992
FAX: 661-276
http://www.nasa.gov/centers/dryden/education/erc.html
Arizona, Southern California (northernmost counties of Kern, San Bernardino, San Luis Obispo)

NASA's Glenn Research Center
NASA Educator Resource Center
21000 Brookpark Road, MS 8-1
Cleveland, OH 44135
Phone: 216-433-2017
FAX: 216-433-3601
http://www.nasa.gov/centers/glenn/education/ERC_GRC.html
Illinois, Indiana, Michigan, Minnesota, Ohio, Wisconsin

NASA's Jet Propulsion Laboratory
JPL Educator Resource Center Village at Indian Hill
1460 East Holt Ave., Suite 20
Pomona, CA 91767
Phone: 909-397-4420
FAX: 909-397-4470
http://www.jpl.nasa.gov/education/
California

NASA's Goddard Space Flight Center
Education Resource Center
Mail Code 160
Greenbelt, MD 20771
Phone: 301-286-8570
FAX: 301-286-1781
http://www.nasa.gov/centers/goddard/visitor/home/erc.html
Connecticut, Delaware, District of Columbia, Maine, Maryland, Massachusetts, New Hampshire, New Jersey, New York, Pennsylvania, Rhode Island, Vermont

NASA's GSFC/Wallops Flight Facility
Visitor Center/RERC
Building J-17
Wallops Island, VA 23337
Phone: 757-824-2297
FAX: 757-824-1776
http://sites.wff.nasa.gov/vc/erc.htm
Eastern Shores of Virginia and Maryland

NASA's Kennedy Space Center
Educator Resource Center
Mail Code ERC
J.F. Kennedy Space Center, FL 32899
Phone: 321-867-4090
FAX: 321-867-7242
http://www.nasa.gov/offices/education/centers/kennedy/erc/index.html
Florida

NASA's Stennis Space Center
Educator Resource Center
Building 1200, Room 208A
Stennis Space Center, MS 39529-6000
Phone: 228-688-1348
FAX: 228-688-7528
http://education.ssc.nasa.gov/erc.asp
Louisiana, Mississippi

Educator Resource Center for
NASA's Johnson Space Center
Space Center Houston 1601 NASA Parkway
Houston, TX 77058
Phone: 281-244-2129
FAX: 281-483-9638
http://education.jsc.nasa.gov/
Colorado, Kansas, Nebraska, New Mexico, North Dakota, Oklahoma, South Dakota, Texas

NASA Educator Resource Center for
NASA's Langley Research Center
Virginia Air & Space Center
600 Settlers Landing Road
Hampton, VA 23669-4033
Phone: 757-727-0900, ext. 757 or 759
FAX: 757-727-0898
http://www.vasc.org/erc/index.html
Kentucky, North Carolina, South Carolina, Virginia, West Virginia

NASA's MSFC Educator Resource Center
U.S. Space & Rocket Center
One Tranquility Base
Huntsville, AL 35805-3399
Phone: 256-544-3922
FAX: 256-544-5820
http://www.nasa.gov/offices/education/centers/Marshall/erc/index.html
Alabama, Arkansas, Iowa, Missouri, Tennessee

APPENDIX 3 – NASA'S MISSION DIRECTORATES

1. Science Mission Directorate (SMD) – devoted to space science studies by means of Earth-orbit and deep space observatories, spacecraft to visit other planetary bodies and robotic landers, rovers and sample return missions. This directorate develops research on Earth[1], heliophysics[2], planets[3] and astrophysics[4]. An example of the work developed under this directorate is the exploration of Mars by Spirit and Opportunity, that have discovered evidence of an ancient wet environments on the planet that may have been suitable for life, and the SOHO mission to the Sun, which helps us understand the Sun-Earth connection, the plasma environment of the Sun and heliosphere, and how it affects Earth by interacting with Earth's magnetosphere (NASA, 2007).

2. Human Exploration and Operations Mission Directorate (HEO) – develops exploration systems, space life sciences research and applications, management of commercial space transportation, and capabilities for human spaceflight, including the servicing of the ISS after the retirement of the space shuttle in 2010; provides support to the Advanced Capabilities program, which focuses on producing the systems and technologies that will support space travelers, such as spacesuits, human medical and support technologies, and power-generation systems (NASA, 2007). This directorate also manages space operations related to launch services, space transportation, and space communications in support of human and robotic exploration programs.[5]

3. Aeronautics Research Mission Directorate (ARMD) – responsible for generating concepts, technologies and capabilities needed to advance aircraft and airspace systems, facilitating safer, more efficient and more environmentally friendly air transportation systems. It also continues to play a vital role in supporting NASA's human and robotic space activities.

[1] Including climate, carbon and ecosystems, surface and interior, atmosphere, weather, water and energy cycles.
[2] Including Sun, heliosphere, magnetospheres, space plasma physics, and space environments.
[3] Including inner and outer solar system, and small bodies of the solar system.
[4] Including the stars, galaxies, interstellar and intergalactic environments, and exoplanet exploration.
[5] Extracted from the NASA's Human Exploration & Operations Web site http://www.nasa.gov/directorates/heo/home/index.html

APPENDIX 4 – "ECLIPSES THROUGH THE CENTURIES" PAPER (ADAPTED)*

Norma Teresinha Oliveira Reis

1. INTRODUCTION

"To the best of our knowledge, our Sun is the only star proven to grow vegetables."

Philip Scherrer[6]

Other than Earth itself, the Sun and Moon are the celestial bodies most closely related to human reality. Although people in ancient times were curious about the stars and the Universe as a whole, the Sun and the Moon have been the main actors in our celestial theater. Since they are most directly associated with quotidian issues such as the day and night cycle as well as the seasons, they influence vital economic activities such as agriculture. Further, in early epochs the Sun was essential to estimate the time of day. The gnomon - a shaft erected perpendicular to the horizon - was used to indicate the passing hours through its shadow projected on the ground.

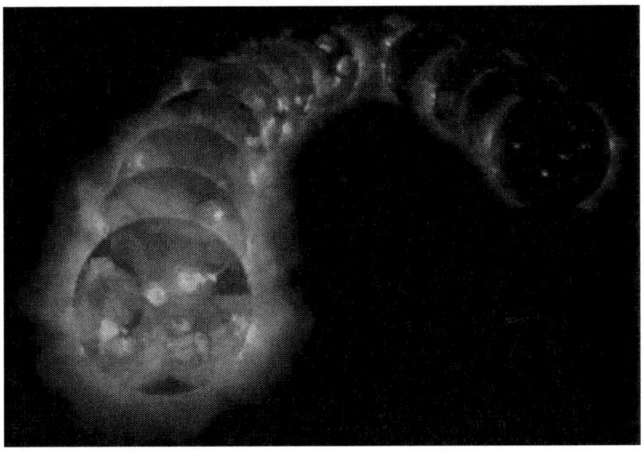

Figure 1. *The changing Sun.* These 13 x-ray images, obtained by the Yohkoh spacecraft between 1991 and 1995, provide a dramatic view of the Sun, highlighting how the solar corona's brightness changes during the waning part of the solar cycle as the Sun goes from an "active" state (left) to a more passive state (right). Source: NASA. Credit: G.L. Slater and G.A. Linford; S.L. Freeland; the Yohkoh Project.

* This paper has been adapted for the purpose of this publication. If you would like more information about the complete 'Eclipses' research I carried out as part of my ISU internship at NASA Goddard Space Flight Center, please send an e-mail to: norma.reis@gmail.com.

[6] Extracted from the Stanford Solar Center Web site, http://solar-center.stanford.edu/sun-on-earth/earth.html.

The Sun is crucial to us all. It is by far the closest star to the Earth and our ultimate source of warmth, light and energy. Could you imagine life on Earth without the Sun?

It must have been most intriguing for early civilizations to be suddenly deprived of the Sun's bright light during the day. They might have asked, "Is this the end of the world?" "Are the Gods angry with our behavior?" Human imagination is very creative at filling in gaps of understanding. Likewise, when a lunar eclipse occurs, the way the full Moon is occulted in a vast area of shadow, and the manner its color changes to a blood red certainly would indicate some disturbance in the cosmic order. So it would not be a large stretch to wonder whether this was a sign of divine anger, or the action of some demonic power [1].

Eclipses are astronomical phenomena that have provoked curiosity, fear, passion, and other feelings throughout the course of history. Humans are sometimes frightened by the unknown and by what we cannot understand or predict. Imagine a spaceship landing on your backyard with creatures from another world – this would be most frightening! If we had no prior knowledge about the mechanism of the Sun's disappearance in broad daylight during an eclipse, we would be afraid that the Sun's light has been truly lost, and that such darkness might last forever. A solar eclipse occurs when the Moon temporarily passes in front of the Sun and obscures it totally or partially. This configuration only exists during the new moon, and then only when the Sun, Moon, and Earth are on a single line, with the Moon in between. Since the Moon's orbit is tilted 5 degrees to the Earth's orbital plane, this configuration will not happen every month.

Figure 2. *The invisible dragon.* People in ancient China and old Southeastern Asian cultures believed that a dragon swallowed the Sun during solar eclipses. Source: NASA.

Though eclipses are well understood today, the apparent disappearance of our star has been attributed to different reasons throughout history. The ancient Chinese, for instance, thought that there was an invisible dragon in the sky. This dragon was angry with them and, for this, would devour the Sun. Whenever this happened, the Chinese followed rituals intended to scare the dragon away and make the Sun reappear. One ancient tradition was to bang drums and pots, and make other loud noises during eclipses in order to frighten the dragon away. Even not long ago, in the nineteenth century, the Chinese navy fired cannons during a lunar eclipse to fright the dragon! [2] The idea of a ferocious animal or beast trying to swallow the heavenly luminaries (the battle between the powers of light and of darkness) was common to many civilizations.

However, eclipses did not frighten some early civilizations. The Eskimo and the Aleuts interpreted them as signs of good fortune. The Sun and the Moon temporarily left their natural places in the sky to reassure themselves that everything on Earth was fine. Some eclipse legends are love stories and many others reflect local beliefs. For example, some Aboriginal cultures had a romantic view of eclipses. The Sun and Moon were husband and wife respectively, pulling curtains in the sky to ensure privacy for their union [1]. In Bolivia, it was believed that dogs chased after the Sun and the Moon to bite their face. It was the Moon's blood that would turn that red! The people howled and wailed to scare the dogs away [3].

Eclipses also used to have political meaning. Empires would rise and fall, kings would be crowned or dethroned, and battles would be lost or won because of chance alignments of the Sun, Earth and Moon. In ancient China, for example, solar and lunar eclipses were regarded as heavenly signs that foretold the future of the emperor [2], thus their prediction was of high importance for the state. From a scientific standpoint, eclipses have often opened doors to important knowledge. For example, in early times, lunar eclipses constituted an important proof of the sphericity of the Earth. Lunar eclipses were essential to debates of Pythagoras, Aristotle, and other Greek philosophers [4]. Thus, if we have an eclipse of the Moon by the shadow of the Earth, the shape of that shadow must represent the Earth's profile [1]. The fact is that humanity has never been indifferent to eclipses. In the course of history, they have been associated to important historical events, and today solar and lunar eclipses keep on having peoples all around the world looking up at the sky to contemplate, without fear, these awe-inspiring encounters between the Sun and the Moon, in their celestial ballet.

2. Eclipses and History

"We had the sky up there all speckled with stars, and we used to lay on our backs and stare up at them and wonder about whether they was made or only just happened."

- Mark Twain (1835-1910)

Since the very beginning of History, people have been amazed by what they see when they look up at the sky. Indeed, looking at the celestial sphere without the unpleasant interference of city lights is magnificent. It can sometimes mesmerize us with a deep desire of traveling out to these celestial spheres to directly experience what our eyes cannot see. Human imagination has no boundaries, but the Universe is infinite! However, most ancient civilizations have viewed changes in the sky with great fear and apprehension. Comets, meteor showers, supernovae, lunar and solar eclipses were viewed as bad omens by most societies.

As we discussed previously, the Sun and the Moon are the main actors in the celestial theater, the former being vital for life on Earth. The Moon, by its turn, has served as special inspiration for poets, writers, and lovers. The Sun and Moon have also been associated with religion and mythology, and sometimes regarded as gods with influence on the destiny of both societies and individuals. In this framework, solar and lunar eclipses were usually regarded as a disturbance in the natural order of the sky - as an indication that something was going wrong. Many historical events coincided with solar or lunar eclipses: battles, crowning or dethroning of emperors, peace treaties, and so forth. It is our nature as human beings to attribute meaning to events by whatever history, tradition or thought is available to us. The same is true for eclipses through the centuries.

Unlike comets, which for a long time have been regarded as unpredictable events, eclipses have been accurately predicted at the earliest stages of history [1]. Chinese astrologers were able to predict them by around 2300 BC [5]. Their predictions were based on empirical relationships governing the recurrence of events in which the geometric configuration of the Earth, Sun, and Moon repeats almost the same manner after 6,585 days. The existence of this regular cycle called *saros*, resulted from coincidences involving complex combinations between the movements of the Moon, Earth, and Sun. The Babylonians and the Chaldeans discovered the *saros* by keeping records of lunar eclipses. They realized that 18 years and 11 days after almost every lunar eclipse, there was another similar one in the same part of the sky and in the same season of its predecessor [6]. However, the general public did not understand these relationships. As governors began to realize how fearful the populace was of eclipses, emperors started to use eclipse-related knowledge as an instrument of power to influence people's psyche. The population would follow rituals and say prayers in order to prevent the supposed dire effects. Governors pretended that they could influence the obscure powers involved. Likewise, astronomers sometimes used their scientific knowledge to manipulate and influence governors, i.e., by establishing a negative correlation between an eclipse and the outcome of a battle.

Only in the last five hundred years or so, or certainly since Galileo's invention of the telescope in 1609, have started to understand these cosmic concurrences primarily in terms of the natural order of the Universe. Further, this invention has allowed the Sun to become an object of scientific investigation [7]. Its corona (extended plasma outer "atmosphere" of the Sun), prominences, and sunspots started to be studied. Solar eclipses are used to photograph and study the composition and dynamics of the Sun's corona, which is only visible when the Sun's bright disk or photosphere is completely blocked out. Today, it is also possible to produce artificial eclipses with the coronagraph – an instrument that uses an occulting disk that covers the Sun's surface (photosphere), and allows one to study the corona at any time. Scientists also take advantage of eclipses to study space weather phenomena such as solar flares and coronal mass ejections (CMEs). These phenomena are important due to the fact that they can directly impact space systems and human activities like telecommunications, navigation, and the highly complex work of astronauts in space, especially while performing extra-vehicular activities.

Currently, most people understand the basic mechanism of eclipses and are no longer frightened by them. They are now usually seen as beautiful and fascinating events. Every year, people of different ages, nationalities and interests join to watch total or partial eclipses of the Sun and the Moon all around the world. These magnificent celestial spectacles are also observed worldwide by students and educators of different levels. These phenomena can capture youthful imagination towards the universe and encourage new understanding about our space environment. But has something been lost? Does knowing how something works reduce its beauty and mystery? Or is the beauty of the world deepened by understanding the things we see? Anyway, eclipses are an effective means of making people curious about the wonders and mysteries of the Universe, and they fan the fires of curiosity and the human desire for exploration and discovery. They are a way of uniting people around the world to look at the starry sky as the place where we all came from long ago, cosmologically speaking. In this article, we present some important eclipses throughout time and their impact on people, societies, and science.

Ancient History

1) Ho and Hi, the Drunk Astronomers, 2137 BCE

Throughout the centuries, Chinese astronomers devoted great attention towards predicting eclipses. However, like all similar efforts before the Renaissance, this could only be empirical [8]. The earliest record of a solar eclipse comes from ancient Chinese history. Identifications of this event have varied from 2165 to 1948 BC [8], though the favored date is October 22, 2137 BC. Ancient Chinese astronomy was primarily a governmental activity. It was the astronomer's role to keep track of solar, lunar and planetary motions, and explain what they meant to the ruling emperor. According to a legend, the royal astronomers Ho and Hi dedicated too much time to consuming alcohol and failed to predict the forthcoming eclipse. Traditionally, the solar eclipse recorded in the *Shu Ching* (Historical Classic) was regarded as from the 3rd millennium BC. *On the first day of the month, in the last month of autumn, the Sun and the Moon did not meet (harmoniously) in Fang'...* [8]. The emperor became very unhappy because, without knowing that there was an eclipse coming, he was unable to organize teams to beat drums and shoot arrows in the air to frighten away the invisible dragon. The Sun did survive, but the two astronomers lost their heads for such negligence.

'Here lie the bodies of Ho and Hi
Whose fate though sad was visible,
Being hanged because they could not spy
Th'eclipse which was invisible (Author unknown). [9]'

Since then, a legend arose that no one has ever seen an astronomer drunk during an eclipse... [1]

2) Thales Eclipse, 585 BC

In the West, the first prediction of a solar eclipse is associated with the Greek philosopher Thales of Miletus. According to Herodotus, he foretold the eclipse of 585 BC. There is a legend referring to the final battle of a 15-year war between the Lydians and the Medes, with ups and downs on both sides, but no decisive victory for either. Also known as the "Battle of the Eclipse," it occurred at the Halys River (Asia Minor), on May 28, 585 BC [10], and was suddenly terminated as the day turned to night, scaring both sides [11].

"In the sixth year a battle took place in which it happened, when the fight had begun, that suddenly the day became night. And this change of the day Thales the Milesian had foretold to the Ionians laying down as a limit this very year in which the change took place. The Lydians however and the Medes, when they saw that it had become night instead of day, ceased from their fighting and were much more eager both of them that peace should be made between them." [12]

The total solar eclipse was perceived as an omen indicating that the Gods wanted the battle to stop. The kings became impressed and stopped fighting. Since the exact dates of eclipses can be calculated, this battle is the earliest historical event for which a precise date is known. Though some argue that Thales used the *saros* cycle of 223 lunations [13,8], it is now agreed by historians that the *saros* was not discovered before the fifth or fourth century BC, therefore Thales could not have used that time system. It is said that Thales was proclaimed a wise man by the oracle of Delphi, possibly due to the eclipse prediction credited to him. However, it is clear that he did not understand the scientific basis of the phenomenon.

3) The Crucifixion Eclipse, 33 AD

Jesus Christ should have been crucified on day of eclipse, during the period when Pontius Pilate was procurator of Judea (26-36 CE). However, there is no consensus on the date. According to the evangelists, Jesus was crucified on a Friday afternoon, some hours prior to the beginning of the Jewish Sabbath. Evidence suggests April 3, 33 AD [16], while others suggest April 7, 30 AD. Another hypothesis is that of a solar eclipse visible in Jerusalem on November 24, 29 AD [1]. The Greek historian Phlegon mentions this eclipse in his *History of the Olympiads*, and says that it has been accompanied by an earthquake [1]. *In the fourth year of the 202nd Olympiad, there was an eclipse of the Sun which was greater than any known before and in the sixth hour of the day it became night; so that stars appeared in the heaven; and a great Earthquake that broke out in Bithynia destroyed the greatest part of Nicaea* In fact, mention is also made in the Christian Bible to the Sun being darkened earlier that day: *The Sun shall be turned into darkness.*

However, there are also various allusions in the Christian Bible to the Moon being dark and turned to blood when it rose in the evening after the crucifixion, which sounds like a lunar eclipse. In *Acts of the Apostles*, Peter also refers to a Moon that is the color of blood and a darkened sky. There is other evidence that on that day the Moon appeared like blood. A New Testament Apocryphal fragment, the so-called Report of Pilate, states *Jesus was delivered to him by Herod, Archelaus, Philip, Annas, Caiphas, and all the people. At his crucifixion the Sun was darkened; the stars appeared and in all the world people lighted lamps from the sixth hour till evening; the Moon appeared like blood.* [17] This may result of a dust storm caused by the *khamsin*, a hot wind from the south. Under these circumstances, a lunar eclipse while there is much suspended dust, one would expect the Moon to appear the dark crimson of blood [4]. The reason why the Moon is blood red is that, though it is geometrically in the Earth's shadow, sunlight is refracted through the Earth's upper atmosphere, where normal scattering prevents blue light from penetrating. But this refracted light would be much weaker than direct light from even a small portion of the Sun, and the blood color associated with the eclipse would not be visible to the unaided eye. However, the Moon would have an amber color from atmospheric absorption, similar to any other occasion when the Moon is low in the horizon [18].

As mentioned, there is controversy among researchers whether that was a solar or a lunar eclipse, and also controversy about the date. In any case, an eclipse occurring in the very same day of the crucifixion would have been interpreted by believers as a supernatural sign and influenced the change of mind of the Jews and Pilate towards the body of Jesus Christ, leading to the placing of a military guard on the sepulcher [19].

Middle Ages

4) Solar Eclipse of the Emperor Louis, 840 AD

This is the story of the eclipse that influenced the division of Europe, as we know it today. Louis of Bavaria, the son of Charlemagne, was head of a great empire when, on May 5, 840 AD, he witnessed a solar eclipse. He interpreted the eclipse as a finger pointed at *him*. Frightened, he did not recover anymore, thinking that he would die very soon. With that "prophecy" in mind, one month later he passed away. Possibly, the emperor Louis associated his father's death in 814 with a set of solar and lunar eclipses that took place between 807 and 810 (and that could be viewed from the kingdom), interpreting the phenomena as a bad omen [4]. After Louis' death, his three sons began to dispute his succession. Their quarrel was settled after three years with the *Treaty of Verdun*, dividing Europe into three large areas, namely France, Germany and Italy.

5) The Black Hour, 1433 AD

One of the most famous eclipses of the Middle Ages was called the *"Black Hour."* It was visible in Scotland and it has been said that a very deep darkness came about 3 pm on June 17, 1433 AD [14]. That eclipse would have been unusually extensive, lasting around one hour. The scientific explanation relates to a specific angle between the Sun and Moon on that occasion: at the time of the eclipse, the Sun was only two degrees from perigee and the Moon no more than thirteen degrees from apogee [20]. Some other interesting facts about this eclipse is that records of the epoch relate that nothing was visible during the height of totality – although it sounds exaggerated – and in the context of a very superstitious time, the pestilence that then prevailed has been attributed to the eclipse.

6) Fall of Constantinople Eclipse, 1453 AD

The Roman Emperor Constantine, in 324 AD, moved the capital of his empire to ancient Byzantium, which he renamed Constantinople. This capital has ruled the eastern Mediterranean and the Black Sea region for more than a thousand years, providing a strong government and assuring the continuation of the Roman Empire after it collapsed elsewhere [21].

Figure 3. *The Siege of Constantinople.* Scene from the battle of defense of Constantinople, painted in 1499, in Paris. Credit: Public domain.

By the 15th century, the expanding Ottoman Empire has decided to conquer Constantinople. The Success and failure can be sides of the same coin, and they usually result of someone's beliefs, here associated with an eclipse of the Moon. Turks attempted assaults in 1402 and 1422 without success, the city protected by its famous impenetrable walls. In 1453, the troops of Sultan Mohammed II returned with a powerful cannon and more than 200 thousand men. However, with only seven thousand men, Constantinople repelled three assaults and repaired their damaged walls each night. They trusted old predictions, according to which Constantinople would never fall. However, the full Moon rose in eclipse on May 22, and Constantinople's invincibility was also eclipsed… The phenomenon has been interpreted as a bad omen by the Constantinople side. Six days later, Mohammed II tried a new assault and succeeded. Accidentally, a postern gate had been left open and some Turks entered the city. As the Sultan's men crossed the walls, the fight became a tumult and Constantinople's defense collapsed. The terrible sack of Constantinople that followed lasted three days and shocked the Western civilization [1].

Modern Ages

7) Christopher Columbus' Eclipse, 1504 AD

After a long trip to the Americas in 1503 AD, in his fourth voyage, Columbus ended up on the island of Jamaica. In principle, he managed to obtain provisions from the Caciques natives in exchange for some trinkets and rubbish. As time went by, novelty and hospitality started to decrease and at the same time the sailors started act aggressively with the natives to obtain food. Unhappy, the Indians communicated to the Spanish that they would not provide any more supplies. Desperate with the threat of famine, Columbus needed to think about a solution. He came up with an ingenious plan. He checked his *Calendarium*, which contained predictions of lunar eclipses for several years. In particular, it informed about a total eclipse of the Moon on the Antilles on February 29, 1504 AD. That evening, Columbus invited the Caciques onboard his *Capitana* for a very serious conversation. He told them that they were Christians and their God did not appreciate the way they had been treating Columbus' crew, and would punish the Indians with famine and pestilence and, as a sign of dissatisfaction, his God would darken the Moon. As he said that, the Earth's shadow started to cover the white disk. Terrified, the natives begged Columbus to bring the moonlight back. According to Ferdinand Columbus (second son of Christopher Columbus), cited by Sinnot [22]:

> *"The Indians observed this [the eclipse] and were so astonished and frightened that with great cries and lamentations they came running from all directions to the ships, carrying provisions and begging (...) and promising they would diligently supply all their needs in the future."*

Columbus replied that he needed to consult his God first. Smartly, he shut himself in a cabin for nearly two hours, and checked his *Calendarium* to see when totality would end. Just before the end of totality, he reappeared and announced that God had given his pardon, and would bring them back the Moon provided that the Christians were given provisions. Immediately, the Moon reappeared. Fearful and impressed with what they witnessed, the natives provided Columbus and his crew their needed provisions until they were able to return to Europe [1].

Figure 4. *Columbus impressing the natives.* This illustration from Camille Flammarion's Astronomie Populaire 1879, shows how Christopher Columbus used an eclipse of the Moon to assert his power over the Indians in Jamaica. Source: Wikimedia Commons. Image published before 1923 and public domain in the United States.

The use of eclipses as a tool to manipulate populations less knowledgeable about eclipses mechanism is also present in diverse works of fiction. In 1889, Mark Twain published *A Connecticut Yankee in King Arthur's Court*, a novel envisioning life in the sixth-century England. The author has Hank Morgan, the yankee in the title, threatening the ignorant folk of that time by invoking prior knowledge of a solar eclipse on June 21, 528 AD. Twain has Morgan, who is jailed waiting for execution, threaten King Arthur with a disappearance of the Sun [4]:

> *"Go back and tell the king that at that hour I will smother the whole world in the dead blackness of midnight; I will blot out the Sun, and he shall never shine again; the fruits of the Earth shall not rot for lack of light and warmth, and the peoples of the Earth shall famish and die, to the last man!"*[23]

This description is accurate in many senses, except for the fact that there was no solar eclipse visible in England in 528 AD. More examples of this theme are available in literature, such as *The Adventures of Tintin* by Georges Remi and *King Solomon's Mines* by H. Rider Haggard.

8) Nat Turner's Eclipse, 1831 AD

The greatest slave revolt in North America was led by Nat Turner (1800-1831 AD). Very clever, Turner learned how to read with his masters' son. Reading was an undesirable skill from the masters' viewpoint for a slave during these times due to fear of rebellion. Turner, afterwards, dedicated himself to religion and became a preacher [24]. In 1828 AD, he had a vision: he would lead his people to liberty, but he should wait for a sign from God – and it supposedly came from the sky. An annular eclipse of the Sun occurred on February 12, 1831 AD. Turner interpreted it as a 'black angel' occulting a white one – the time had arrived for blacks to overcome whites. In other words, it was time for rebellion. Several months later, after having murdered his original masters, Turner and his band of insurgents headed for the small town of Jerusalem where militiamen interrupted their march. Most slaves, including Turner, went into hiding for seventy days, and ended up hanged [1]. Many people have died during this revolt, and in no other episode in U.S. history have so many slave owners perished, a reason why Turner is regarded a hero of the resistance to oppression against black people in the United States.

9) The Great Eclipse of 1878 AD

Considered one of the principal eclipse events of the century, it is known as the great eclipse of 1878 for its large path of visibility. The Moon's shadow took some 20 minutes to cross Wyoming and Colorado, in the United States, on July 29, 1878 AD. Many tourists filled hotels to see the spectacle - even the famous inventor Thomas Edison was there! Extensive preparations were made by officers in charge of the National Observatory to observe the phenomenon. Five expeditions were in charge of relevant scientific investigations, such as making drawings of the corona, and studying the physical constitution of the Sun [25]. Though the corona was photographed in 1851 AD, the results were not satisfactory and, in 1878, drawings provided the best information about its size and shape [26].

The 1878 eclipse was observed by the American astronomer Maria Mitchell, the first woman astronomer to join the prestigious American Academy of Arts and Sciences. She led a team of five of her female students on a cross-country journey to Denver, Colorado, to observe and scientifically report a total solar eclipse [27]. They traveled by train at a time when ladies did not travel unescorted. Her students were fascinated with the trip, though a little frightened. Maria took care of all logistics of sending telescopes to an observational site, and provided students with instructions of the observations they should make. *You will see Nature as you never saw it before – it will neither be day nor night – open your senses to all the revelations*, she pointed out. *Let your eyes take note of the colors of Earth and Sky. Observe the tint of the Sun. Look for a gleam of light in the horizon. Notice the color of the foliage. Use another sense – notice if flowers give forth the odors of evening. Listen if the animals show signs of fear – if the dog barks – if the owl shrieks – if the birds cease to sing – if the bee ceases its hum – if the butterfly stops its flight – it is said that even the ant pauses with its burden and no longer gives the lesson to the sluggard.* [28] She described also the most glorious moment, that is, the observation of the corona! *"As the last rays of sunlight disappeared, the corona burst out all around the Sun, so intensely bright near the Sun that the eye could scarcely bear it; extending less dazzlingly bright around the Sun for the space of about half the Sun's diameter, and in some directions sending off streamers for millions of miles..."* [28]

The young ladies were enthusiastic about the experience and harbored a brave attitude at a time when women were not supposed to be inside scientific circles. At an epoch when men's colleges rarely engaged science students with direct field experience like this, Mitchell's students were entering a new era of learning for women. This event represented significant scientific, societal, and pedagogical advancement promoted by a pioneering woman.

10) Einstein Eclipse, 1919 AD

This eclipse has definitively revolutionized the history of science in the twentieth century, as it helped to confirm the General Theory of Relativity by Albert Einstein (1879-1955). Indeed, one of Einstein's most remarkable contributions to science was his General Theory of Relativity (GTR), formulated between 1913 and 1916 AD. This theory contradicted Newton's and provided fundamental new understandings of time and space measurement. GTR was extremely complex to understand though and, in order to be accepted, it was necessary that Einstein's theory predicted or explained some observed phenomenon that the Newtonian theory could not [4].

In this scenario, Eddington came up with the idea that a total eclipse of the Sun would provide such a unique opportunity to quantitatively test Einstein's theory. How? If it were correct, the light from stars would be bent by the strong gravitational field of the Sun. And what circumstance was necessary to observe this? Totality and stars appearing close-by because the test required several bright stars close to the limb of the Sun during the eclipse. The eclipse of May 29, 1919, offered these conditions. The path of totality crossed Brazil, in South America, and Principe, an island owned by Portugal in the Gulf of Guinea, just to the north of the equator and 150 miles from the African coast. In northeastern Brazil, the city of Sobral, Ceara state, was the best post of observation of the phenomenon, and two expeditions of U.S. and English scientists joined the Brazilians to observe the eclipse. Their purposes were distinct. The Brazilian commission focused on studies of the solar corona (its form and shape), and carried out spectroscopic analysis of its constitution. The U.S. and the England scientists planned to verify experimentally the consequences of GTR [29].

After analysis of the eclipse results, the royal astronomer Frank Dyson announced, in November 1919 AD, that the results confirmed the theory and it was made public: Einstein was right! In fact, what provoked such commotion was precise measurement of the deviation of starlight passing close to the Sun. The value of such deviation agreed with the prediction of Einstein's GTR (1.75 arc seconds), but was almost double the value predicted by Newton's gravitational theory (0.87 arc seconds). This is one of the most dramatic events in the history of science and was front-page news around the entire world. The London Times featured the headlines, *"Revolution in science. New theory of the Universe. Newtonian ideas overthrown"* and The Washington Post, with *"New theory of space: has no absolute dimensions, nor has time, say Savants."* The president of the Royal Society, J. J. Thomson, described the general theory as *"the greatest discovery in connection with gravitation since Newton... Our conceptions of the fabric of the Universe must be fundamentally altered."*[30] Although people were still mystified by Einstein's theory and did not understand that, Einstein's worldwide popularity as a legendary scientist [24] increased exponentially almost overnight [1] thanks partly to the fanfare that followed the eclipse. Einstein was also very charismatic and became famous for his equation $E=mc^2$.

Acknowledgments

Writing a paper is such a special act - it is an act of love. We always wonder how to make that better, but at some point we just have to stop, and hope that readers will enjoy that. Few significant accomplishments can be completed alone, so that I would like to express here my gratitude to a few exceptional people who have been fundamental to help me make this project a beautiful reality.

First, I want to express my deep and special gratitude to *Dr. Timothy E. Eastman* (Perot Systems at NASA Goddard), my mentor and supervisor, who provided full and unique guidance and support to make my NASA internship possible. Dr. Tim is a brilliant plasma physicist, and a wonderful person. He has made a tremendous difference in my life, teaching me fundamental professional, academic and life lessons.

Figure 5. *Dr. Timothy E. Eastman.* Plasma physicist and group manager at NASA Goddard. Creator of the Web site *plasmas.org*. Source: NASA.

I am very thankful to *Dr. Louis A. Mayo* (SP Systems at NASA Goddard), my mentor, for our countless pedagogical and scientific discussions, which largely enriched the research. He is a great astronomer and space science education specialist, and a special person who has greatly contributed to my internship experience at NASA.

Figure 6. *Dr. Lou Mayo.* Astronomer, space science education specialist at NASA Goddard. Source: NASA.

Many thanks to *Ms. Constance Carter*, Head of the Science Reference Section of the Science, Technology and Business Division of the Library of Congress - LOC, who assisted me in finding several outstanding sources, which greatly contributed to the quality of this paper.

I am deeply thankful to the guidance of *Dr. Fred Espenak* (NASA Goddard) who, for very good reasons, is known as "Mr. Eclipse." Dr. Espenak kindly provided singular guidance and assistance during this research, indicating valuable sources, and making available various images from his eclipse-dedicated Web sites, books and papers.

Other distinguished NASA specialists were exceptionally generous by collaborating to the enhancement of this project: *Dr. Troy Cline*, for his support in the production of this paper Web site and podcasts; *M.Sc. Rita Johnson*, for her partnership in the production of the Portuguese podcast; *Dr. Robert L. Kilgore*, and *Dr. Glen A. Asner*, for their critical reading and suggestions to the paper; and *Dr. Jay Friedlander*, who provided design and artistic support. Thanks to all NASA people who made my internship experience just wonderful and unforgettable!

My special thanks to my parents and masters, *Mrs. Antonia Reis* (in memory) and *Mr. Antonio Oliveira Reis*, whose unconditional love, continued support and encouragement have been essential to help me reach my dreams.

Foremost, I want to thank God for the unique combination of gifts, opportunities, and for introducing these so special people in my life. He has been opening doors to turn my "pathway to the stars" into such a wonderful reality – the lifetime passion I always dreamed about: NASA!

REFERENCES

[1] Brunier, S., Luminet, J. P., *Glorious Eclipses: Their Past Present and Future*, Cambridge University Press, 2000.
[2] NASA – Sun-Earth Day – Space Weather Around the World, http://sunearthday.gsfc.nasa.gov/2008/TTT/63_eclipse.php
[3] Andrews, T., *Wonders of the Sky - Wonders of Nature: Natural Phenomena in Science and Myth*, Westport: Libraries Unlimited, 2004.
[4] Steel, D., *Eclipse: the Celestial Phenomenon that Changed the Course of History*, Washington, DC: Joseph Henry Press, 2001.
[5] NASA Web site, "Eclipses Through Traditions and Cultures," http://eclipse99.nasa.gov /pages/traditions_Calendars.html.
[6] Littman, M., Willcox K. and Espenak F. *Totality: Eclipses of the Sun*, New York: Oxford University Press, 1999.
[7] NASA Web site, "Historical Aspects of our Sun," http://www.nasa.gov/vision/universe/solarsystem/sunhistory_prt.htm.
[8] Needham, J., *Science and Civilization in China*, New York, Oxford, Cambridge University Press, 2008.
[9] NASA Web site, "NASA - Solar Eclipses of History," http://eclipse.gsfc.nasa.gov/SEhistory/SEhistory.html.
[10] Ionides, S. A. and Ionides, M. L., *Stars and Men*, Indianapolis, New York: Bobbs-Merril, 1939.
[11] Solar Eclipse Newsletter, Volume 6, Issue 5, 2001. eclipse.gsfc.nasa.gov/SENL /SENL200105A.pdf
[12] Wikipedia.
[13] Wolfram Research Web site, http://scienceworld.wolfram.com/astronomy/ Lunation. html.
[14] Chambers, G. F., *The Story of Eclipses*, New York: D. Appleton and Company, 1902.
[15] Brewer, B., *Eclipse*, Seattle: Earth View, 1991.
[16] Humphreys, J. C., Waddington, W. G., "Dating the Crucifixion," *Nature*, v. 306, n. 5945, December 22, 1983, p. 743-746.
[17] Humphreys, J. C., Waddington, W. G., "The Date of the Crucifixion," *JASA* 37 (March 1985): 2-10.
[18] Schaefer, B., E. Q., J R. Astr. Soc. (1990), 31, 53-67.
[19] Humphreys, J. C., Waddington, W. G., "Dating the Crucifixion," *Nature*, v. 306, n. 5945, December 22, 1983, p. 743-746.
[20] Littell, E. and Littell, R. S., *The Living Age*, Boston: Son & Company, 1868.
[21] Schaefer, B. E., "Lunar Eclipses That Changed the World," *Sky and Telescope*, December 1992, p. 639-642.
[22] Sinnott, R. W., "Columbus and an Eclipse of the Moon," *Sky and Telescope*, October 1991, p. 437-440.
[23] Twain, M., *A Connecticut Yankee in King Arthur's Court*, New York, London, Harper & Brothers, 1889.
[24] Schaefer, B. E., "Solar Eclipses That Changed the World," *Sky and Telescope*, May, 1994, p.36-39.

[25] Scientific American, "The Total Eclipse of the Sun," *APS Online*, v. XXXIX, n. 04, July 27, 1878, p. 56.

[26] Eddy, J. A., "The Great Eclipse of 1878," *Sky and Telescope*, June 1973.

[27] The Vassar Encyclopedia online Web site, maintained by the Vassar's College Historian, http://vcencyclopedia.vassar.edu/

[28] Wright, H., *Sweeper in the Sky: the Life of Maria Mitchell First Woman Astronomer in America*, New York: The Macmillan Company, 1949.

[29] Soares, M. N. M., *Eclipse de 1919: Múltiplas Visões*, ed. 2, Sobral: Edições UVA, 2003.

[30] Boston, A., "*At Another Eclipse, a Star Was Born*," *New Statesman*, v. 128, August 9, 1999, p. 16.

PART II

SPECIAL INVITED CHAPTERS

In: The Space Education Phenomenon at NASA ...
Editor: Norma Teresinha Oliveira Reis
ISBN: 978-1-62618-188-5

Chapter 1

THE RADIO JOVE PROJECT

James R. Thieman*
NASA/Goddard Space Flight Center,
Greenbelt, Maryland, US

ABSTRACT

The Radio Jove Project is a non-profit education and outreach project begun more than a decade ago with the intent of acquainting students and other interested groups and individuals with the science of radio astronomy.

By purchasing an inexpensive kit participants can build a radio receiver and antenna and use the equipment to monitor the radio signals of Jupiter, the Sun, the galaxy, as well as Earth-based emissions. Radio Skypipe software supplied in the kit or through a download on the internet allows participants to show the output of their radio telescope on a simulated chart recorder displayed on a personal computer. The observer can also make the signal available to other Radio Jove participants monitoring through Radio Skypipe so they can compare what they are seeing to what others are receiving at the same time. Over 1700 kits have been distributed throughout the world.

INTRODUCTION

Astronomy has long served as a source of knowledge and inspiration for humankind given the vast reaches of the Universe and our minuscule part of it. Since the time of Galileo the use of the telescope has revolutionized our understanding of the Universe and the discoveries continue to grow exponentially. This is due not only to the increasing sophistication of our optical telescopes, but also to the realization that the explanation of the phenomena ongoing in the Heavens is revealed through means other than what we see with light observable by our eyes alone. The information contained in the non-visible parts of the electromagnetic spectrum is overwhelming in comparison to what we learn from just the small slice of the spectrum that is revealed to us by our eyes [1]. As the study of the spectrum

* E-mail address: james.r.thieman@nasa.gov

increased in the 20th century we learned that all parts of the electromagnetic spectrum are to be found in the Universe, but only the optical, and some of the radio, infrared, and ultraviolet parts of the spectrum penetrate the Earth's atmosphere and are detectable on the Earth's surface.

We have used optical telescopes at ground level for 400 years to discern quite a bit about the physics of the Cosmos, but only in the last three quarters of a century have we understood how to use the long wavelengths of the radio part of the spectrum to expand the science of astronomy. The optical telescope is well known to the general public as a tool for the astronomer, but the radio telescope as an additional tool for discovery is relatively unfamiliar. It is the purpose of the Radio Jove project to give students and other interested individuals and groups a basic knowledge of what can be learned through the different realm of radio astronomy. The project is a building block leading to the understanding that the Universe has much more to tell us not only in the radio portion of the spectrum but also in the entire spectrum which can be monitored through different types of telescopes both on the ground and flown in space.

The Origins of Radio Jove

Radio Jove began in the late nineties when a group of scientists, engineers, etc. with interest in radio astronomy met at NASA's Goddard Space Flight Center and planned a project to promote the science of radio astronomy for students. Many of the group had ties with the University of Florida radio astronomy program so Radio Jove put emphasis on the decametric (approximately ten meter wavelength) radio observations of Jupiter which had been the main emphasis of the Florida radio research. The group set a goal of creating a radio telescope kit for $100 or less so it would be inexpensive for participants to be introduced to radio astronomy. Interest in the subject could be carried further by setting up more sophisticated antennas or using commercially available receivers or even spectrographic receivers.

Radio Jove was fortunate enough to have a proposal to the NASA Goddard Space Flight Center Director's Discretionary Fund accepted for a multi-year start to the program. The grant provided supplementary funding so that the first hundred radios could be sold at half price, $50, and allowed a quick start to the effort. The first ten kits were actually offered free of charge to ten "beta testers" to assure that the kits would work and to provide feedback for any problems in the kit makeup. In that initial period Radio Jove was also helped greatly by a similar program called INSPIRE [2] which had preceded Radio Jove and successfully distributed VLF radio kits for the monitoring of Earth-based radio emissions. The two founders of INSPIRE, Bill Taylor and Bill Pine were also among the founders of Radio Jove and the project benefitted tremendously from their knowledge and support. Bill Pine was the first distributor of the Radio Jove kits. Bill Pine is now retired and Bill Taylor passed away a few years back, but Jove is in debt to the project very aptly named "INSPIRE".

Over the years Radio Jove did benefit from a number of other small grants, but it has continued as a self-sustaining non-profit program in which the sales of previous kits provides the funds necessary to continue the generation of new kits. The members of the core staff of Radio Jove are mostly volunteering their time for the benefit of the program. The cost of the kit has doubled to $200, but that seems quite reasonable for the duration of more than a decade of parts cost increases. In that time period more than 1700 kits have been distributed to more than 70 countries. Radio Jove has been used in schools, science fairs, museums and science centers, projects for organizations, and as an amateur scientist tool for individuals and groups. We hope that this successful approach can be continued well into the future.

How Does Jove Work?

The details of the Radio Jove project are found on the associated website [3], but some of the aspects of it will be described in this chapter. Those who are interested in participating in Radio Jove are invited to purchase the radio telescope kit by going to the Radio Jove website and clicking on the "Kit Ordering" link. The link leads to an order form that provides the details about the items that can be ordered, including the costs for purchase and for shipping.

The usual approach is to purchase the standard kit that includes all the parts necessary to build the radio receiver as well as the parts for the standard dual dipole radio antenna.

Figure 1 is a top-down view of the receiver as it appears after construction. Normally there is a metal lid on the receiver, but that has been removed so the parts inside the receiver can be seen. Figure 2 is a picture of the typical setup for the dual dipole antenna. What is not included in the kit is the support structure for the antenna. The antenna consists of two copper wire dipoles approximately 25 feet in length that are suspended parallel to the ground and usually running in an East-West direction. The two wires are suspended parallel to each other at a height of 10 to 20 feet and at a distance of about 20 feet from each other. In Figure 2 support for the wires has been provided by lengths of PVC pipe but many other types of materials have been used to suspend the wires. Figure 3 is a schematic of the antenna construction.

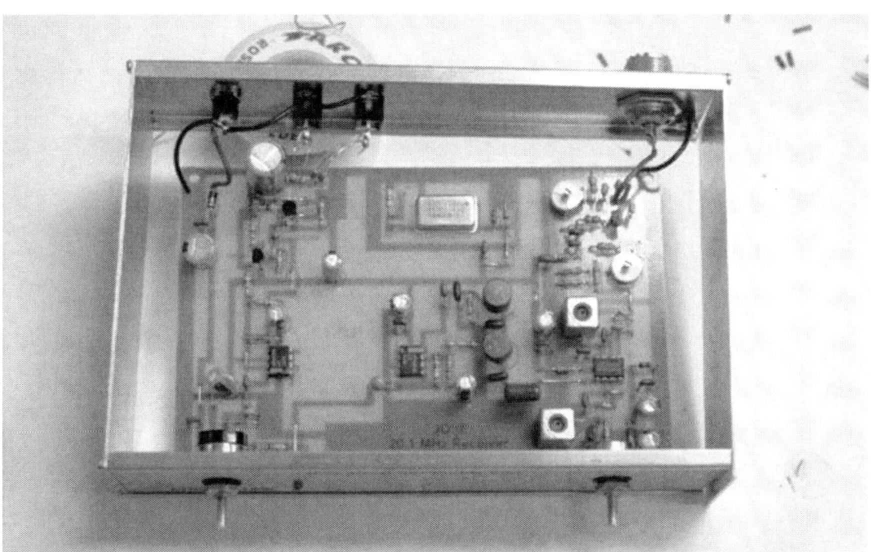

Figure 1. Radio Jove RJ1.1 20.1 MHz receiver. View is from the top.

Figure 2. Radio Jove dual dipole antenna at a height of 10 feet using PVC pipe, guy-ropes, and stakes to support the antenna wires.

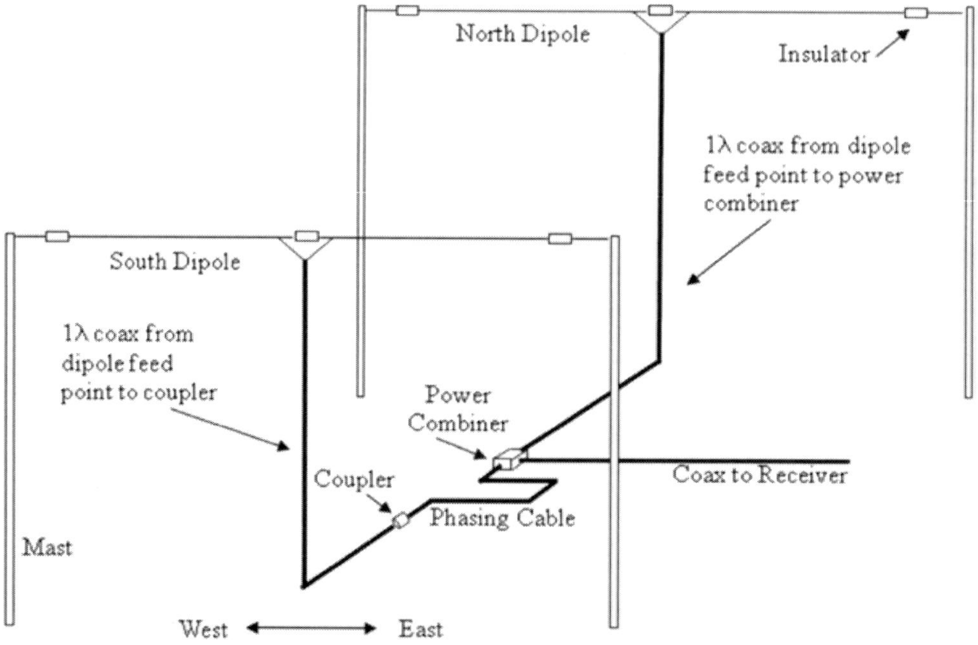

Figure 3. Schematic for setup of the Radio Jove dual dipole antenna.

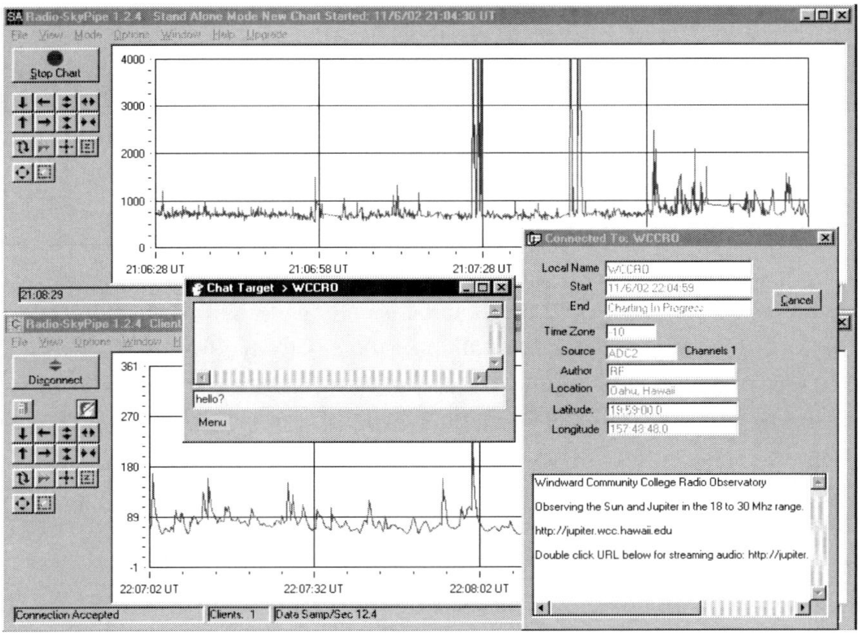

Figure 4. Radio Skypipe software emulates a chart recorder showing the signal strength from the antenna and can send signals to and receive signals from other antennas through an internet connection.

Also in the antenna kit is a CD that contains Radio Skypipe software. The software emulates a chart recorder on a Windows-based Personal Computer (PC). When the antenna is connected to the receiver, the receiver converts the electrical signal to an analog sound signal that can be input to the sound card of a PC. The PC digitizes the signal and the digital signal is displayed using the Radio Skypipe software. Figure 4 shows a sample display using the software. The vertical scale of the graph represents signal strength and the horizontal scale indicates the time, usually in Universal Time. The Radio Skypipe software is very versatile and has capabilities for variation of the vertical and horizontal axis displays, magnification, offsetting, color variation, etc. More importantly, if the PC is connected to the internet, the software allows the signal to be sent through the world wide web to other computers also running Radio Skypipe and they can open a new chart emulation window and display the signal almost simultaneously. Thus, the Radio Jove participants can compare the signals they are receiving with other's signals side-by-side. They can also open a chat window and send messages back and forth to other Skypipe users.

Also on the CD is a copy of the book "Listening to Jupiter" by Richard Flagg [4]. This is a valuable source of information for all aspects of the Radio Jove project.

WHAT CAN I OBSERVE?

As mentioned previously, the standard Radio Jove dual dipole system provides an entry point into the mysteries of the Universe revealed through radio waves. We know there are solids, liquids, and gases around us on Earth and these have been observed to be present beyond Earth as well. A fourth state of matter, plasma, is not very prevalent on Earth, but is a main component of the rest of the Universe. Indeed, in our solar system the Sun contains by far most of the matter and it is almost entirely in the plasma state. Such is true for all the stars that we see at night as well. Plasma is somewhat like a gas in that the particles in a plasma are not tightly bound to each other, but, unlike the gases we often experience on Earth, the particles are electrically charged and the forces that result from the movement of these charged particles cause the plasma to behave in very different ways from a non-charged gas. Moving charged particles can generate magnetic fields and these magnetic fields can, in turn, affect the motion of other charged particles. When the charged particles move through a region where there is a magnetic field they can be accelerated and this acceleration will cause them to emit electromagnetic waves. The nature of the electromagnetic waves will depend on the amount of acceleration and ultimately the energy received by the charged particles. Acceleration to low energies will lead to the emission of low energy electromagnetic waves, which are radio waves. As the energy increases the type of wave emitted may be infrared, visible, ultraviolet, X-ray, or Gamma-ray radiation. Radio waves, however, are common since it does not take much energy to generate them.

As indicated in the introduction, visible light and portions of the ultraviolet and infrared radiation spectrum are able to penetrate the Earth's atmosphere and reach us on the Earth's surface. Likewise, portions of the radio spectrum also penetrate. The Radio Jove radio telescope is made to receive radio waves that have a frequency around 20.1 MHz (MegaHertz) since this frequency does reach the Earth's surface. This is a useful observing frequency because several of the strong natural radio sources can be detected with the Radio Jove equipment tuned to this frequency. The main natural sources observed by Radio Jove are the Sun, Jupiter, the galaxy, and the Earth. The equipment also receives many man-made radio signals that can be of interest as well.

When the radio receiver is turned on the operator hears the radio waves converted to sounds that can be played on the speaker and fed into the computer sound card. The normal sound is a constant hiss of "white noise" or static. At the observing frequency of 20.1 MHz this is actually the constant emission coming from our Milky Way galaxy. The charged particles moving in the magnetic field of the galaxy are constantly emitting this level of background hiss that forms the baseline of radio noise that the telescope detects. How do we know this is from the galaxy? One experiment that helps to confirm this is to listen to the background over an entire day or longer. The background noise will grow stronger when the center of the galaxy passes through the area of the sky where the antennas are most sensitive.

The plasma in the Sun is in a constant state of turmoil and therefore constantly giving off low levels of electromagnetic radiation including radio waves. This is a natural result of the heat and energy that are generated by the Sun. The Radio Jove equipment is not sufficiently sensitive to pick up this constant level of emission from the Sun called thermal emission. The background level of emission from the galaxy is stronger than the Sun's normal emission and "drowns it out". Every so often, however, the magnetic fields generated by the Sun's turmoil are concentrated into the areas we usually associate with sunspots (see Figure 5) and the magnetic fields may break and reconnect suddenly releasing a much larger amount of radiation called a solar storm. These solar storms can be strong enough to be easily detected by the Radio Jove equipment. They are also strong enough that similar signals will be obtained by many Radio Jove observers scattered over the sunlit side of the globe. It is hard to predict when exactly the Sun will have radio storms so it is best to set the equipment in operation and patiently wait for a storm to appear. This may take hours and sometimes several days of operations, but when successful the results can be quite fulfilling.

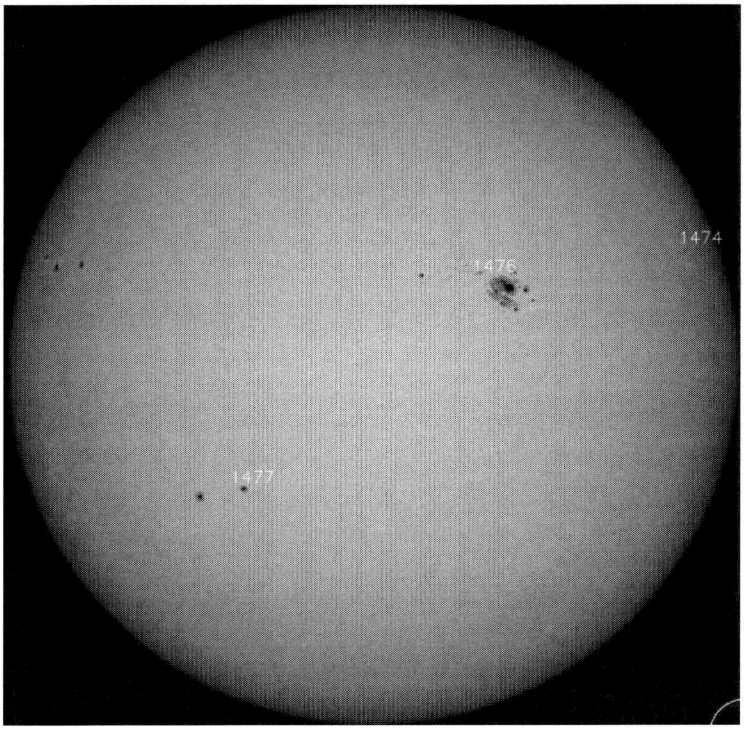

Figure 5. Solar Dynamics Observatory image of the Sun with sunspots labeled by sunspot group number.

The planet Jupiter can also be a very powerful radio source since it has a large number of charged particles moving in the big magnetic field generated by the planet. Jupiter radio emissions are not usually as strong as the Sun, but they can be quite intense. For reasons which we will not discuss here (see *Listening to Jupiter* [4] for background information) Jupiter can be a very predictable radio source and it is not hard to find periods of several hours when Jupiter storms are very likely to occur. Again, software, called Radio Jupiter Pro (Figure 6), is included on the CD that comes with the Radio Jove kit and this software enables the prediction of when would be a good time to observe Jupiter for any location on Earth. The software also determines the sensitivity pattern of the radio telescope as a map on the sky that also helps to determine the best arrangement of the antenna for particular targets.

The natural question is to ask if radio emission from planets other than Jupiter can be detected. Although we now know from spacecraft flybys of the other planets that every planet that has a significant magnetic field also emits radio waves, all of the other planets except Earth do not have strong enough emissions to be detected by the Radio Jove equipment. Since we are so close to some of the radio emission sources on Earth it is quite easy to detect them with the Radio Jove telescope. Earth-based lightning storms are easily detected from hundreds of miles away and sometimes farther. These would be too far away and too weak, however, if we were on Jupiter with our equipment and trying to hear radio emissions from the Earth.

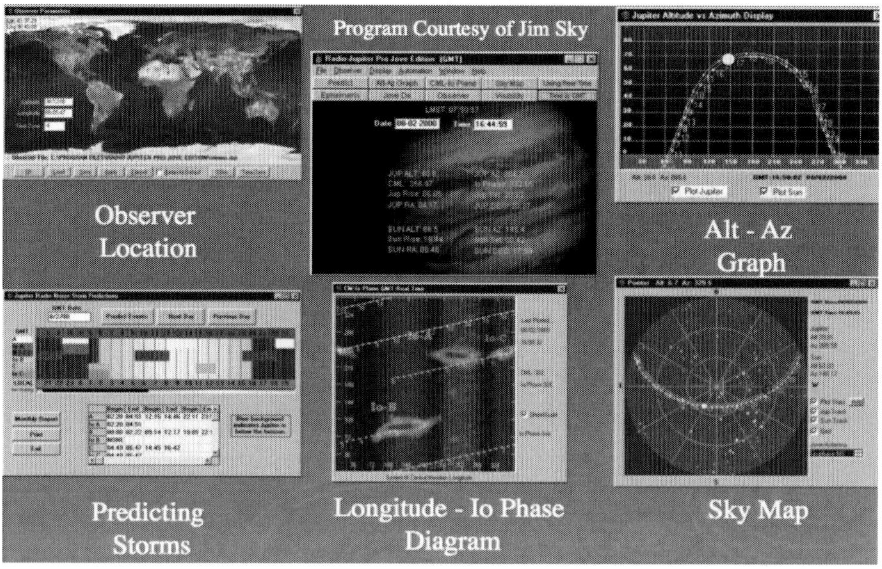

Figure 6. Radio Jupiter Pro software predicts times to listen for Jupiter radio storms and how best to set up the antennas for particular targets.

One other source of radio emissions from the Earth is that made by man. Radio stations are sometimes detected by the Jove antennas. The radio receiver was designed to receive natural radio emissions so it is not a particularly good station receiver. Nonetheless, stations sometimes are received, and one has to tune the radio away from them if you want to hear the natural radio emissions described before. So, the stations are considered interference for the purpose of the observations of natural radio sources.

How Will I Benefit?

Radio Jove was begun for the purpose of giving students and other interested individuals and groups an introduction to the science of radio astronomy. It is more than just information about radio astronomy. It is a "hands-on" experience of the scientific method. Those who wish to learn about astronomy and, in particular, radio astronomy, do so by repeating the types of observing that first resulted in the discoveries of extraterrestrial radio emission. So, for example, the discovery that Jupiter is a radio source happened in 1955 by radio astronomers Kenneth Franklin and Bernard Burke [5} using a Mills Cross radio antenna which is simply a large number of connected dipole antennas laid out in a cross shape (see Figure 7). Their equipment was more elaborate and took longer to create, but they used the same process that Radio Jove observers do today. They built their own instrument and learned a lot about the behavior of it in the process. They overcame the problems that inevitably appear. They carefully made observations and noted strange things that were not expected. They guessed what the unknown phenomena represented and then tried to determine how to prove or disprove their guesses. They carried out further experiments. When they felt they had the correct explanation and had learned something that was of benefit for others to hear they published their results.

The same process can be done by Radio Jove observers. In building the radio telescope there is much to be learned about electronics, about careful attention to detail, and about overcoming difficulties when things go wrong. With reasonable persistence the participant has a working instrument that displays signals and gives off sounds both anticipated and unexpected. Trying to understand what the signals and the sounds are telling you about their source is all a part of the huge number of puzzles that are presented to the scientist. Answers always seem to lead to more questions.

Figure 7. Antenna used by Bernard Burke and Kenneth Franklin to discover that planets (Jupiter) could be radio emission sources.

It is hoped that the initial Radio Jove experience will lead to a wish to go deeper and try still more sophisticated approaches to understanding the world. To that end Radio Jove will have accomplished its purpose.

CONCLUSION

Radio Jove is just one example of an interactive pedagogical approach to science education. The project is intended for high school level students, but has been done by college students, middle school students and even elementary grade children. The main problem is that it does take time to do the project properly and the higher grades have less and less available time to spend on such projects. Higher level students often become involved in Radio Jove as an after-school or extracurricular project. Many students have done quite well using Radio Jove equipment as part of science fair projects. There are about ten people who help to manage Radio Jove on a day-to-day basis and who are quite willing to help answer questions from any of the participants. Usually this is done by email through the Radio Jove email distribution listserver. Periodically the Radio Jove management staff advertise "Coordinate Observing Teleconferences" through the email distribution. The teleconferences are usually scheduled at a time when the Sun or Jupiter have a higher than normal probability of activity. During that time a toll free teleconference number is advertised and all are welcome to call and compare/discuss what they are seeing on their equipment, ask questions, or just relate their thoughts about the project. Non-U.S. observers are encouraged to join the teleconferences through calling methods such as Skype or its equivalent. Radio Jove tries to make sure that help is always available and there is no reason for not being able to participate. As mentioned before, the main source of further information may be found at http://radiojove.gsfc.nasa.gov.

REFERENCES

[1] See http://missionscience.nasa.gov/ems/index.html

[2] See http://theinspireproject.org/

[3] See http://radiojove.gsfc.nasa.gov/

[4] R. S. Flagg, *Listening to Jupiter,* 2nd Edition.

[5] Burke, B. F. and K. L. Franklin, Observations of a variable radio source associated with the planet Jupiter, *Journal of Geophysical Research,* vol. 60, pp 213-217, 1955.

In: The Space Education Phenomenon at NASA ...
Editor: Norma Teresinha Oliveira Reis
ISBN: 978-1-62618-188-5

Chapter 2

MOBILE DEVICES FOR STEM LEARNING TO GO: GUIDELINES FOR SELECTING MOBILE DEVICES AND APPLICATIONS TO ENHANCE STEM LEARNING

Laurie F. Ruberg*, Kendyl Morris Kimberly and Charles A. Wood

Center for Educational Technologies, Wheeling Jesuit University, West Virginia, US

ABSTRACT

This chapter describes how mobile devices such as smartphones, tablet computers, and other portable electronic devices can support and enhance science, technology, engineering, and mathematics (STEM) education. The authors organize their research and exemplary cases of mobile device applications according to five key questions: 1) What unique qualities do mobile digital devices offer for supporting open, distributed learning experiences that inspire or advance student learning in STEM areas? 2) How do mobile devices offer context-appropriate teacher professional development that can be adapted for STEM content, pedagogy, and technology training? 3) How are National Research and Education Networks (NRENs) made available through mobile technologies, and what are the costs and benefits of this access? 4) What opportunities for data collection and dissemination are available through mobile technologies? 5) What opportunities and challenges are there for using mobile devices and smartphones to access learning materials and collaboration platforms? The authors discuss each question in terms of related educational research, cases to illustrate exemplary practice, and guidelines for future applications in these areas. The appendix provides a rubric that STEM educators and mobile device app developers can use and adapt. The conclusion links to an ongoing online discussion related to the ideas and examples in this chapter.

* E-mail: lruberg@cet.edu

INTRODUCTION

Give most students a list of facts, worksheets, or set of computational problems without also providing them some context, and those students are unlikely to be very inspired. Students are more likely to be motivated to learn when a teacher creates a context that makes the subject matter inviting or when they're motivated simply by their own interests. In this chapter we look at how science, technology, engineering, and mathematics (STEM) educators can use mobile devices such as smartphones (e.g., iPhones), tablet computers (e.g., iPads), portable media players (e.g., iPods), and similar technologies to inspire students to learn.

Mobile devices are rapidly entering educational settings. They can be great tools because they can follow students wherever they go. In this chapter we discuss the emerging possibilities for using mobile devices and apps to improve STEM education from the kindergarten through high school (K-12) level in both formal and informal learning contexts. We provide a rubric to help STEM educators select devices and applications (apps) that use the affordances of mobile technologies, provide access to tools and resources that can improve student learning, and include design features that help teachers address student needs. We also provide examples of mobile apps that address one or more of the criteria in our rubric.

What makes mobile devices and applications so appealing? According to a Pew Research study (2010), three key features draw users to mobile devices. First and most importantly, mobile devices and apps provide 24/7 access to the web and all of its resources. Science and technology are among the top six categories of interest that mobile device users listed as the most frequent subjects that young adults report searching for online. And when asked what subjects online news seekers would like more information about, 44 percent chose science and technology.

Further Pew Research (Purcell et al., 2010) analysis of user behavior finds three features that attract news seekers to using their cellphones for information: *portability*: 33 percent of respondents accessed news on their smartphones; *personalization*: 28 percent of Internet users customized their home page to include news from sources and on topics that particularly interest them; and *participation*: 37 percent of Internet users contributed to the creation of news, commented about it, or disseminated it via postings on social media sites such as Facebook or Twitter. The research we conducted for this chapter suggests that STEM educators look for mobile devices and apps that feature these same affordances in their design and features.

A Pew Research survey of 2,277 U.S. adults (Aug 16, 2011) found the following smartphone usage statistics: downloading apps (69%), using their smartphone to play games (64%), listening to music (64%), accessing social networking sites (59%), recording videos (59%), and posting photos or videos online (45%). This same survey found that only 15 percent of respondents reported using Twitter, and 13 percent used their smartphone to make video calls. Of course, reporting smartphone usage stats is a snapshot of a point in time—like taking a photo from the window of a speeding train. Figure 1 (below) displays the amount of time 12- to 24-year-olds spend in various types of online venues as compared with the total number of Internet users. Most significantly, we see that teens and young adults spend much more time on social media and multimedia Internet sites, which are both highly active areas of teen and young adult mobile device use.

The smartphone usage we observe and count today is part of a pattern that is rapidly changing across all population groups, but especially among young adults. This trend is evident around the world among all population groups as reported by an Education Sector Report published by the World Bank and African Union (Adam, Butcher, Tusubira, and Sibthorpe, 2006). A UN study (Lynn, 2008) found that use of mobile phones has spread faster than any other technology. Lynn also found that mobile phones can improve the lives of the poorest people in developing countries by providing access to information where landlines or the Internet are often unavailable.

In this chapter we look at some ways teachers can use mobile devices to further engage and inspire their students in STEM topics and processes. We are particularly interested in free applications from NASA and other federal agencies and nonprofit groups that educators, parents, and students can download to extend their STEM learning and interests.

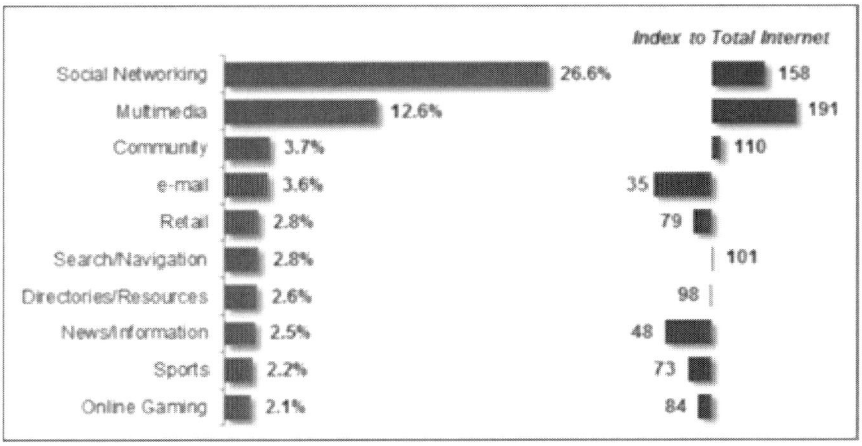

Figure 1. Top Categories by Share of Time Spent Online for 12- to 24-year-olds. Sites visited most by those between 12-24 years of age included Google sites, which ranked first as the most visited website, reaching 90 percent of the demographic and accounting for 16.3 percent of all time spent online. Facebook, ranked second and accounted for the greatest share of time spent online at 23.2 percent. 12- to 17-year-olds spent an even greater share of their time on both Facebook (25.1 percent) and YouTube (13.9 percent). Source: comScore Media Metrix, U.S., October 2011. Downloaded 4 June 2012 from http://blog.comscore.com/2012/02/.

The mobile technologies and applications we feature entice students to electronic venues that support STEM learning. Our first section outlines several key strategies for integrating mobile devices and applications into classroom or afterschool learning environments. We present examples of educators using mobile devices to enhance their STEM learning. We also offer guidelines to help educators choose mobile devices that meet their needs. The guidelines can be applied to the continually changing set of mobile devices available.

In this chapter mobile devices refer to small, handheld computing devices, typically having a display screen with touch input and/or a miniature keyboard and weighing less than 2 pounds (0.91 kg). A mobile phone (also known as a cellular phone, cellphone, and a hand phone) is the most common device in this category. Its purpose is typically distinguished by its capability to make and receive telephone calls over a wireless network, usually a radio frequency interface that keeps a connection even while the user moves around a wide geographic area. The phone works by connecting to a cellular network provided by a mobile phone operator, allowing access to the public telephone network. Early pocket-sized telephone devices are now frequently supplemented by tablet computers such as iPads, personal digital assistants (PDAs), and portable readers, all of which typically include input and output interfaces combined in a touch screen design. Although tablet devices are not designed to place phone calls, this chapter focuses on those mobile devices that have Internet/WIFI connectivity. Smartphones, PDAs, and tablet computers offer a subset of the affordances of a personal computer in a mobile package that's easy to hold, start, use, carry, and store.

Our goal is to evaluate how mobile devices can engage students and educators in STEM-related learning activities. Here are our research questions:

1. What unique qualities do mobile digital devices offer to support open, distributed learning experiences that inspire or advance student learning in STEM areas?
2. How do mobile devices offer context-appropriate teacher professional development that can be adapted for STEM content, pedagogy, and technology training?
3. How are National Research and Education Networks (NRENs) made available through mobile technologies, and what are the costs and benefits of this access?
4. What opportunities for data collection and dissemination are available through mobile technologies?
5. What opportunities and challenges are there for using mobile devices to access learning materials and collaboration platforms?

METHOD

To conduct this research, we reviewed literature, trade journals, and industry white papers about mobile devices; we examined surveys and interviews involving K-12 teachers; we identified relevant government and professional society research reports; and we reviewed case study research of three NASA-funded applications for mobile devices. Our review of research and trade journal articles on educator use of mobile devices over the past five years helped us compile a taxonomy of mobile device features, functions, and affordances that is useful for comparing and contrasting mobile devices and apps. We identified 10 key features of mobile technologies that educators commonly referenced as benefits and concerns to look for when selecting and integrating use of mobile devices and apps into STEM curriculum.

Table 1 summarizes the benefits and concerns we identified. In the Data Analysis section we analyze selected cases that represent these 10 key features.

What measures effectively gauge the impact of mobile device implementation in STEM education settings? We reviewed apps and devices according to the five conditions required to invite students and adults to engage in learning: affirmation, contribution, purpose, power, and challenge. The examples we selected illustrate one or more of the key affordances of mobile devices and had documented evidence of meeting one of those five conditions. Many of the examples cited are from early classroom pilot studies with mobile devices, and the classifications presented are more projections of potential than final recommendations for broad application. We do suggest areas for further research and expanded pilot testing.

Table 1. Summary of Mobile Device Features, Benefits, and Concerns

AFFORDANCE	BENEFITS	CONCERNS
VISION	leads to a coherent digital learning environment	requires advance planning and agreement by all stakeholders on the idealized digital learning environment
CUSTOMIZATION	allows adaptive teaching and student choices for instructional media and learning styles	increases complexity of support and variability in curriculum quality and depth
PORTABILITY	offers conveniences such as increased battery life, screen size, flexibility, and decreased weight and startup time	carries risks of a short history with individual products and trial and error uses in the classroom setting
TIME	allows fast, flexible deployment of technology as well as new or modified curriculum; provides better service desk response time	requires ongoing maintenance of device compatible/incompatible issues that is time consuming and involves frequent updates with feature changes
BUDGET	shifts funding toward classroom technology tools and instructional innovations rather than generic infrastructure tools	requires frequent upgrades which may cause confusion as operating system features for mobile devices change more rapidly than for personal computers and network servers
INDEXING	allows learning objects within digital instructional component to be tagged with curriculum content, resources, content standards, and/or learning styles; allows teachers and students to choose interfaces with greatest appeal and efficiency	needs a standard system for index criteria and tools that is consistent and easy for teachers and students to access and use
DATA COLLECTION	makes individual access patterns accessible so that student/class participation can be quantified	involves administrative balance of access, protection, social collaboration, and storage requirements
ASSESSMENT	allows both ungraded, self-assessment and graded assignments and quizzes to be easily embedded into the mobile curriculum	increases concerns regarding identity issues for distributed learning, which are not easily resolved
STRATEGIES	gives educators using mobile technologies opportunities to update and even radically reform school vs. out-of-school learning	requires documentation of implementation strategies to be able to understand how mobile devices support existing and new instructional approaches and strategies
ACHIEVEMENT	appeals to student interest in using mobile devices as part of their learning toolset and stimulates student interest	requires controlled research settings to identify contexts in which integration of mobile technologies are effective in improving student achievement

Successfully integrating mobile devices into an existing STEM education context requires investments in:

- Affordable and reliable connectivity.
- Relevant applications for retrieving and managing educational information and electronic content (teaching and learning materials for both school and health contexts).
- Professional development of all involved, including government officials, school principals, administrators, teachers, and learners.
- Support and maintenance strategies.
- Legal regulatory policies and procedures.
- Effective governance and operational systems and processes.

Appendix 1 applies the list of mobile device features, benefits, and concerns in Table 1 to a rubric that educators can use to guide their reviews of devices and/or apps for possible integration into classroom, afterschool, or out-of-school enrichment activities.

Research Question 1: What unique qualities do mobile digital devices offer for supporting open, distributed learning experiences that inspire or advance student learning in STEM areas?

Pearce (2011) proposes seven key beneficial reasons for using the iPad in educational settings:

- Learning styles—audio, kinesthetic, and visual aspects.
- Creativity—various apps to meet specific needs for specific individuals.
- Technical—minimal maintenance and installation of software.
- Accessibility—light load and accessible anywhere with Internet access.
- Battery life—seven hours.
- Users—any age.
- Motivation—gets students more active and excited in learning.

The advantages Pearce proposes, though, depend on the context of the learning activities. The context must be an invitation to learning instead of simply a way to present an electronic version of traditional worksheets, drill and practice, or listing of facts to be memorized.

Technical Features of Mobile Devices and Mobile Apps

As many mobile device users can testify, relying on the devices for all the users' needs is risky when bandwidth requirements are not always available. The following technology issues also may sometimes limit mobile device use, but with the rapid growth of this industry, problems cited today may already be resolved in some areas:

- Security standards: Working mobile means depending upon public networks, requiring careful use of WIFI, but not all free wireless networks are equal. The public networks offered at many commercial locations such as McDonalds, Barnes and Noble, and similar venues are not secure. This type system functions much like the old analog party line phone system. If you picked up your phone in the party line while someone else was using the same line, you could overhear the conversation. The same is true for public wireless networks. When your iPad, smartphone, or other WIFI-enabled mobile device connects to a public network, it shares the airwaves with other devices in the vicinity. Unless your computer transmissions across the WIFI are encrypted, other network devices have the ability to piggyback onto your data line.
- Power consumption: When a power outlet is not available, mobile computers must rely on battery power. Combined with the compact size of many mobile devices, this often means unusually expensive batteries must be used to obtain adequate battery life.

- Transmission interferences: Weather, terrain, and the range from the nearest signal point can all interfere with signal reception. Reception in tunnels, some buildings, and rural areas is often poor.
- Human interface with device: Screens and keyboards tend to be small, which may make them hard to use. Alternate input methods, such as speech or handwriting recognition, require training.
- Potential health hazards: Using mobile devices while driving distracts from driving, increasing an assumed likelihood of being involved in traffic accidents; many states now ban cellphone use while driving. Cellphones also may interfere with sensitive medical devices and cause planes to crash.

Many are familiar with apps for iPhones and Android phone devices. These "native" apps are written and programmed in languages specifically for mobile devices and will not run on personal computers. Programmers must learn the mobile device operating system code to take advantage of unique smartphone features. Mobile devices equipped with Internet-compatible data capabilities can access web-based programs, just as personal computers do.

Mobile App Systems: What's the difference between a native application and a mobile web browser-based application? Those who favor use of mobile web-based app systems argue that these systems are easier to maintain and run because they are all built to link to the web, a uniform platform. This means that essentially all mobile web systems are built, launched, and maintained on a single platform. On the downside, mobile app systems offer users less control and a slower experience, no single point of distribution, and few quality or security standards across mobile browsers.

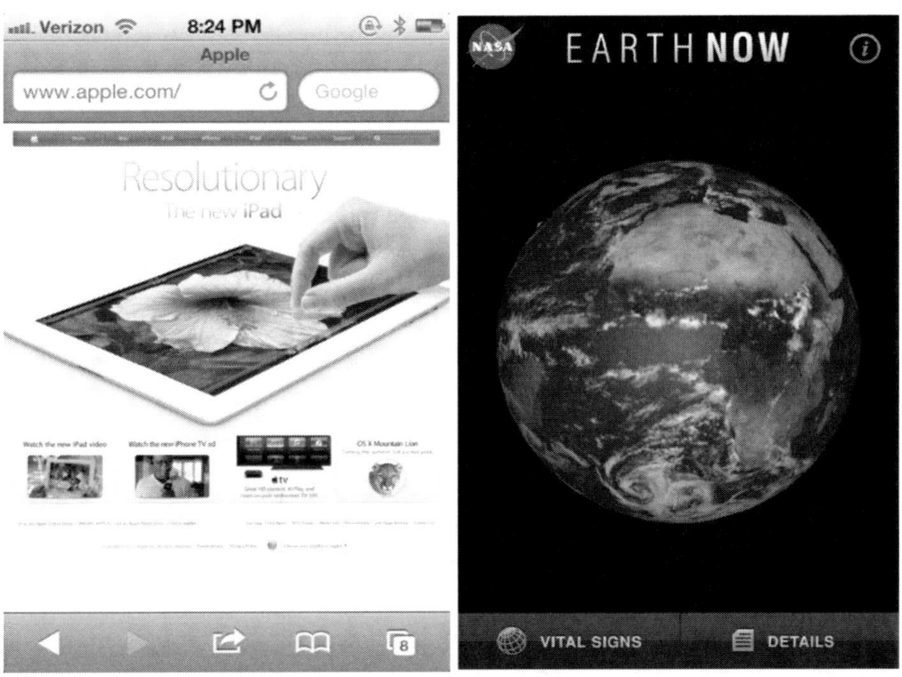

Figure 2. A screen shot of a mobile browser (left) and a native app (right).

On the other hand, native apps offer more media functionality (such as the touch control to zoom in or out of the display) and a more controlled, content-customized, faster running application. The native apps are also easier to find, as they are made available through app portals. Native apps are limited when it comes to running on multiple platforms because they have to be reconfigured. Each native app includes its own build, test, maintenance, and distribution costs as well. In addition, the added functionality that native apps offer is not always needed or integrated into the app design.

Research Question 2: How do mobile devices offer context-appropriate teacher professional development that can be adapted for STEM content, pedagogy, and technology training?

The same qualities that inspire students to learn also motivate and engage teachers to pursue professional development. The inspiration to learn varies across individuals as their interests and needs change. A learning environment that addresses individual needs through innovative and customizable design features draws learners into the content of learning. While there are many ways to categorize and explain inspiration and motivation, the following set of constructs posed by Tomlinson (2002) is useful because this approach builds on the beneficial features of mobile devices. The five strategies adapted from Tomlinson's model support teacher professional development activities that foster student-centered instruction, embedded reflection, and self-regulated learning practices.

Affirmation: Mobile device tools and apps can help teachers more readily manage and enjoy the demands of the classroom setting. Using cell phone cameras and free apps like *Dragon Dictation* can make the first few days of setting up a classroom a lot more fun. The personal aspects of using photos and audio along with text files to set a class roster reinforce students' need to know that they are accepted, safe (physically, emotionally, and intellectually), cared for, listened to, and that there are people who know how they are doing and that it matters that they do well. With guidance and scaffolding from experienced mentors, teachers can make use of the mobile device built-in tools like cameras as integral tools to improve and enhance their pedagogical strategies and classroom management techniques. Many smartphones include digital and video camera capabilities that are more than sufficient for classroom media activities. Building on these and other ideas gives teachers a chance to be creative and innovative while addressing routine classroom management and initial rule- and routine-setting activities.

Contribution: While every school district has the responsibility and opportunity to establish its own criterion for curriculum mapping, each individual teacher should have an avenue to demonstrate their contribution to the district's overall excellence. Mobile devices and apps can provide an easy way for teachers to show each other how to communicate to parents great things that happen in their classroom. Tools such as Facebook, Twitter, Linked In, TeacherTube, school websites, e-mail, and even text messaging let teachers connect easily with parents. Learning how to make effective use of these communication tools takes practice and persistence. The focus of these frequent messages is most effective when teachers highlight achievements of children from varied perspectives and abilities. Teachers need to find ways to recognize every child and to show that individual success is connected to the entire group and contributes to the success of the whole class through mutual work on common goals. Teachers who learn to use the communications features of mobile devices to support frequent exchanges with students' parents gain significant knowledge worthy of sharing with other teachers.

Purpose: All teachers bring individual gifts and talents to the classroom and broader school setting. Mobile devices and apps can be useful tools to help teachers find ways to integrate their talents and special interests in ways that enhance instruction. For example, a biology teacher who is able to share his or her interest in ornithology by showing students apps used to track and document bird sightings by observation and birdcalls would make zoology a much more authentic study. Other apps that would engage students when demonstrated as extracurricular activities by their teachers include music, drawing, translation, and plant identification apps (like *Snap!*). Students should see that their teachers are personally invested in the subjects that they teach and that the work their teachers do has purpose and absorbs them.

Teachers may also use mobile devices to give students new motivations and methods for STEM learning. Brown, Prendergast, and Woodfin (2011) describe how the multisensitivity of the iPad can open up new venues for meeting the needs of students who are limited by one or more learning disabilities. The easy device accommodations for enlarging text, changing fonts and background, and individually customizing the iPad and other mobile devices give teachers quick and manageable ways to adapt instruction for students with special needs. To help teachers manage the process of adapting their instruction for students with special needs, the iPad app, *Catalyst*[1], allows teachers to collect, manage, and apply data to suit the needs of individual students. The app is also designed to move the differentiated instruction process from paper to mobile device format—saving teachers and school systems time and money.

Power: The great thing about mobile devices is that they indeed are mobile and can easily be integrated into every aspect of one's daily routine. This portability and convenience are not always welcomed, but the potential for continuous access and contact and uninterrupted links with areas of interest offers a sense of power that can be brought into the classroom. For example, a teacher who is willing to share his or her cellphone number with students has the power to tell students that they can call if they need help. Suddenly, there is no excuse for students to say I didn't know what to do, I lost the assignment, or I couldn't solve that math problem on my own. Mobile technologies can offer a connection between teachers and their students that contributes to quality work and responsibility for every learner's journey in and beyond the classroom. For example, the MIT Media Lab *App Inventor*[2] is a free browser-based program that teachers can use to engage their students in simple Android app development. Imagine the power that teachers can share with their students when they offer students opportunities to show what they've learned by having them develop games, share data through apps, or complete quizzes to assess peer-to-peer learning.

Many more suggestions for using iPads and other mobile devices to improve classroom and schoolwide management of school activities can be found in the *iPads in Schools* blog[3]. As this blog points out, various educators—even school bus drivers—are using the iPad. One bus driver, for example, uses it to take attendance. Through an app containing the bus route management system, the location and time of when a child entered the bus can be logged automatically.

---

[1] More information about this app can be viewed at Catalyst: Data Collection Solution for Autism—Compass Solutions for Autism: http://www.compassautism.com/products/catalyst.

[2] *App Inventor* is in beta form at the time of this writing. The tools for app creation are published by the MIT Center for Mobile Learning and are available in open source format on the web at http://www.appinventor.mit.edu/.

[3] The *iPad in Schools* blog is an active resource highlighting unique and innovative educational uses of Apple's iPad and can be found at http://www.ipadinschools.com/.

Challenge: In an April 2012 survey conducted with K-12 teachers in Western Pennsylvania attending a daylong, district professional development in-service workshop on iPads, 92 percent of the teachers reported that they were not currently doing any activities in the classroom where students used technology to share data (Ruberg, 2012). In addition, 75 percent of the teachers reported that they do not use any type of computer games in their classroom curriculum. However, teachers also reported that they want technology resources to complement and enhance their curriculum, and they see potential in mobile devices to "greatly engage my students and enhance their technology skills."

A recent study of iPad use among kindergarteners conducted in Auburn, ME, randomly assigned half of the district's 16 kindergarten classes to use iPads for nine weeks.

This study showed that classes using the iPads outperformed the non-iPad students in every literacy measure they were tested on. Further, Sue Dorris, principal at East Auburn Community School, in reflecting on this study conducted at her school, reported, "We are seeing high levels of student motivation, engagement, and learning in the iPad.

"The apps, which teach and reinforce fundamental literacy concepts and skills, are engaging, interactive, and provide children with immediate feedback. What's more, teachers can customize apps to match the instructional needs of each child, so students are able to learn successfully at their own level and pace."[4]

Emerging from the movement to greater integration of mobile technologies and customization of instruction to provide differentiated instruction are strategies for rethinking how schools can change from the present outdated Industrial Age, assembly line structure to a mass customized learning structure with the capacity to meet the individual needs of every learner (Schwahn and McGarvey, 2011).

Teachers should have learning experiences that stretch them so that they are on their own learning journey and can share some aspects of their learning experiences with their students. Like their students, teachers should have opportunities for success that reinforce accountability for personal growth that contributes to the growth of others. This is an area that changes constantly. Teachers who integrate mobile devices and apps into their instruction will have to be willing to undergo unexpected app updates, device upgrades, and system upgrades, all of which can lead to periodic frustration and incompatibilities. Finding apps and activities that deeply engage students will make teacher professional development and investment of their time worthwhile.

Research Question 3: How are National Research and Education Networks made available through mobile technologies, and what are the costs and benefits of this access?

Using mobile devices for data collection is practical and engaging for students and teachers. This process can be managed in several ways. We selected the following examples to represent different data collection, analysis, and interaction functions available on a variety of mobile devices.

The *DataAnalysis* app allows users to quickly generate a plot of all types of data that can be formulated as x, y pairs. Data can be entered in a variety of formats, including text file (.txt), comma delimited values (.csv), gzipped cs file (.csv.gz), or by simply dragging and dropping within iTunes. The data entered by hand or by imported file are plotted and can be fit to a variety of math equations by linear regression or more complex equations.

---

[4] Read more about the kindergarten iPad program at Auburn High School on the web at http://www.auburnschl.edu/education/components/whatsnew/default.php?sectiondetailid=4&itemID=2884.

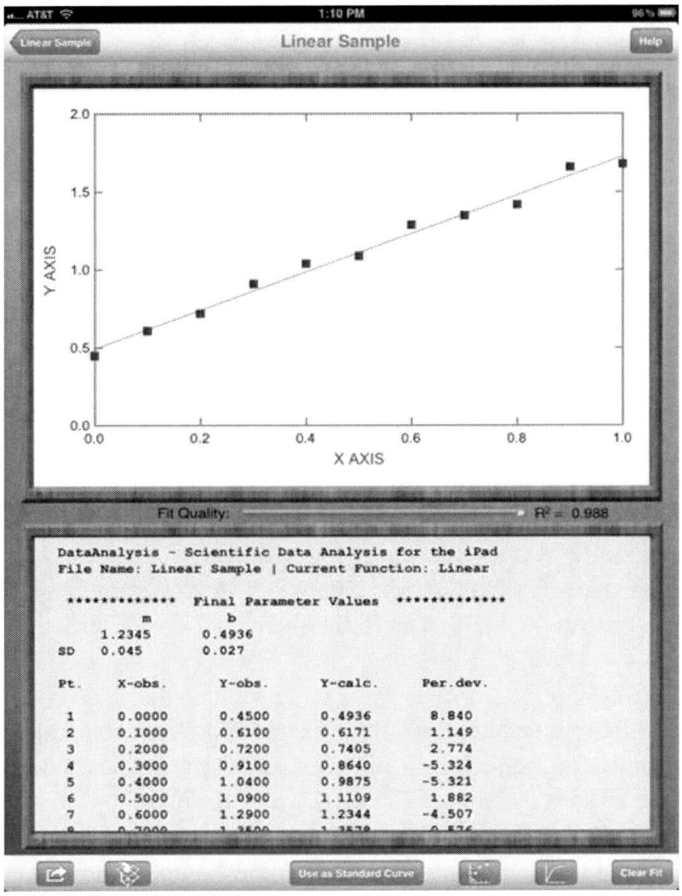

Figure 3. A screen shot of a linear graph dynamically created by the *DataAnalysis* app based on the x, y data presented.

As the *DataAnalysis* authors explain, more equations are being added all the time. This tool features an easy-to-use interface for scientific data analysis on the iPad.

The app also includes a series of tutorial movies, which help introduce program functions with simple exercises. This incredibly powerful and easy to use program is free. Figure 3 shows a screen capture of a sample mathematical function graph fit to data.

The app includes text, .png, or .pdf file save and export functions. Graphs created can also be copied and pasted directly into other iPad apps, such as *iPages*. The My Documents folder includes four sample documents that can be used to show four cases of data functions.

The *BLiSS Sim* is an example of a data simulation app. It is available as a free download for the iPad from the iTunes Store. *BLiSS Sim* is built on data generated from NASA research into how plants could be grown in an engineered, hydroponic environment to provide oxygen, potable water, and food to support humans on lunar or Mars colonies or on long-duration space missions.

The bioregenerative life support system (BLiSS) data used to create the *BLiSS Sim* includes four crops—wheat, potatoes, soybeans, and lettuce. In the iPad sim players face a series of challenges that build on crop-specific questions that players can answer by interacting with the simulator. The screen shots and captions provided in Table 2 overview the five major pathways from the menu trees, introduce the scientific concepts and scenario context behind the sim, and show samples of the game play screens.

When reviewing the *BLiSS Sim,* consider: How closely does the sim relate to the original data and research? What type of cognitive scaffolding is provided to link the sim interactions to the science in the original research? Are there links to websites or publications where students and teachers can read more about the source research projects?

The related links, wiki, and blogs associated with the BLiSS Sim app provide educators with hands-on lab activities, related plant science teaching resources, and opportunities for posting student work and collaborating with other teachers.

The science context of the BLiSS Sim is designed to engage students in interacting with plant science data in iterative designs that model scientific and engineering practices. The sim thus addresses Dimension 1 of *A Framework for K-12 Science Education*[5], which blends science and engineering practices, cross-cutting concepts, and disciplinary core ideas. Readers can participate in educator discussions about the NASA science behind the BLiSS simulations at http://bliss-wiki.wikispaces.com/BLiSS+Sim+Home.

Data Display Apps offer a convenient way to have instant access to data in areas of great interest on a personal or professional level. Here are a few examples of data display apps developed by NASA that are available for free download. Most of these apps will work on Apple or Android mobile device platforms.

NASA'S EARTH NOW: Vital Signs of the Planet science app is available on the iPad or iPhone or other smart mobile technologies. It is distributed by the Jet Propulsion Laboratory (JPL). This NASA-sponsored free app is available through the iTunes Store. JPL and Caltech have operated robotic spacecraft funded by NASA that have visited all of the planets from Mercury to Neptune, including Earth. JPL is a world leader in science and technology.

The tools JPL developed for Earth and space exploration have provided new discoveries and insights that have expanded scientists' understandings of Earth, its atmosphere, climate, oceans, geology, and the biosphere.

The ongoing invention of ever more sensitive space sensors has also resulted in a myriad of technology applications widely used for medical, industrial, and commercial uses on Earth. What's exciting about this app is that a summary portion of the data that NASA satellites and sensors are collecting is viewable on the *EARTH NOW* app.

The app offers differentiation of views of each of the seven vital signs: air temperature, carbon dioxide, carbon monoxide, gravity field (used for finding water underground), ozone, sea level, and water vapor. There is a short overview description for each vital sign, a data key, a description of what the data map displays, and a selection of other data mappings. When we have shown this app to students to rate its user friendliness, users were attracted to the simplicity of each page view and easy navigation, which enabled a quick startup and use. Most students will not seek out this app on their own, but they will use it without hesitation when a teacher recommends it. Student motivation to use it will flourish if teachers suggest that the data presented can help students get a quick overview of key indicators to track in climate science research. The data are available electronically as dynamic, interactive 3-D map displays on the iPad, iPhone, or iTouch device.

[5] http://www.nextgenscience.org/sites/ngss/files/How%20to%20Read%20Next%20NGSS%20%20PUBLIC%20January%20Draft%20-%20FINAL.pdf

Table 2. *BLiSS Sim* Screen Captures and Descriptions

The menu guides navigation through the *BLiSS Sim*. It includes the Introduction at right, and the Instructions and Play game interfaces. The Learn More button takes you to related links. The BLiSS Wiki button links to an open collaborative for activity sharing among game players, educators, scientists, and *BLiSS Sim* developers.

The Introduction illustrates how the simulation closely ties to NASA plant science research and human exploration needs for biological life support systems.

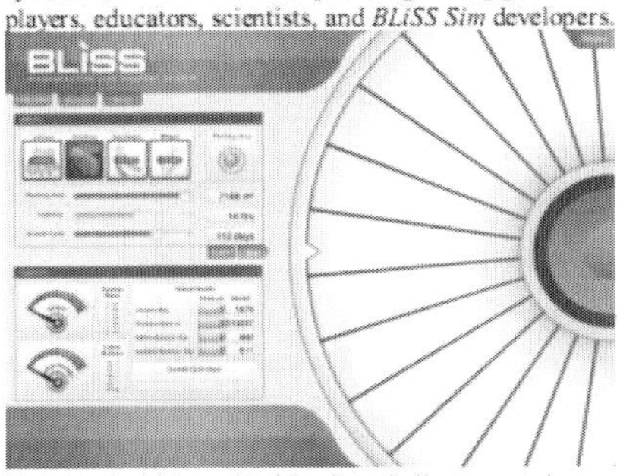

Each level within each of the four challenges requires players to select a crop, planting area, lighting level, and crop schedule.

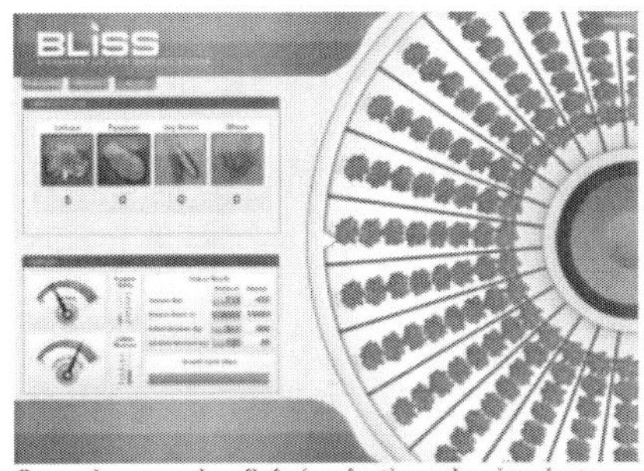

Once players make all their selections, the sim plants, grows, and harvests their crops.

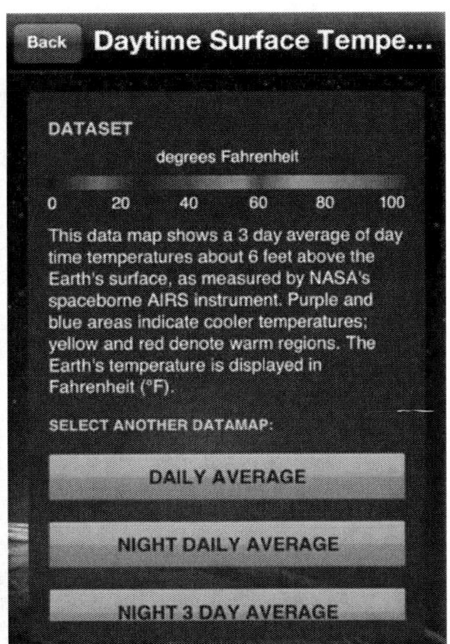

Figure 4. Two screen shots from the display of surface temperatures (one of the seven vital signs data sets) presented in the Earth Now app. The image on the left shows the 3-D globe interactive display that the user can rotate to view data from all over the globe. The image on the right shows the data key and other data maps available.

The graphical data reports can be recorded as screen captures or accessed on the related source websites, which are referenced for each vital sign. The presentation of the data (the left image in Figure 4) is offered in a 3-D globe that can be rotated, which makes the summary data easy to view and fun to access.

Below are some other Earth and space science data view apps:

Mars Globe HD: This virtual globe combines a high-resolution satellite map with laser altimeter data and advanced topographic lighting to present Mars as it appears from above. Users can choose to view topographical features through a guided tour function, or they can choose to explore the Martian surface independently. The visualization tool includes more than 1,500 surface features with annotated information and links. This app provides a large database of information about Mars. Through its interactive exploration, those who download this app can enjoy exploring the many and varied landscapes of Mars through high-density images. *Mars Globe HD* is available for iPhone and iPad at a cost of 99 cents.

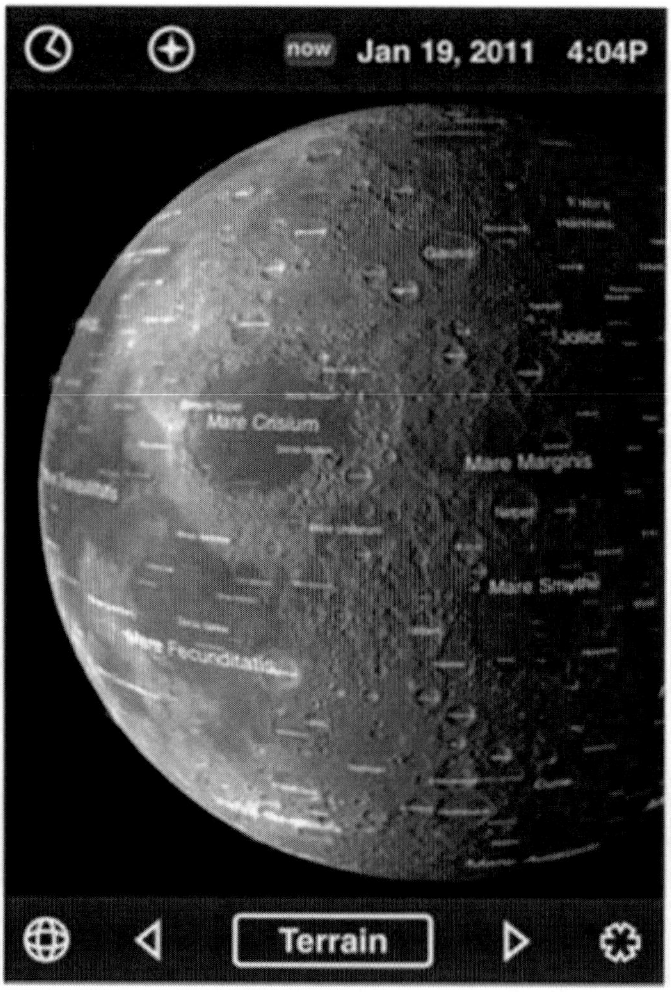

Figure 5. A screen shot from *Moon Globe*.

The author of *Mars Globe HD,* Midnight Martian, has also created *Moon Globe* and *Moon Globe HD*. Figure 5 is a screen shot of the lunar surface using the *Moon Globe* app's Terrain view. These apps provide the tools to transform an iPhone, iPod Touch, or iPad into a kluge of scientific instruments that depict the surface of Earth's Moon. As the app description explains, "Satellite imagery and topographic laser altimeter data are combined to render the Moon with realistic lighting in real-time 3D."[6] The multitouch screen allows users to interact easily with the view as if to spin above the surface, or by switching to telescope mode, the view changes to a perspective of the Moon as it appears from the viewer's location on Earth. Adjustments to the display time allow dynamic changes in the sunlight reflections across the lunar landscape. A tag cloud floats just above the surface, labeling terrain features and spacecraft; touch any label to find out more about that feature. *Moon Globe* also shows you where the Moon and sun are in the sky with its dynamic sky compass. The *Moon Globe* astronomy simulation is free; *Moon Globe HD* is 99 cents.

3D Sun: This app created by Dr. Tony Phillips, LLC, displays 3-D views of the sun based on digital reconstruction of satellite images accessed from NASA's STEREO satellites. This app allows 3-D-rendered views of the sun from any angle with touch-enabled capability to zoom in and out and across the sun's dynamic surface. This app includes a solar event alert system, which when enabled notifies users of solar flares or other solar weather events. Science educators are using the alert system provided by this app to incorporate dynamic, real-time solar events into their solar system and space weather curriculum. This app encourages students to engage in authentic, current space science inquiry and technologies. *3D Sun* is free through iTunes.

ArcGIS is a free app from ESRI that provides a convenient way to discover and use maps with portable devices. These GIS maps are beautiful to view (as shown in Figure 6) and are tactile or location-coordinate interactive. Options for geo-information include map queries, search, measures of distance, hyperlink to locations of interest, and map sharing through embedded mobile device social media, web, or mail services. Access via mobile devices allows links to community hosted maps from ArcGIS Online–ESRI's online GIS.

More experienced *ArcGIS* users can integrate the ArcGIS.com authoring tools to create personal maps that can be exported and shared. *ArcGIS* maps can extend to iOS devices using ArcGIS Server.

Features available through the mobile *ArcGIS* include being able to:

- Navigate map galleries.
- Use maps authored in ArcGIS.com.
- Access personal GIS data.
- Display and zoom to current location.
- Collect and edit data, including using the built-in mobile device GPS or the map to collect and update GIS data, attach photos and movies, create intelligent form-based data entry driven by ArcGIS data models, and access industry specific collection sample maps.

[6] Read more about *Moon Globe, Moon Globe HD,* and *Mars Globe HD* on the Midnight Martian: Apps with Depth website: www.midnightmartian.com.

Educators will be interested to know that the app includes activities that require map viewers (i.e., students) to perform linear and area measurements based on a current location or by interacting with the map. The mobile device *ArcGIS* also allows retrieving detailed map metadata, viewing feature attribute information, performing keyword search and accessing relevant information, executing predefined searches, changing the visibility of layers, customizing access to popular maps by creating a personal favorite maps list, and sharing maps with other iPhone/iPad users. The ESRI education community[7] website houses a wide range of lessons for elementary through postsecondary audiences.

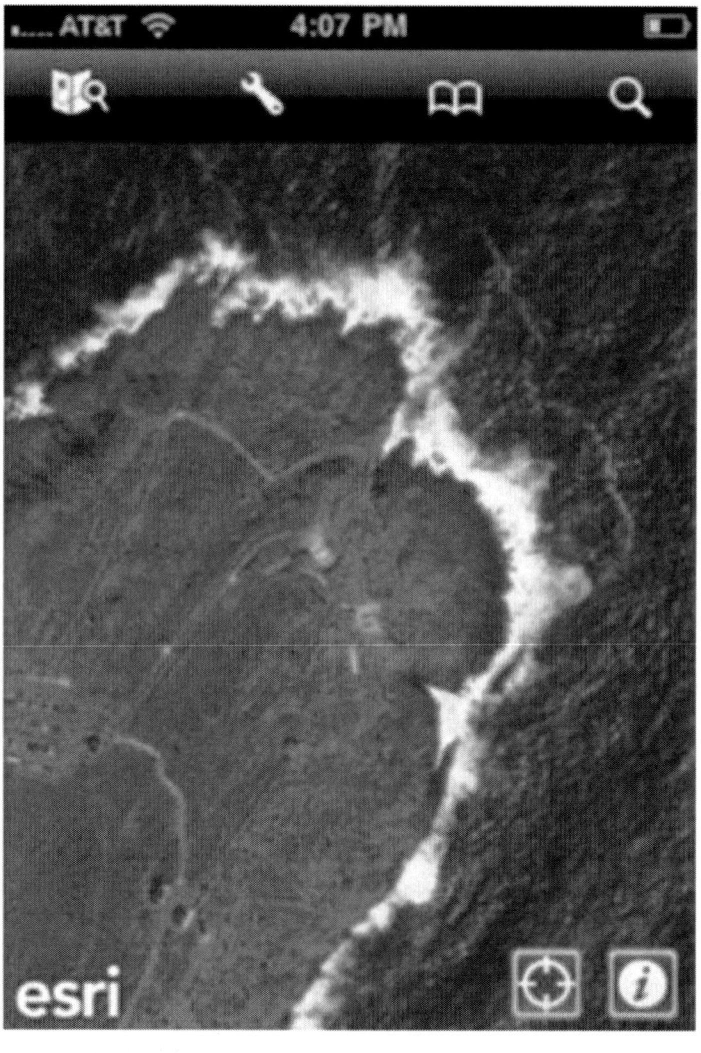

Figure 6. A screen shot from *ArcGIS*.

A section on the education community website called *Most Recent Additions to ArcLessons* includes such instructional topics as Exploring 10 Landscapes w/ArcGIS Explorer Online; Investigating the Harrisburg, Illinois, 2012 Tornado; Analyzing

[7] The *ArcGIS* education community lessons are published at http://edcommunity.esri.com/arclessons

Demographic Components and Implications; Core Elements of GIS: Create, Analyze, Represent; and Essential Skill Sets for Conducting GIS Training.

Research Question 4: What opportunities for data collection and dissemination are available through mobile technologies?

According to the *New York Times* (Hu, 2010), New York City; Chicago; Scottsdale, AZ; four major cities in California; and many school systems around the United States have ordered iPads for teachers and students to use as a way to motivate students and rejuvenate school curriculum. In some cases schools are using iPad apps as an alternative to textbook curriculum. For example, several urban schools systems in California are using an iPad-only algebra course developed by Houghton Mifflin Harcourt.

Some technology enthusiasts, however, caution that integration of mobile devices does not require purchasing the latest, greatest, most expensive tool. Smartphones, which are offered in many different models with varying features, can offer similar benefits as the iPad at a greatly reduced cost but with smaller screens. Acknowledging that nearly every teen has a smartphone that they bring to school and rather than punishing them for use of these devices, some school systems have adopted a policy of BYOT—bring your own technology. Below is a list of innovative ways that feature mobile device features and capabilities that have not device model or app-specific requirements.

Carolyn Skibba's blog, The Power of a Camera[8], discusses how students can use the iPad camera to document what they're learning and create movies, blogs, or other multimedia presentations. Using the digital camera embedded in most mobile devices allows students to capture their world at a moment's notice and share their image or movie with their peers, parents, or teachers.

The camera in a mobile device allows picture-taking from dual perspectives—the camera can be pointed outward or inward—engaging students in considering the perspective that best fits the message they want to convey. The digital camera embedded in the mobile device can also be used as part of student assessment, giving teachers physical evidence of student interpretations and understanding of the assignment. Other learning activities in which students can incorporate the camera in a tablet computer include electronic posters, online experiments, and digital demonstrations. Thus, the camera can help educators integrate multimedia assessment and documentation into their instruction—if all students have access to their own device.

Data collection apps are useful for managing grades, assignments, classroom roster/attendance, and several other measurable factors in a classroom environment. Such apps can even assist in measuring the goals and effectiveness of integrating mobile devices into the classroom.

Jermey Mathews[9] discusses ways of utilizing the sensors and software inside cellphones and mobile tablets to conduct scientific measurement. Students have access to the same tools inside their pocket device as scientists who use their mobile devices as microscopes and have been working on an attachment to allow cellphones to work as a wide-field fluorescence microscope for water-quality screenings. NASA, for example, has used mobile devices and

/arclessons.cfm.

[8] This blog is available on the web by Augustus H. Burley School, Chicago, at http://ipadsatburley.blogspot.com/2011/10/power-of-camera.html.

[9] This article can be accessed at Harnessing Consumer Mobile Devices for Science: http://www.physicstoday.org/resource/1/phtoad/v64/i8/p24_s1?bypassSSO=1.

applications in space such as on the International Space Station for research and collecting data. As mobile devices offer more technical proficiency and capabilities, the uses for mobile devices in science can be enhanced and further explored.

Ziemke Fastabend[10] gets her students involved by leading a project called, App Reviews: For Students, By Students, at the Augustus H. Burley School in Chicago. In her online article Fastabend discusses the process of students' using iPad apps and sharing their learning by compiling app reviews.

Research Question 5: What opportunities and challenges are there for using mobile devices and smartphones to access learning materials and collaboration platforms?

This section features several examples of how integration of mobile devices can be economically advantageous, serve the vision for technology integration and student-centered learning, or serve a school budget by decreasing the need for printed materials by substituting electronic versions of textbooks and applets that address lab activities and homework exercises. However, as shown by the rubric (provided in Appendix 1) and the itemized list of affordance, benefits, and concerns in Table 1, educators, technology coordinators, administrators, parents, and students need to reflect on all aspects involved with integration of mobile devices.

Use of mobile devices in the classroom can be a good choice for economic and logistic reasons.

For example, Steven Matthews writes that iPads in classrooms save money and time and put information "at our fingertips."[11] Students have easy access to their virtual textbooks and assignments. Teachers save time spent photocopying, and using less paper saves schools money.

Along this same theme of mobile devices economies, Wesley Fryer[12] suggests that schools are wasting their money on expensive whiteboards and, instead, should connect an iPad to an LCD projector through a computer. Educators have to use things that are appealing to the students, and according to Fryer, technology can provide a common ground for meeting students and for getting their attention. The cost of whiteboards is significant, and Fryer suggests that the iPad offers greater capabilities anyway. His article also provides step-by-step instructions and support on how to use an iPad as an interactive whiteboard.

Our last example in this chapter looks at the research study of Project K-Nect[13], which involved students at Southwest High School in Jacksonville, NC. In this study students were given cellphones equipped with programs to help with algebra studies (see Figure 7).

Richtel and Stone (2009) conducted the evaluation of the Southwest High School cellphone study funded by wireless technology company Qualcomm. "This is a device kids have. It's a device they are familiar with and want to take advantage of," said Shawn Gross, director of Digital Millennial Consulting, which received a $1 million grant from Qualcomm to conduct the research.

[10] More information about how to get students involved in app reviews is available at http://burley.cps.k12.il.us/.

[11] Read Matthews article at http://www.middletownjournal.com/news/middletown-news/ipads-in-classroom-save-money-time-and-puts-information-at-our-fingertips-1272741.html.

[12] Read Fryer's article at http://www.speedofcreativity.org/2011/03/24/ipad-as-an-interactive-white-board-for-5-or-10/.

[13] Read more about Project K-Nect at http://www.fastcompany.com/magazine/140/ cellphonometry.html.

In Project K-Nect students used smartphones to videotape their explanation of math processes and post them online to help others having trouble with similar math problems.

Instead of passively absorbing teachers' lectures, students in this curriculum had to participate intensely.

They began by taping up big poster boards, writing out how they got the solution to a particular problem, videotaping their explanations of their solution, and posting the videos online for other students to view. As one middle school student explained, "We actually ended up teaching our classmates" (Svoboda, 2009).

Tools like QR readers allow teachers to build in interactive messages to students, parents, board members, supervisors, and colleagues. A mobile device online collaborative[14] allows educators to share ideas about the topics presented in this chapter after the book is published.

Figure 7. In Project K-Nect and similar school projects, students use smartphones to make videos of one another explaining math concepts. Students also collaborate on cellphones to get help with math homework. Photograph © Jake Hellbach - Fotolia.com.

GUIDELINES AND RECOMMENDATIONS

In addition to the guidelines provided throughout this chapter, the recommendations in Table 1, and the rubric in Appendix 1, the following points summarize key ideas about use of mobile devices to support and enhance STEM education:

---

[14] Mobile Devices in Education at http://www.nasatalk.com.

- Establish a plan or method of how to incorporate mobile devices to enhance STEM education in the classroom.
- Set goals. Know what you want to achieve and list possible methods using mobile devices.
- Track measurable findings. Find out what works and what does not.
- Become knowledgeable. Find out how other educators are successfully incorporating mobile devices into the classroom. Discover which types seem to work best for a specific student audience.
- Recognize both the positives and negatives. Stay open-minded; find what devices offer and what they don't to enhance the subject matter.

CONCLUSION

Adults and youth will continue to use mobile devices as a way to communicate, seek and share information, and find immediate, portable opportunities for entertainment.

The challenge for STEM educators is to find clever and useful ways to integrate mobile device technologies and apps that accomplish the achievements outlined in this chapter.

When done thoughtfully with a context-based practical plan, integration of mobile technologies can save money as alternatives to traditional, paper-based instructional systems. When educators and administrators work together to plan integration of mobile devices and apps, both partners can save time and work more efficiently. As the examples provided illustrate, assessments, lab activities, homework, communication between schools and parents, and enrichment activities can be disseminated and managed through mobile technologies in ways that can save paper, can be programmed to go directly to the intended audience, can immediately update individual and districtwide databases, and can be customized by the recipient.

Technology is enhancing education at an extremely rapid rate. This places a major importance on conducting research on how mobile devices play a role in education. Educators who are actively using mobile devices to promote STEM education need to collect and report on how they are incorporating them. They should contribute discussions on their successes and possibly failures of doing so. The constant changes in mobile device capabilities make it difficult to collect hard data, but if there is a focus on key areas of mobile devices that occur across the board, then that is where educators need to place emphasis.

For years education theory has promoted a movement from teacher-directed to student-centered instruction. With strategies in place like BYOT and direct access for educators and students to open source programming tools, applications, and online courses, the teacher's role has clearly shifted from conveyer of knowledge to designer and mentor for goal-oriented learning experiences.

APPENDIX 1. COPY OF MOBILE DEVICE AND/OR APP FEATURES, BENEFITS, AND CONCERNS RUBRIC

BASIC APP FEATURES		CHECKLIST	COMMENTS
	Download app from:		
	Latest version:		
	Release date:		
	App size:		
	Requirements:		
	Review comments available from:		
	Ratings:		
	Cost to download:		
AFFORDANCE	**BENEFITS (+1) & CONCERNS (-1)**	**SCORE**	**COMMENTS**
VISION			
	BENEFITS: Supports your school's vision for a coherent digital learning environment.		
	CONCERNS: Requires advance planning and agreement by all stakeholders on the idealized digital learning environment.		
CUSTOMIZATION			
	BENEFITS: Allows adaptive teaching and student choices for instructional media and learning styles.		
	CONCERNS: Increases complexity of support and variability in curriculum quality and depth.		
PORTABILITY			
	BENEFITS: Offers conveniences such as increased battery life, screen size, flexibility, and decreased weight and startup time.		
	CONCERNS: Risks of a short history with individual products and trial and error uses in the classroom.		
TIME			
	BENEFITS: Allows fast, flexible deployment of technology as well as new or modified curriculum; better service desk response time.		
	CONCERNS: Requires ongoing maintenance of device compatible/incompatible issues that is time consuming and involves frequent updates with feature changes.		
BUDGET			
	BENEFITS: Shifts funding to be more directed to classroom technology tools and instructional innovations rather than generic infrastructure tools.		
	CONCERNS: Requires frequent upgrades, which may cause confusion as operating system features for mobile devices are changing more rapidly than for personal computers and network servers.		

Appendix 1. (Continued)

INDEXING			
	BENEFITS: Allows learning objects within digital instructional component to be tagged with curriculum content, resources, content standards, and/or learning styles; allows teachers and students to choose interfaces with greatest appeal and efficiency.		
	CONCERNS: Needs a standard system for index criteria and tools that is consistent and easy for teachers and students to access and use.		
DATA COLLECTION			
	BENEFITS: Makes individual access patterns accessible so that student/class participation can be quantified.		
	CONCERNS: Involves administrative balance of access, protection, social collaboration, and storage requirements.		
ASSESSMENT			
	BENEFITS: Allows both ungraded self-assessment and graded assignments and quizzes to be easily embedded into the mobile curriculum.		
	CONCERNS: Increases concerns regarding identity issues for distributed learning that are not easily resolved.		
STRATEGIES			
	BENEFITS: Gives educators using mobile technologies opportunities to update and even radically reform school vs. out-of-school learning.		
	CONCERNS: Requires documentation of implementation strategies to be able to understand how mobile devices support existing and new instructional approaches and strategies.		
ACHIEVEMENT			
	BENEFITS: Appeals to student interest in using mobile devices as part of students' learning toolset and stimulates student interest.		
	CONCERNS: Requires controlled research settings to identify contexts in which integration of mobile technologies improves student achievement.		
SUBTOTAL SCORE	**BENEFITS (+1) & CONCERNS (-1)**		

Student Needs	AppDesign Facilitates Teaching Strategies that Address Student Needs. Yes (+1); No (0)	Score	Comments
AFFIRMATION	Facilitates student-centered instructional approach and requires student contributions.		
CONFIRMATION	Supports multiple learning styles through use of the mobile device media tools such as images, text, activity completion, and embedded assessments to demonstrate understanding.		
PURPOSE	Gives learners access to device capabilities in an applied context that demonstrates how their data collection and use of the mobile device has purpose and absorbs them.		
POWER	Empowers students to take ownership of their work quality while providing access to dependable support for their learning journey.		
CHALLENGE	Offers learning experiences that stretch learners, provide opportunities for success, and reinforce accountability for personal growth that contributes to the growth of others.		
SUBTOTAL SCORE	**APP SUPPORT FOR MOBILE DEVICE AFFORDANCES**		
TOTAL SCORE	**SUM OF ALL CATEGORY SCORES**		

References

Adam, L., Butcher, N., Tusubira, F. F., and Sibthorpe, C. (Revised August 2006). *Transformation-Ready: The strategic application of information and communication technologies in Africa: Education Sector Study: Final Report. Second Decade of Education for Africa (2006-2015), Plan of Action: World Bank, African Union, and ict Development Associates, Ltd.* Retrieved April 15, 2012, from http://www.nepad.org /system/files/Second%20Decade%20of%20Education%202006-2015.pdf

A commitment to America's future: Responding to the crisis in mathematics and science education. (2005). Business-Higher Education Forum: Washington, DC. Retrieved Dec. 11, 2011, from http://www.bhef.com/about/index.asp

Brown, T., Prendergast, L., and Woodfin, L. (2011) *Use of the iPad for students with significant intellectual disabilities.* Chesterfield County Public Schools, VA. Retrieved Dec. 15, 2011, from http://www.ttacnews.vcu.edu/2011/09/use-of-the-ipad-for-students-with-significant-intellectual-disabilities.html

Callaos, N. (2008). *The essence of engineering and meta-engineering.* Retrieved Nov. 6, 2011, from www.iiis.org/Engineering-and-Meta-Engineering

Cohen, E. (1999). Reconceptualizing information systems as a field of the discipline informing science: From ugly duckling to swan. *Journal of Computing and Information Technology*, *7*(3), pp. 213-219. Republished in T. G. Gill and E. Cohen (Eds.) (2009), *Foundation of informing science*: *1999-2008,* pp. 7-12. Santa Rosa, CA: Informing Science Press.

Davis, B. G. (1999). *Tools for teaching.* Jossey-Bass. Reprinted with permission, University of California, Berkeley. Retrieved Dec. 12, 2011, from http://www.ware.k12.ga.us

/superintendent/Goal%20Presentations/
0195/WCHS_Motivating_Students_article_to_teachers.htm

Dell. (2012). *Connected classroom: Learning through technology*. Retrieved Jan. 2, 2012, from http://dell.triaddigital.com/connectedclassroom/

Gill, T. G. (2010). *Informing business: Research and education on a rugged landscape*. Santa Rosa, CA: Informing Science Press. Retrieved Jan. 2, 2012, from http://dell.triaddigital.com/connectedclassroom/

Hu, W. (Jan 4, 2011). Math that moves: Schools embrace the iPad. *New York Times*. Retrieved Dec. 12, 2011, from http://www.nytimes.com/2011/01/05/education /05tablets.html?pagewanted=all

Lynn, J. (2010, Oct. 14). Mobile phones help lift poor out of poverty: U.N. study. *Reuters*. Retrieved April 2, 2012, from http://www.reuters.com/article/2010/10/14/us-telecoms-poverty-idUSTRE69D4XA20101014

National Science Board. (2004). *9 science and engineering indicators*. Retrieved Jan. 4, 2012, from http://www.nsf.gov/statistics/seind04

Pearce, S. D. (2011). *Why use iPads in education?* Retrieved Jan. 2, 2012, from http://innovativescholar.com/2011/05/26/why-use-ipads-in-education/

Purcell, K., Rainie, L., Mitchell, A., Rosenstiel, T., and Olmstead, K. (2010, March). *Understanding the participatory news consumer: How internet and cell phone users have turned news into a social experience*. Pew Internet and American Life Project: Pew Research Center. Downloaded April 29, 2012, from http://www. pewinternet.org/ ~/media//Files/Reports/2010/PIP_Understanding_the_Participatory_News_Consumer.pdf

Pew Research. (2011, Aug. 16). *Pew Research reveals latest U.S. smartphone usage stats*. Retrieved April 23, 2012, from http://www.mobile-ent.biz/news/read/pew-research-reveals-latest-us-smartphone-usage-stats/015165

Quillen, I. (2011, June 15). Educators evaluate learning benefits of iPad: With the release of Apple's iPad 2, educators are still determining best practices for the classroom. *Education Week: Digital Directions*. Retrieved Jan 2, 2012, from http://www.edweek.org /dd/articles/2011/06/15/03mobile.h04.
html?tkn=XUYFQPVXFQlB54nFbqKJ%2F6fJs9R2x9o2EGGoandcmp=clp-edweek

Richtel, M., and Stone, B. (2009, Feb. 15). Industry makes pitch that smartphones belong in classroom. *New York Times*. Retrieved Nov. 6, 2011, from http://www.nytimes.com /2009/02/16/technology/16phone. html?ref=technology

Ruberg, L. F. (2012). Learning with iPads. Unpublished manuscript, Center for Educational Technologies, Wheeling Jesuit University, Wheeling, WV.

Schwahn, C., and McGarvey, B. (2011). *Inevitable: Mass customized learning: Learning in an age of empowerment*. ISBN: 1453748628.

Svoboda, E. (2009, Nov. 1). Cellphonometry: Can kids really learn math from smartphones? Schools are partnering with mobile-phone companies to help kids conquer math. Are smartphone-learning initiatives more than a corporate gimmick? *FastCompany Online Magazine*. Retrieved Feb. 15, 2012, from http://www.fastcompany.com/magazine/140/ cellphonometry.html

Tomlinson, C. A. (2002, Sept.). Do students care about learning? *Educational Leadership*, *60*(1) 6-11.

uTest (2010). eBook: A guide to mobile app testing. Southborough, MA: uTest, Inc. Retrieved April 21, 2012, from http://www.utest.com/landing-page/crowdsource-your-mobile-app-testing?ls=Banner%20Adandmc= DISPLAY-MobileAppTesting-Mobile-200x200-Rebrand-Static

Vielle, U. (2011, Nov. 28). *Mobile computing*. Retrieved Nov. 30, 2011, from http://www.AMC.org

Wikipedia. (2011). *Mobile technology tools.* Retrieved Nov. 6, 2011, from http://en.wikipedia.org/wiki/Mobile_computing

Ying, Z., and Qunli, S. (2011, Sept.). *Design and development of mobile learning resources based on learning object.* Electrical and Control Engineering (ICECE), 2011 International Conference, Yichang, Coll. of Educ., Zhejiang Univ., Hang Zhou, China. ISBN: 978-1-4244-8162-0, pp. 6486-6489.

In: The Space Education Phenomenon at NASA …
Editor: Norma Teresinha Oliveira Reis
ISBN: 978-1-62618-188-5

Chapter 3

SCIENTIFIC EDUCATION IN BRAZIL

Marcos Cesar Pontes*
United Nations Industrial Development Organization
Goodwill Ambassador, Brazil

ABSTRACT

This chapter brings discussions and analyses of the scientific education phenomenon in Brazil. The education system faces several challenges currently, and fostering students' interest in Science, Technology, Engineering, and Math (STEM) is one of them. Many students are not interested in these themes, as they seem not to be popular, and because the way that these contents are taught – as if they were very distant from the youth's reality – in many cases are not adequate to assure learning and engagement. Actually, students are more interested in themes such as soccer and other forms of entertainment, though sciences are not that popular among children and teenagers. In order to assure students' interest for STEM education, it is insufficient only to promote improvements at scholarly level. It is also critical that researchers devote part of their time to education and outreach activities. The teaching profession is the most important one that helps construct a solid and sustainably developed nation. It is the teachers the main agents in the process of encouraging students to consider careers in STEM. However, in order to assure that teachers could fulfill their mission, it is necessary that the government work towards assuring proper salary and working conditions. Finally, in order to assure students' success, it is also critical more involvement of parents in their children's education. Discipline is an important ingredient to assure not only that students learn more, but also for the construction of a responsible society, committed with the development of every citizen.

Keywords: scientific and technological education; discipline; teachers

* E-mail address: astronauta@marcospontes.com.br

SCIENTIFIC EDUCATION IN BRAZIL

"I hate Physics!" How many times have you heard this sentence?

Perhaps even you yourself have already said that!

Anyway, the fact is that many people in Brazil (I confess that it concerns me to imagine that it could be the majority) are averse to science-related subjects, i.e., Physics, Chemistry, Biology, Mathematics, and their "derivatives."

My concern with this finding is to be aware that the development of any solidly built country depends upon science and technology careers. One cannot construct a first-world country only with soccer players, journalists, philosophers, lawyers, sociologists, anthropologists, politicians, economists, artists, and other humanities' careers. Certainly, these professions are essential to the entire development scenario. Our own career, as educators, belongs to the humanities. However, the process of transforming knowledge into products of high added value, thus encouraging research, innovations, the generation of companies and employment in the country, mandatorily requires an expressive percentage of professionals in science and technology. We need engineers, physicists, chemists, biologists, medical doctors etc.

In spite of this, the fact is that the youth definitely demonstrates aversion to science classes! And this idea, brought and increased year after year in primary and secondary school, generates reduced demand by science and technology careers in higher education!

The first necessity of scientific dissemination is clear: we have a concerning reduction in search for science and technology careers.

We have a problem in our hands that in national scale can potentially compromise the national development itself in a near future. Actually, this effect of lack of interest by careers in science and technology has also been noticed in the United States, which encouraged the establishment of *Science, Technology, Engineering, and Math* – STEM programs, bringing more dynamism and interest in education in these areas.

As in every problem, we should start by the reaffirmation of the objective, followed by the identification of factors that contribute to the deficiency and, finally, identify factors that contribute to generate the deficiency, and then develop and apply corrective actions (solutions).

Our purpose is to increase the interest of students in science-related classes and, in doing so, contribute to amplify search for careers in science and technology. At the moment, we live with the fact that most youth is averse to science subjects.

It is certainly part of factors that contribute to this noticed problem: insufficient scientific dissemination on the means of communication, and discredit of the teaching profession, which contribute to the insufficient preparation of these professionals in the scientific areas, especially among those who work with precollege education, and the lack of more adequate tools and methods to introduce science-related subjects inside classrooms, laboratories... and outside them!

The solution to this problem may comprise several pathways and stages, and compose extensive, frequent discussions in circles of researchers in education, currently. Thus, what I present here are only a few considerations to "put some wood on the fire" of ideas and contribute somehow to reach our purpose. These considerations have been raised as part of studies for implantation of programs of incentive to sciences in elementary and middle education of the Astronaut Marcos Pontes Foundation.

Initially, note that it is unnecessary expensive equipment or complex methods to capture the youth's interest in some science and technology subjects.

Teachers have, as always, enormous influence in the process.

The stile of classes based on transcription of theory and equations on the blackboard, even when accompanied by songs or other "memorization techniques," may work for preparatory courses to university entrance examinations, but they are not at all efficient for learning. It is necessary to have the subjects (including equations) leave the 2D of the blackboard to the 3D of students' lives and hands. The subject needs to "be embodied," justify its existence and limits as theory that attempts to model principles and natural quotidian phenomena. Each part of an equation, each concept involved, has a story behind, a reason for being, a possible emotion to be associated with, and this relates more to our quotidian life than to books and mathematical expressions.

"Imagination is more powerful than knowledge." This sentence has been said by a person of great wisdom! Do you remember?

We can complement with something like this: "Imagination is an essential part of motivation for knowledge."

Observe that imagination is related to two emotions that are in the root of motivation of all our activities: fear and pleasure. Imagination relates to fear when we realize a "necessity" and need to "imagine" alternatives, and to pleasure when we seek manners of winning challenges to satisfy our "curiosity."

Interesting! If we combine these ideas to the fact that we usually learn by association, that emotion intensifies these associations, and that often finding pleasure in a same activity generates habit, we could start identifying manners of transforming science learning into something more pleasurable, interesting, remarkable and frequent in students' lives.

Considering that fear affects global learning, as focus attention and inhibits the extrapolation of concepts (nobody truly learns under pressure, and each one learns one's own way), perhaps the key that we are looking for would be in "curiosity." Note that curiosity, accompanied by overcame challenges generate pleasure, which generates imagination, which generates more curiosity, which generates more pleasure at each discovery... and so on. Exactly! It is possible that this circle generates the good habit of research that, ultimately, will encourage students to search for careers in science and technology!

But… how can we generate curiosity?

Again: teachers have, as always, enormous influence on the process.

Do not forget that we have three "levels" or "kinds" of subjects related to science and technology:

1. Basic knowledge, such as Physics, Chemistry, Biology etc.;
2. Tools, such as Mathematics, technical drawing, artistic drawing, history of science, scientific methodology etc.;
3. Applications, such as electronics, aerodynamics, neurology, meteorology, programming, agronomy etc.

Have you already noticed that the majority of youth "live" technology (applications) and rapidly learn how to use them (even without reading the manual)? Have you noticed that part of motivation comes from the pleasure of "discovering" and sharing with friends, as a type of emotional affirmation? All of this can be explored!

Observe that the subjects related to applications comprise systems and equipments that are "cool" of assembling, disassembling and operating, such as radio amateurs, robots, sensors and so forth, which are all multidisciplinary with regard to matters of basic knowledge and tools. Thus... summing all this...

Perhaps we could have fewer *"reality shows"* and more TV programs targeting science and technology; perhaps we could encourage more the establishment of science clubs, competitions and fairs; perhaps we could provide students with more challenges of constructing, enhancing and modifying some of these "cool applications", even in a rudimentary way and without the complete base of knowledge; perhaps we could better foster the utilization of research resources (including teachers, books and internet) in order to have students finding their answers and solutions to overcome challenges; perhaps we could better prepare ourselves and make the correct "connections" with subjects and histories "behind" equations and concepts as students look for us to help them find their solutions; perhaps we could invert the usual direction: instead of going from the blackboard to the laboratory, go from the laboratory to the blackboard.

Maybe we could?

No! We can! And fortunately I already see some educators doing just that.

It is an advance, but this group still serves a very small percentage of Brazilian students. It is necessary a much larger adhesion.

In this scenario, we need to state something: the problem is much more complex than it seems to be.

In order to obtain more adhesion, we need better-motivated and prepared teachers. Further, it is also necessary transformations in Brazil's politics, especially with regard to public education. In summary, we need to improve our education system.

To construct a robust educational system is not a matter of option for a country that aims at belonging to the so-called "first world;" it is an indispensable condition.

Currently, there is no solid development without a strong base, constructed on the foundation of knowledge and sustained by the columns of technology.

More companies, more employment, more health, more safety, fewer incompetents and more aware citizens, which in first place fulfill their duties with society and, after, require their rights in an efficient way, demanding results and behaviors of their elected public servants (politicians): these are only some of the good consequences that an education with quality can generate.

On the other hand, a people deprived of education, unaware of its potential as human beings and of its power as citizens, is a people that can be easily manipulated, and votes for a dish full of promises and crumbs put on the dirt floor of the backyard, without knowing that they could sit to the main table and enjoy the banquet offered to all citizens of a developed and professionally qualified society.

In other words, a nation of success is composed of educated citizens.

And how to accomplish this?

Answer: With good teachers and good politicians.

You could be thinking: "How does one thing relates to the other?"

Keep reading...

We certainly have some politicians who are honest and willing to contribute to the process of transforming Brazilian education. The question is: Would it be possible that this small "group of warriors" could win the resistance of those unscrupulous and corrupt who use the people's ignorance as electoral tool?

It is essential to increase the group of the good elected representatives; it is necessary that you, who are competent and honest, stop complaining and do your part. Your participation is crucial! And one does not accomplish this by sticking his head in the sand and moving away of the "dirt of politics." White vote, null vote or "protest" vote are synonyms of a combination of cowardice and ignorant behavior. And this is not what I expect from our compatriots! I hope that they have courage and maturity to face the challenges ahead; to search the facts and expel from public service all of those bad people; to take their positions in decisions, i.e., in politics, and show how things should be done; give the example!

Observe that the complete national development process depends directly on education and that it should include a great "cleaning," followed by considerable restructuring in political activity. If we really intend to live in the country of our dreams, these areas need to be integrated and treated with much more seriousness by all of us Brazilians.

Now, realize that in order to allow these cultural changes and knowledge acquisition to occur, there is a central figure that needs to "definitely enter the combat:" the teacher! Currently, this is a profession whose image and importance is so decadent that makes us feel pity, but be aware that this is the professional category that will change the country.

Whether, on one hand, there is no real development without education, also there is no solid education without good teachers. They are the fundamental pieces in this structure. They are the ones who, actually, construct development. Think: without teachers there is NO other qualified professional. Further, they have the power of molding and constructing the ideas of tomorrow. And this is VERY interesting...

The logic of this affirmative is not difficult to realize, correct? Especially if you are not one of those alienated who are only interested on soccer, shows and appellative TV programs, without content or signal of intelligent life.

However, being so important professionals, how are our teachers being treated in Brazil?

In the press, from time to time we see some manifestations, sometimes expressed by the despaired complaining of a master captured in short videos on the internet, other times in strips and strikes on the streets, other times... we did not know.

It is evident that, in response to these events, some people get heartened, touched, and post beautiful messages on social networks, we listen to well-trained speeches, some agreements are signed and... Everything stays the same! Many ideas and words. Of real and permanent results, not so much.

In some private schools, educators are treated as if they were service providers to their young clients (students). When there is difference of opinion between the immature client and the teacher's experience, the "client" soon requires his " right," and parents, instead of supporting an attitude of more respect towards the teacher, threaten their job.

In public schools, it is not the job that is at risk, it is the physical integrity of the teacher!

Good times were those of my primary education! At that time, teachers always had the last word in every situation. Who was I to complain or question the word of one of them with my parents! My mother's reaction would be immediate: "Son, you are at school to learn to become someone in life. This is not easy and certainly it will not be done the way that you, without knowing almost anything in life, imagine. Stop complaining, be more attentive in classes, do your part well done, obey the teacher and LEARN. It is this, ONLY, which you have to do in this phase of your life. Who commands the classroom is the teacher!"

And currently? Where is due respect to the master? In which part of the "great educational theories" it has been included?

OK! Think about this later.

Now I would like to pose you another question: And what about the teacher's salary?

You have stopped to see the values and do some calculations. Is this the way you expect that those who are responsible for the education of your children be remunerated? I hope not!

Teachers have families, responsibilities and bills to pay, such as all of us! Have you ever seen some supermarket attendant ask "teacher's card" to someone and let him bring the purchases home for free? Of course not! Teachers usually are professionals passionate by their pupils and live for the satisfaction of seeing them succeeding in life. For this reason, they are abnegated and in many cases even naïve, as they do not require what would be their rights.

Right! Very beautiful! However, the reality is that they get "ridiculous" salaries and as many of them do not feel free to complain, I can speak for them: Brazilians, let us WAKE UP in this country! We need to realize the relative importance of the activities and pay the salaries that our masters deserve and, of course, require from each one of them the corresponding performance! It is just like that, simple, without hypocrisy! Nothing is for free! If we want good education to our children, we need to pay the educators well!

Apparently, our dear compatriots do not think strange the huge expenses with soccer stadiums, shows, propagandas, and political campaigns. On the other hand, it seems that investing in educational infrastructure and to increase the minimum wage of the most important professional to the country's development – the primary and secondary school teacher – to one level that could be at least considered more "decent" is a great absurd. It seems that this (only) would cause "an enormous hole in the public expenses." Would it? Really? Just to give you an idea, the first Brazilian human space mission (which has already generated scientific and social results that paid many times the investment done) cost R$0,07 (seven Brazilian cents) per Brazilian; while the corruption cases (only the known ones) cost more than R$15,00 (fifteen Brazilian "reais" – the name of our currency) per year, per Brazilian (the equivalent to more than 200 space missions!). If this annual corruption value, pocketed by criminals, would be converted in teacher salaries, several good things could be done! Do you agree?

With regard to priorities among activities, pay attention; please do not take me bad. I am not against soccer, shows, and so forth. Actually, I am a big fan of these activities. I love watching a good game; some TV programs, and listen to my favorite singers. The work of these professionals is very good to our emotions. Fun is good to our soul! But, remember that your success can only be constructed by your own work, not by the work of those who provide you entertainment.

Likewise, you should agree with me that the country development requires professionals in science and technology prepared to create and provide high added value products and services. It is impossible to prepare these professionals without a robust education system. The importance of these professionals in your life cannot be compared with the pleasure of entertainment. In summary, it is boring when you cannot watch a show, but when there is not a competent medical doctor to take care of you in sickness it is tragic! Think about it! What is necessary to turn a young person into a good physician? Shows and stadiums or good schools and teachers?

In any case, note that the essential question of the country's social and political development has very profound roots. They reach the realm of culture and family values themselves.

At this point, I believe that it should be clear that there are many things to be done in this process that goes far beyond the dominium of scientific dissemination. The official solution will surely comprise several analyses, beautiful studies, proposals, expensive and beautifully shielded reports and - well - meetings and speeches. I hope that one will not forget to put everything into practice! "Action" has to stop being something rare in this area.

Yes, we need much action to bring our education to a satisfactory level. It means an enormous transformation with regard to what exists today. A long pathway has to be followed with patience and determination. A journey that surely starts with the good and old family education, with parents' commitment towards the development of their children's basic and essential values with regard to life. Things such as respect by elder people, the masters and the established laws, cordiality, honesty and so forth. All of this needs to be retaken and emphasized.

This pathway continues with the power of schools in establishing and requiring the observation of fundamental behavior rules from everyone involved in the educational process: employees, teachers, parents and, of course, students! In one single word: discipline. Establishment and observation of limits, with the normal frustration associated to "we cannot do everything we want/ the way we want," have been always essential for learning and for healthy emotional development. Creativity has never been diminished by discipline, and real life does not follow theories in paper. It is part of schools' obligation to participate in the preparation of the youth for real life conditions and events!

However, currently, what happens inside schools?

Student fights, teachers beaten, classes interrupted, attention lost... The main purpose, which is human development, becomes increasingly distant.

The news brings, almost every week, serious cases of violence inside scholarly institutions. Which factors contribute to this problem? What can be done?

In order to start addressing this subject, we need some reference. Will it be possible that someday, somewhere, this problem was (or still is) inexistent or extremely reduced?

Yes! In Brazil, for 20 or 30 years, we did not listed about this kind of behavior inside schools. Currently, in several countries, such as Japan, South Korea, and others, "education," in the widest meaning of the word, does not mix with violence.

As we consider only these two references, first we can assume that there is surely some cultural component involved in the results observed; second, though teaching theories and techniques have advanced considerably in the last decades, in Brazil, for some reason, the attitude of everyone involved in the process – students, parents, teachers, employees, authorities and so forth – has changed, for worst.

Considering "culture" as a result constructed gradually through "collaboration" of everyone that compose our society, it would be logical to imagine that it is not something that "changes from one time to the other." Thus, maybe the attitude change observed here in Brazil is caused by factors that are more transitory, less profound in our community, but that with due "permission of culture," has *also* been influencing negatively behavior inside schools.

Anyway, before we reach the scenario that we have today in our schools, the last stop in this train of factors is called *indiscipline*.

In the past, discipline was regarded as an essential element of learning for life. It started at home, with values taught by parents and continued at school, with due respect to the master, conscience of fulfilling duties before demanding rights, awareness and acceptance of rules and norms of conduct required in order to have the right to attend classes, the mandatory use of uniform, care with schools' patrimony, strict observation of the social etiquette for the living together inside school, punishment, including expulsion due to disrespect of rules or inadequate behavior, and a greater participation of parents in values formation, school activities, and support to the authority of teachers. All strictly required by schools principals, observed by teachers valued by society and always supported by laws and by authorities who are not afraid of "losing votes" by using words such as discipline, obligations, rules, behavior, mandatory, punishment, exclusion, and so forth.

"Your right ends where the right of others begin;" "Treat others as you would like to be treated;" "If everyone observe the rules, everyone shares the success." These were simple sentences that my parents used to say to teach me about the necessity of respecting other people, of learning the importance of discipline to overcome challenges in life, to realize that the right of the group is always more important than my personal right, and that following the rules is the manner to assure the entire group's success, including mine.

In practical terms, it is simple: many people scream by "their rights," claiming by freedom of doing everything one wants to. If the person who claims for rights also fulfills his or her obligations, all established rules and does not interfere in the group's rights, excellent! This is positive for everyone. On the other hand, if this person only considers his or her rights, this person does not deserve to be part of the group.

Thus, with regard to the current scenario in our schools, which would be some possible actions to reduce the problem of indiscipline?

Authorities could give the example and establish norms of conduct in a stricter and more detailed manner, to be observed by all schools and by all participants in the process, including more participation of parents in school, and in responsibility for the results. They could also re-empower and value the teaching career, offering due support for their decisions and authority inside classroom, with regard to observation of established norms.

The school management could assure the legal acquiescence of parents, at the moment of enrollment, of all rules to be observed by their children in order to allow them to remain in school, supervise the strict observation of these rules in schools' quotidian and punish according to norms those that did not observe the foreseen behavior. Further, school's management could provide guidance to parents for better accompaniment of children and to address questions common to education in values at home.

Small conflicts and natural adjustments of living together at scholarly age are normal, have always existed and will always exist. They are part of the development and can be treated case by case among teachers, students and parents, always guided by clear rules that everyone knows.

Constant presence of policemen especially trained to deal with students in all schools would certainly provide more safety and would enhance social integration, especially in regions with more criminality incidence.

Further, with regard to the discipline aspect we need everyone's participation!

The creation of a fair, honest, disciplined, and develop society requires more dedication from all.

Thus, encourage discipline in schools, support the obedience of established norms, do not turn your back to the problem, "regarding normal" that schools "reflect the current violence present in society," participate in solutions, visit the schools of your neighborhood, help in some way, even if your children are not still in scholarly age, even if you do not have children. Do not wait for government. Your attitude makes a lot of difference. Schools cannot be the reflex of the bad things that exist in today's society; they should represent the seeds of good things that we want for tomorrow's society.

Finally, considering the importance of teachers in practically all actions above, I hope that all of them realize their importance and struggle in a clever way by the valuing of their profession. Politics and education are DIRECTLY related. Include citizenship in your classes, in all subjects. The real master is not only a "knowledge transmitter," but also an example and the lighthouse to attitudes and behaviors of their pupils' lives. If you want real changes in the country, use your classes and conversations to emphasize the importance of education for the success of everyone in the country; teach your students how to vote in a conscious way; and encourage these young students to participate actively in politics, not to stick their head in the sand cowardly and not to run away of that such as something "dirty." It is in politics that are defined the laws that promote or devaluate the teacher's profession!

In summary, "dear teachers," the essential part of the problem's solution in education is inside classroom everyday: your students. Mold in them the future that Brazil needs! Only you, teachers, have this power! Have this in mind!

All of these actions in schools, working together, should certainly help the development of science and technology in the country.

Problem solved?

Unfortunately, not. School work can certainly increase search for careers in science and technology. Through that work, we have a promise of future: to bring a new attitude to our students. However, in addition to our youth, there is still a large part of the population that does not know the importance of science and technology and that, the same way, can influence negatively our politicians against investments in the research area.

We, who work with science in Brazil listen very frequently the following question: "How can our country invest in science and technology when there are many people in poverty?"

This type of questioning shows clearly the lack of awareness of a vast part of the population with regard to the importance of long term vision towards research and development to increase employment opportunities, to reduce the roots of basic problems such as health, violence, hunger etc., that is, to the betterment of the quality of life for all.

Thereby, in addition to all actions towards education in Brazil, the second necessity of scientific dissemination becomes evident: we need to bring science and technology closer to the general public, providing logic and practical meaning to the activities and investments carried out in the areas of research and development.

Again, remember that research and development usually involve public resources that are approved by politicians. These ones, as elected representatives of the people, in addition to have, by definition, the same level of preparation and culture of their electors, carry a constant concern with their own political career, that is, with votes for the next election.

If those who elect them, for insufficient knowledge and reduced strategic vision, do not associate social priority to investments in science and technology, certainly politicians will follow the same booklet, and the projects of research and development will always need budget and "constant justification" for their existence. This is exactly what we observe in Brazil.

How can we turn the science and technology subjects popular?

Events such as the "National Week of Science and Technology," carried out annually by the Ministry of Science, Technology and Innovation, have an interesting result, but they reach part of the population who, someway, already consider the subject interesting and, for this reason, attend events promoted that week.

During the rest of the year, it brings difficulty the fact that, usually, researchers get so entertained "struggling" to keep budgets to their own projects, or sometimes so in love with their productions, that they do not consider important to invest some time disseminating their ideas and results in a language accessible to the public. It is obvious that there is nothing bad or anti-professional in disseminating their work (indeed it would be exactly the contrary), but apparently there is some fear of "less recognition inside the category" if the scientist also becomes a good communicator. A belief that is not healthy to sustain projects as a whole.

Note that the problem is self-sustained (in a negative way): the smaller the scientific dissemination, less public recognition by the public and the politicians, less investment, less incentive to researchers, more "struggle" for budget, less desire or availability to scientific dissemination, and so forth.

It is necessary that we adopt more effective awareness actions among science professionals with regard to scientific dissemination, as well as more activities that directly connect science and technology applications to the quotidian of the common people. For that, for example, projects related to the sustainability of cities, involving energy, water, transportation, pollution, and so forth, things that are daily in houses and lives of all people in all urban areas, should be encouraged by authorities, not only by the fact that they contribute in the dissemination of science, but also by the urgent necessity for these solutions.

Likewise, the efficiency in business related to rural area activities can be tremendously improved through the use of science and technology. It means more production, more companies, and more employment.

If well presented to the public, with due care of emphasizing where the scientific solutions that today improve quality of life "came from," certainly the culture of "we should not invest in science to focus uniquely and constantly in the symptoms, such as bad health, violence, and so forth" will be gradually eliminated.

Thus, it is necessary an intense work of scientific dissemination inside schools (in addition to other actions to turn our education system more robust), and in activities of substantial popular reach.

In: The Space Education Phenomenon at NASA ...
Editor: Norma Teresinha Oliveira Reis
ISBN: 978-1-62618-188-5

Chapter 4

BRAZILIAN OLYMPIAD OF ASTRONOMY AND ASTRONAUTICS

João Batista Garcia Canalle*
State University of Rio de Janeiro, Brazil

ABSTRACT

We have been organizing the Brazilian Olympiad of Astronomy and Astronautics since 1998 and in this Chapter we will show the results of the event of 2011. The main goal is the popularization of Astronomy and Astronautics, but also equally important is the capacitation of the teachers of elementary and high school who teach these sciences. We have been successful in our goals because there are over one million students that are taking part in the Olympiad, distributed over all states of Brazil. The number of teachers involved (64,000) in the process is also very important. In fact, it is the most important aspect of all, because we wish to teach them Astronomy and Astronautics. The Olympiad takes place in the middle of May and it lasts one day, but the most important part is everything that those teachers do before that. With our help they teach a little more of Astronomy and Astronautics to their students, who observed the sky and made didactic experiments during the day using, for example, the Sun, or the shadow of a stick. After the Olympiad day we collect all tests and organize the lists of participants. We send certificates to everyone and about 33,000 medals. We invite a few hundred to take part in special events, such as the Space Journey, the Energy Journey, the Brazilian Olympiad of Rockets, the Latin American Olympiad of Astronomy and Astronautics, or in the International Olympiad of Astronomy and Astrophysics.

INTRODUCTION

The Brazilian Olympiad of Astronomy and Astronautics (OBA in Portuguese) began in 1998[1]. Initially the subject was Astronomy only and it was organized by the Brazilian Astronomical Society. In 2005 we included the Astronautics and since then we have been

* Email: joaocanalle@gmail.com

working together with the Brazilian Agency of Space. The main point, however, is that since 1998 the main goal of the OBA is not just the competition between students or schools or cities, but the teaching and popularization of Astronomy and Astronautics. How can we do that? We use the natural interest of young people for competitions. Everyone likes competitions, especially when it comes to sports, but young people appreciate competitions of any kind, not just sports. Every human being loves to be the "winner", the champion. Well, we are using this special interest of young people to attract them to the OBA. If we motivate the students, we can teach them, and more, we can also get their teacher involved, and we can also help them to better understand Astronomy and Astronautics. This is the main point. We teach the teachers so that they can teach their students. Unfortunately, the teachers that teach Astronomy and Astronautics in elementary schools and also in high schools have only a superficial understanding of Astronomy and Astronautics. Therefore we need to help them to better understand these sciences, so that they can better teach their students. We are organizing the OBA, but we have also been organizing many others activities in order to reach our goals.

Besides our main goals with all our activities, which includes OBA, Brazilian Olympiad of Rockets (OBFOG in Portuguese), Workshops, International Olympiads and Regional Meetings on Teaching Astronomy and Astronautics, we also have the following goals: 1) To train and inform the ordinary citizen of the importance of the scientific and technological apparatus existing in the country, 2) To awaken interest in the study of science, 3) To serve as a mobilizing agent community; 4) To emphasize the importance of education for personal development, 5) To motivate teachers to teach astronomy, astronautics, mathematics, physics, science, geography and chemistry as well as promoting the creation of study groups within schools to enable the dissemination of information and knowledge; 6) To establish innovative forms of learning; 7) To discover and encourage new talents towards a career in science in general and in astronomical research or aerospace science, in particular; 8) To allow the development of disseminating agents that often, for lack of support, do not carry out projects of interest to students and their communities; 9) To use the formulations of the questions and their solutions to bring correct and updated knowledge on astronomy and astronautics to students and , indirectly, to their teachers, because they will grade the tests according to a template provided by the National Organizing Committee; 10) To encourage the student population in the study of science using the OBA as a motivator; 11) To reward with medals about 30,000 students who presented the best performances; 12) To issue certificates of participation for all students as well as for their teachers who contributed to the event; 13) To distribute educational material for teachers, who are representatives of OBA; 14) To use their questions and their feedback to change misconceptions about Astronomy traditionally found in textbooks and among teachers; 15) To encourage teachers responsible for teaching the astronomy in elementary and high school to be able to upgrade their knowledge of the content to better meet the needs of their students; 16) To encourage the creation of astronomy clubs or clubs of amateur astronomers; 17) To involve teachers, their pedagogical coordinators and school principals in a nationwide joint effort to support the teaching of Astronomy; 18) To establish closer contacts between Brazil's professional and amateur astronomers with teachers of elementary and high school, responsible for teaching content of Astronomy; 19) To encourage the participation of those teachers in the Regional Meetings on Teaching of Astronomy and Astronautics and in the National Symposium on Teaching Astronomy; 20) To encourage visitation of the fixed and portable Planetarium; 21) To disseminate and discuss the reasons why Brazil should possess, maintain and encourage a space program; 22) To present the post-war historical context in which took place the Space Race between the United States and the former Union of Soviet Socialist Republics; 23) Show the benefits of the Space Race, such as remote sensing, weather forecasting, the controlling of deforestation, satellite communications, among others; 24) To promote the development of aerospace activities and emphasize its importance for the Brazilian scientific development; 25) To explain the problem of global warming and the changes that the global population will have to implement in order to reverse the situation.

The Brazilian Olympiad of Astronomy and Astronautics proved to be an excellent tool for the interaction between teachers responsible for teaching this content in elementary and high school levels and the Brazilian astronomers and aerospace engineers. We use this event as a pedagogical vehicle to teach astronomy and the fundamentals of aerospace science to the teachers responsible for those contents. Although the Olympiad involves the students, we are much more concerned in cooperating with the proper training of the teachers, after all they are the ones who have to teach astronomy and basic concepts of aerospace, year after year, class after class. Therefore, our interest lies in the completion of teacher training. In order to achieve this we have sent educational material to schools participating in the OBA. We believe we are the only Olympiad that has this concern and uses most of its resources to print educational material for teachers who are our representatives within their respective schools. In addition, between 2010 and 2011 we distributed, about 16,000 telescopes called Galileoscopes, rotating celestial planispheres for the Southern Hemisphere, books, etc.

Each school that wishes to take part in the OBA has a teacher representative of the School. They receive the OBA regulation with all instructions on how to publicize the event among their students and how to prepare them. They also receive a registration form to register the students who volunteer to participate in the OBA. The teachers teach additional classes of Astronomy and Astronautics, guide the studies, watch the sky and launch homemade rockets and other activities.

All students take the tests on the same day in their own school, which has been previously registered, under the supervision of the teacher representative of the school, because we wish to effectively involve teachers in the whole process. The teachers grade the tests from a template provided by the National Organizing Committee of the OBA and then mail only the top 10 tests of each level, as well as the list of names and grades of all participating students (not only the best of each level) to the National Commission. We type names and grades and ranks in descending order of students within each grade level. The ranking is done on a national level, as opposed to state, city or school.

Overall, during this process we are teaching Astronomy and Astronautics to both students and their teachers, although they have never studied those contents in their regular courses. As the textbooks were not written or reviewed by experts, they still have mistakes[2][3], despite the efforts of the Ministry of Education to improve the quality of those books. In the OBA tests we have the opportunity to provide the correct information. We donated printed materials or DVDs with the contents of Astronomy and Astronautics to all schools that effectively took part in the OBA. Below we are going to show the results and actions of the XIV OBA.

Annual Participation of Students in the XIV OBA

The enthusiasm of the students participating in the Olympiads is contagious, because the participation rate is rising as we can see in Figure 1. In 2009, the International Year of Astronomy, there was a giant leap in the number of participants: we achieved to 868,171 participating students! Therefore, there was an increase of 95,1% over the previous year! In 2010 we expected a small reduction in the number of participants. In fact 786,000 students participated, distributed over 9,152 schools. But in 2011 the number of participants rose again, as we reached a total of 803,218 students, as show in Figure 1. We also advise teachers not to force their students to participate, because what interests us are the ones who are volunteers to participate. We are not interested in quantity but quality, i.e., in really interested participants.

For the foregoing reasons it is evident that this type of event brings together teachers of elementary and high school and professional astronomers, as well as specialists on Brazilian aerospace. In the following sections we will present the main results of participation of students, schools and teachers, during the XIV OBA.

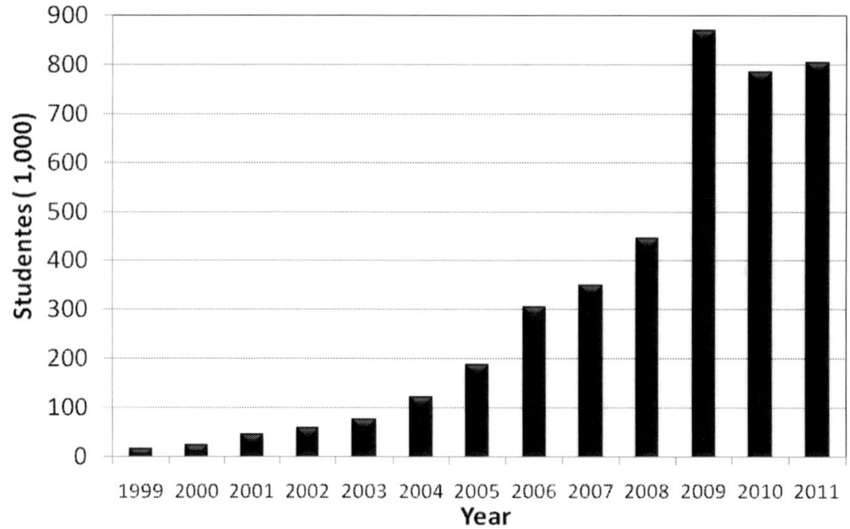

Figure 1. Total number of students participating in the OBA throughout the years.

State Distributions of Students and Schools Participating in the XIV OBA

Our goal is to have participating students in all Brazilian states. We believe that talent can be anywhere in Brazil, in any school. We hope with this Olympiad to discover them and encourage them to continue their studies, to believe in their potential, to have confidence in themselves and show that achieving success is a matter of individual effort, and that there are people and institutions who are willing to help them.

Figure 2 depicts the distribution of students participating in the XIV OBA (2011) by Brazilian states. As we insist to have participants from all Brazilian states our publicity always involves all Municipal Secretaries, State and Regional Education. It can be seen from Figure 2 that the State of Sao Paulo has the largest participation, with 193,000 students, followed by Ceará with almost 94,000 students, and in third place comes Minas Gerais, with approximately 85,000 students. But the most important thing is that we have interested students in all states and in 13 of them we have more than 20,000 students participating. As the Brazilian states have huge differences in size, population and therefore population density, Figure 3 shows a graphic of the percentage rate of schools participating in the OBA.

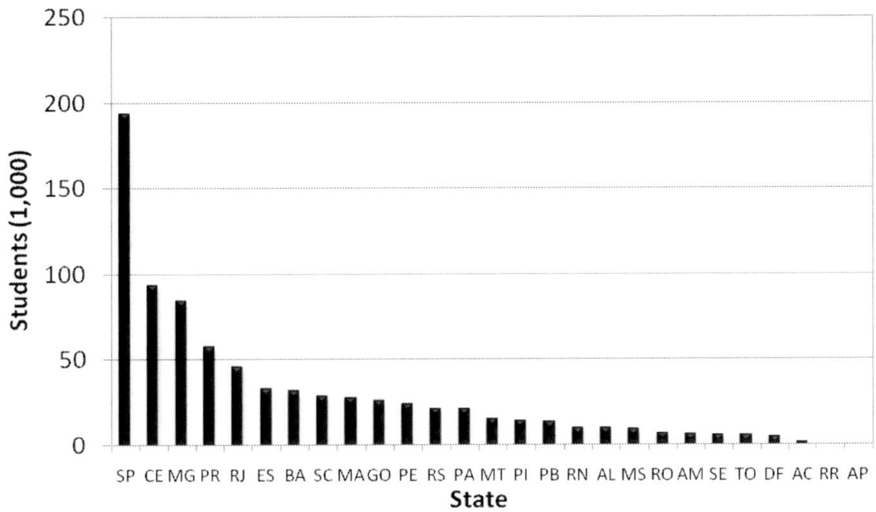

Figure 2. Distribution of students among the Brazilian states that participated in the XIV OBA.

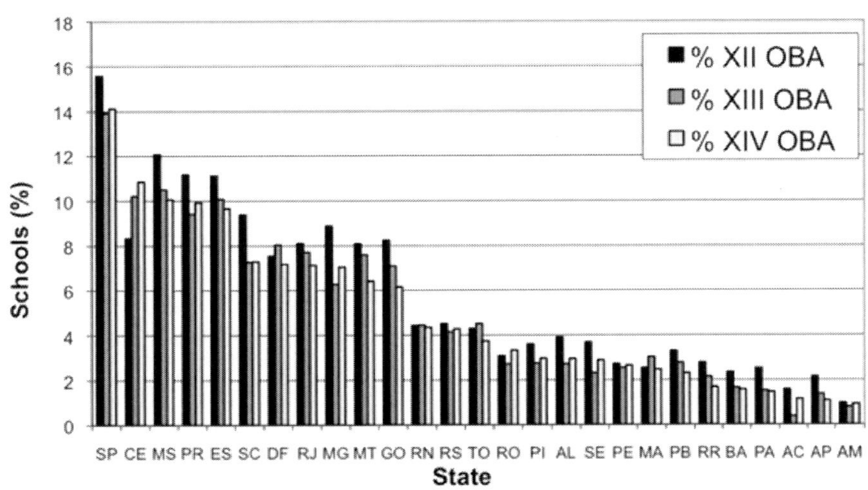

Figure 3. Percentage distribution, per state, of schools that have already participated in the OBA in the years 2011 (XIV OBA), 2010 (XIII OBA) and 2009 (XIIOBA) for purposes of comparison.

The graph thereby eliminates the need for the number of schools in each state. Figure 3, also shows the participation rate of schools in the last three years for comparison purposes. It is very satisfying to see that we have states such as São Paulo, Minas Gerais, Ceará and Paraná with more than 10% of all schools participating in the OBA and several other states already have about 8% of their schools participating in the OBA. With the exception of the Ceará, all the states in the North and Northeast of Brazil have less than 4% of its schools participating in the OBA. In those states we find a high concentration of rural schools, which makes it very difficult to communicate with them.

The Olympiad has also stimulated many schools to organize small groups of astronomy that are the embryos of future astronomy clubs, who organize themselves to study various astronomical topics, ranging from the practice of astronomical observation to the more theoretical issues. Some schools are even committed to buying telescopes for their students, due to the same motivation.

We encourage the realization of practical activities, because every year we send out to every school registered some detailed proposals for low cost experiments concerning astronomy and astronautics, such as educational rocket launching, astronomical observations during daytime or nighttime, construction of solar and lunar clocks, determining the mass of the Earth, comparing the volumes of the planets and the Sun, the correct determination of the cardinal points, observation of the celestial sphere, recognition of stars of different temperatures (colors), etc.

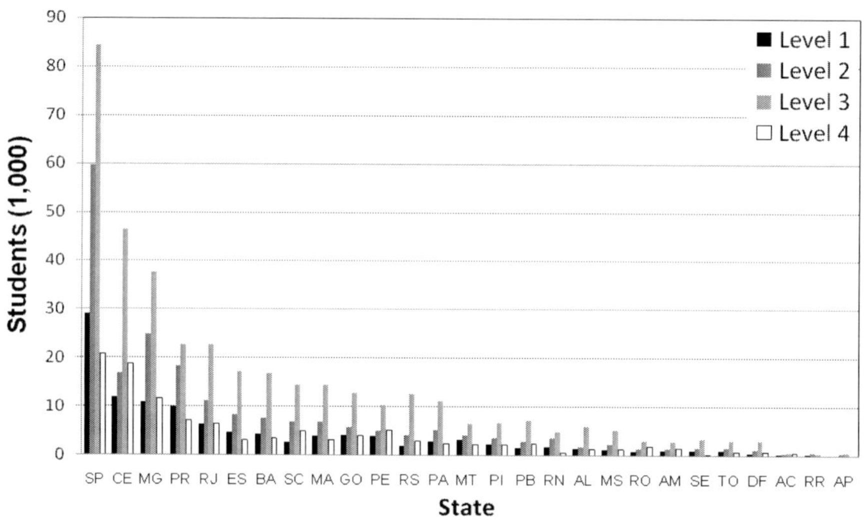

Figure 4. Distribution of students according to level and state.

Thus, there is no doubt that the Olympiad is an extremely efficient vehicle for enhancing student motivation in their studies. And as it is necessary to study physics, mathematics and even geography in order to better understand Astronomy, certainly one who intensifies their studies in astronomy will also be studying several other sciences. And as anyone who studies always improves his/her knowledge, this is an Olympiad in which each attendee is a winner.

Figure 4 shows the number of students participating in the XIV OBA separated by four levels. From Figure 4 we see that we have approximately the same percentage of students at levels 1 and 4. The level 3 has the majority of participants followed by the level 2.

1.1 Distribution of Test Scores of Levels 1, 2, 3 and 4

The Olympiad is open to the first year students of elementary school till to the last year of high school but separated in four levels, namely:

Level 1 - for students from 1st to 3rd grade of elementary school,
Level 2 - for students from fourth to fifth year of elementary education;
Level 3 - for students from sixth to ninth years of elementary school and
Level 4 - for high school students.

Figure 5 shows the frequency distribution of grades at level 1. These graphics are important in order to know whether the test will be feasible to be solved by the students.

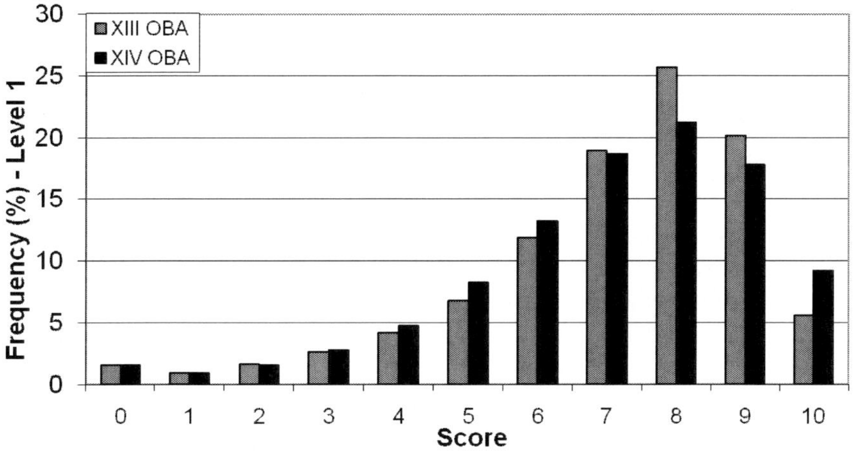

Figure 5. Frequency distribution of test scores of the level 1 of XIII and XIV OBA.

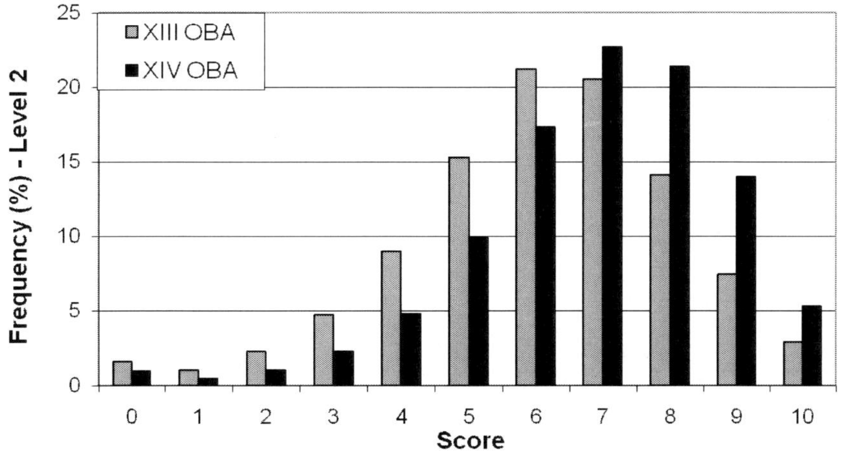

Figure 6. Frequency distribution of test scores of level 2 for XIII and XIV OBA.

Figure 5 compares the frequency distribution of grades, including the distributions of the XIII and XIV OBA to the same level 1. It can be observed that the test of XIV OBA was simpler than the earlier, because much more students got score 10 in XIV OBA than the previous one. However, the two peaks of the distributions are near the score 8. Figure 6 shows the frequency distribution of student scores at level 2 of the XIII and XIV OBA. It can be said that the test of XIV OBA has a score rate much higher than that of the XIII OBA. XIII OBA had scores 6 to 7 at the peak, whereas there was a higher percentage of scores 8, 9 and 10 in the XIV OBA. This tells us that we can ask questions a little more complex for level two than we did in 2011, but not as much as we did in 2010. Similar results were obtained for levels 3 and 4.

2.1 Distribution of Medals

The medals are highly motivating elements for the students. Therefore, there is no sense in distributing only a few tens or even hundreds. This would configure a group of highly talented young people extremely focused on the Olympiads who are in general already studying in excellent schools. They will have success independent of our actions. Therefore, we distributed 33,500 medals in XIV OBA, which represented 4.2% of rewarding students. The medal also honors, though indirectly, the teachers of these students and their school as a whole. Many of these students become news in the newspapers of their cities, which in turn highlights the works of his school.

The medals are distributed at intervals of grades. The ranges of notes do not change much from one year to another, except for the lower limit for the bronze level 4, which rose one point in relation to the XIII OBA. This is explained by the change introduced in this competition in 2011.

Events Resulting from the XIV OBA

It is undeniable that every year the OBA becomes an increasing event, not only in terms of the number of schools, students and teachers, but also due to events arising from it and involving students and their teachers. In 2011 we organized the following events: 1) VII Space Journey; 2) IV Energy Journey; 3) V Brazilian Rocket Olympiad; 4) III Rocket Journey; 5) Organization and participation in the III Latin American Olympiad of Astronomy and Astronautics and 6) Participation in the International Olympiad of Astronomy and Astrophysics.

1) VII Space Journey

We organized the VII Space Journey together with the Brazilian Space Agency, in São José dos Campos, from October 30th to November 5th, 2011, with students who had excellent answers to the questions of Astronautics in the XIV OBA. Each year we select a group of 60 high school students and their teachers, among those with the best grades in Astronautics and invite them to participate in the Space Journey. During one week, students and teachers visit the Integration and Testing Laboratory of the National Institute for Space Research and the Brazilian Aerospace Memorial. They take part in workshops, courses and lectures about the space sector. The topics were: climate change, rockets, satellites and space exploration. The Brazilian astronaut, Marcos Pontes, gave a lecture to the participants. In 2011 six of these students received a scholarship, as a reward from Fundação Educar. The scholarship consists of a notebook, an English course and vocational guidance.

2) IV Energy Journey

We performed the IV Energy Journey in the cities of Rio de Janeiro and Penedo, from the 19th to 21st of October, 2011, with students who had excellent answers for the energy questions of the tests of the XIV OBA.

Figure 7. PET bottle rocket being launched.

Each year a select group of 25 students and their 25 teachers are invited to participate in the Energy Journey. On this occasion they visit the Planetarium of Rio de Janeiro, the National Organizer System, the company that controls the generation and transmission of energy throughout Brazil and also visit the hydroelectric power plant in Itatiaia.

3) V Brazilian Olympiad of Rockets

We promote annually the realization of the Brazilian Olympiad of Rockets, with students from elementary and high schools. At the elementary school level they are asked to launch a rocket using a simple blowing of soda straw, for example.

At high school level we ask them to launch rockets using, for example, vinegar and baking soda in a PET bottle. The two levels are given some preliminary guidelines and safety rules. It is asked for attendees to discover which improvements are necessary to launch their rockets as far as possible. In 2011 we had the participation of 19,130 students distributed over 640 schools and 2,729 teachers involved. All of them received certificates of participation. In Figure 7 we show a PET bottle rocket being launched.

Figure 8. Picture of participants of the III Rocket Journey in Passa Quatro, MG.

4) III Rocket Journey

We performed the III Rocket Journey in Passa Quatro, MG, from October 31st to November 2nd, 2011. One hundred and fifty people attended, divided into 30 teams, one from each school, with five people on each team. Participants were selected because their rockets, which were constructed from PET bottles, reached the longest distances. All attendees that took part in the III Rocket Journey came from high school. During the event they showed their rockets and launched them in front of a jury, and attended lectures by experts in rocketry. The evaluation committee was composed of teachers from all teams and they had to judge the following items for each team: 1) Completion of the rocket; 2) Originality of the rocket; 3) Completion of the base; 4) Originality of the base; 5) Safety (general aspects) and 6) Team presentation. The best six teams in the jury evaluation were rewarded with trophies. The 12 teams who launched their rockets the farthest won trophies. The trophy was a replica of the Satellite Launcher Vehicle. Figure 8 shows the participants in front of the tower of the Pousada do Verde, in Passa Quatro, MG.

5) III Latin American Olympiad of Astronomy and Astronautics (III OLAA)

The I OLAA took place in Brazil in 2009. The II OLAA took place in Colombia and the III OLAA was set to be in Chile. However, problems caused by the earthquakes in Chile, made it impossible for the III OLAA to be organized there. No other country offered itself to host the III OLAA on such short notice, therefore, we decided to have the III OLAA in Brazil again. It took place in Rio de Janeiro, where occurred the opening ceremony and tests of Planetarium. After that we went to Passa Quatro, in Minas Gerais State, from October 23rd to the 30th, 2011. The advantage of Passa Quatro was the price of lodging, the unpolluted sky and the proximity to the National Laboratory for Astrophysics, which we were able to visit.

Eight countries took part in the III OLAA, namely, Argentina, Bolivia, Brazil, Chile, Colombia, Mexico, Paraguay and Uruguay. Each country came up with four or five students and one or two teachers. The event was a huge success in the opinion of all present in the event and had great media coverage, including TV and radio. We must emphasize that the focus of this Olympiad is on the integration between participants and countries rather than to stimulate competition.

Figure 9. The picture shows the participants of the III OLAA during their visit to National Laboratory for Astrophysics.

To encourage interaction among them all three tests were done in multinational teams, except for individual tests. Figure 9 shows a photo of the participants on their visit to the National Laboratory for Astrophysics in Brasópolis, MG. The III OLAA assembled 34 high school students from eight countries in Latin America, who achieved the best grades in the Olympiads of Astronomy and Astronautics in their respective countries.

6) Participation in the International Olympiad of Astronomy and Astrophysics (IOAA)

We played an active part in the founding of the IOAA in 2007 and we have been participating in the IOAA on an annual basis with a team of one or two team leaders and four or five students selected after a long process of training a big group of high school students. We coach about two hundred students for a few months after the OBA accomplishment on May. Then, on April of the following year we select two teams. One will represent Brazil in the IOAA e the other in the OLAA. Every year our teams were awarded with medals on both Olympiads.

REGIONAL MEETINGS FOR TEACHING ASTRONOMY

The Regional Meetings for Teaching Astronomy (EREA in Portuguese) was created in the International Year of Astronomy (IYA), in 2009, as a subprogram of the celebrations of IYA. The EREAs focus is to promote the training of teachers of elementary and high school presenting practical methods for teaching astronomy and also astronautics, giving to the participants educational material for teaching astronomy. We donate telescopes (Galileoscope) to each school that participates in the event. The EREAs have been coordinated by OBA. We organized three of them in 2009, eleven in 2010, twelve in 2011 and eleven in 2012. We intend to continue organizing about a dozen EREAs each year. In each event we have the participation of about 100 teachers and each event takes three or four entire days and sometimes even five days.

CONCLUSION

The OBA is a great opportunity for professional and amateur astronomers and scientists of aerospace sciences to collaborate with teachers of elementary and high schools that teach Astronomy and Astronautics. During the whole process we have been distributing didactic material, such as telescopes, planispheres, books, DVDs, as well as proposing daytime and nighttime experiments related to Astronomy. In the Astronautic we have been organizing the Brazilian Olympiad of Rockets which is a great success because students have the opportunity to put in practice their knowledge on rocket launching. The students that launch their rockets the farthest are invited to take part in a Journey of Rockets, and we award junior scholarships to the best students of this event.

We send the best students in Astronomy to represent Brazil in the Internationals Olympiads. The teachers have been receiving proper training in the EREAs.

Many more students and teachers have been visiting the Planetariums and Observatories and, most importantly: they are looking a lot more to the sky and having fun.

ACKNOWLEDGMENTS

We thank the financial support from Conselho Nacional de Desenvolvimento Científico e Tecnológico (CNPq/MCT), Brazilian Space Agency, Universidade Paulista (UNIP), State University of Rio de Janeiro, Fundação Estudar, National Institute of Studies of Space (INespaço/MCT) and the Museum of Astronomy and Related Sciences (MAST).

We also thank all Planetariums and Observatories of Brazil that have been supporting the students and teachers to take part in the OBA, as well as all groups of amateur astronomers that have been helping students and teachers to improve their knowledge of Astronomy, especially when it comes to observation.

We thank all the people that have helped us prepare the tests of the OBA and train the students to take part in the International Olympiads. We also thank the hundreds of astronomers and engineers of aerospace science who have helped us organize the Regional Meetings for Teaching Astronomy.

Special thanks to the Brazilian astronaut Marcos Pontes for kindly participating in many of our activities.

We thank the secretaries of the OBA for the efficiency and team spirit, especially in periods of high workload: Giselle Bayer do Amaral, Pamela Marjorie Correia Coelho and Marcela Pereira Barreiros.

References

[1] D.F. Lavouras and J.B.G. Canalle, *Boletim da Sociedade Astronômica Brasileira* 18 (3), 39 (1999).

[2] J.B.G. Canalle, *Física na Escola* 4 (2), 12 (2003).

[3] Also available in http://www.sbfisica.org.br/fne/Vol4/Num2/ v4n2a06.pdf.

[4] J.B.G. Canalle, R.H. Trevisan and C.J.B. Lattari, *Caderno Catarinense de Ensino de Física* 14 (3), 254 (1997). Abstract available in http://fsc.ufsc.br/ccef/port/14-3/a2.html.

In: The Space Education Phenomenon at NASA …
Editor: Norma Teresinha Oliveira Reis
ISBN: 978-1-62618-188-5

Chapter 5

SPACE TECHNOLOGY IN EDUCATION - INDIAN PERSPECTIVE

Bhushit G. Vaishnav*

Physical Research Laboratory, Ahmedabad, Gujarat, India

ABSTRACT

From the very inception, the use of space technology for education, communication and training has been a major focus of the Indian Space Programme. This chapter chronicles this aspect of the Indian space programme and discusses various innovative and successful practices followed in the Indian space programme for Education. Discussions on various pathways in which students and teachers can participate in education is also presented with a triadic model for space education. This chapter broadly discusses the instigation of the Indian Space Programme, the efforts put by India in implementing and imparting the education through the powerful medium of space technology.

PROLOGUE

Evolution of Space Programme in India

The space age in India, just over 55 years old, has witnessed some astonishing achievements. Space technology enabled communications capabilities, now indispensable parts of our modern life, are generally taken for granted. The space technology in India has been growing at ever accelerating pace, in terms of developing indigenous satellites, new spectral windows, improvised launch vehicles and commercial launching facilities and the development of a Geoportal named “Bhuvan” to augment the Indian Earth Observation Capabilities from Indian Remote Sensing satellites. Today India stands sixth in the International space techno arena.

* E-mail address: bhushit@prl.res.in.

The Indian space programme was envisaged by Dr. Vikram Sarabhai, who believed, *"There are some who question the relevance of space activities in developing nations. To us, there is no ambiguity of purpose... But we are convinced that if we are to play a meaningful role nationally, and in the comity of nations, we must be second to none in the application of advanced technologies to the real problems of man and society*". Prof. Sarabhai's vision for the space programme was a one page of vision statement in 1970 which stated "*India with her mighty scientific knowledge and power house of young, should build her own huge rocket systems (satellite launch vehicles) and also build her own communication, remote sensing and meteorological spacecrafts and launch from her own soil to enrich the Indian life in satellite communication, remote sensing and meteorology. Space programme will transform our developing society into a developed society*". Looking back after 50 years, we are overwhelmed to see the results of this statement. Today India builds numerous types of satellite launch vehicles, satellite payloads, and spacecrafts and launches them from Indian soil. Societal aspects such as communication, resource mapping, disaster prediction, disaster management and entertainment have substantive inputs of space technology.

In early 1960s', Prof. Vikram Sarabhai located a place technically most suited for space research. This was Thumba, near the magnetic equator and was ideally suited for ionospheric and electro-jet research in upper atmosphere. The major challenge then was to get this specific area for space activities. Initial reaction for the designated land was negative as this will affect the lives of fishing folks who lived there and affect an ancient St. Mary Magdalene Church, Bishop's House and a school. However "Rev. Father Peter Bernard Pereira" the Bishop of the region when approached by Prof. Vikram Sarabhai, met with the statement, *"Oh Vikram, you are asking my children's abode, my abode and God's abode. How is it possible?"* However he invited Sarabhai to a prayer meeting and introduced him as, "*Dear children, here is a scientist, Prof. Vikram Sarabhai. What do sciences do? All of us experience, including this church, the light from electricity. I am able to talk to you through the mike which is made possible by technology. The treatment to patients by doctors comes from medical sciences. Science through technology enhances the comfort and quality of human life. What do I do, as a preacher? I pray for you, for your well being, for your peace. In short, what Vikram is doing and what I am doing are the same - both science and spirituality seek the Almighty's blessings for human prosperity in body and mind. Dear Children, Prof. Vikram Sarabhai says, he would build within a year, near the sea-coast, alternative facilities to what we are having. Now dear children, can we give your abode, can we give my abode, can we give the God's abode for a great scientific mission?*" There was a total silence, a pin drop silence. Then all of them got up and said 'Amen' which made the whole church reverberate.

This church was the first design centre for space technology where India started rocket assembly. The Bishop's house was the scientists' working place. Later the Thumba Equatorial Rocket Launching Station (TERLS) led to the establishment of Vikram Sarabhai Space Centre (VSSC) and, the space activities grew into multiple space centres throughout the country. Dr. Kalam summarizes thus, *"When I think of this event, I can see how enlightened spiritual and scientific leaders can converge towards giving reverence to the human life.*" Of course the birth of TERLS and then VSSC gave India the capability for launch vehicles, spacecraft and space applications that have accelerated social and economic development in India to unprecedented levels. (Incident quoted from: www.abdulkalam.com)

INDIAN SPACE EXPLORATION

Indian space programme initiated space exploration with its first unmanned lunar probe 'Chandrayaan-I' in 2008, which carried 11 payloads including Indian payloads and payloads from other countries. The mission was a model of International collaboration with many new results including the discovery of water. New mission to Mars and Chandrayaan- II are being planned now.

India has played an active role in international bodies such as, the United Nations Committee on the Peaceful Uses of Outer Space (UN-COPUOS), United Nations Economic and Social Commission for Asia and the Pacific (UN-ESCAP), the international COSPAS-SARSAT system for search and rescue operations, the International Astronautical Federation (IAF), the Committee on Earth Observation Satellites (CEOS), the Committee on Space Research (COSPAR), the Inter Agency Debris Coordination Committee (IADC), the Space Frequency Coordination Group (SFCG), the International Lunar Exploration Working Group (ILEWG), the Coordinating Group on Meteorological Satellites (CGMS), the International Global Observing Strategy (IGOS), the International Space University (ISU), the Asian Association for Remote Sensing (AARS), the International Society for Photogrammetry and Remote Sensing (ISPRS), etc. As a partner in the International Charter for Space and Major Disasters along with CNES, ESA, the Canadian Space Agency (CSA) and NOAA, Indian Space Research Organization (ISRO), is actively partnering in programmes that will provide satellite data for the management of natural disasters.

POTENTIAL USES OF SPACE TECHNOLOGY

The major objectives of the Indian Space Programme are to establish self-reliant operational space services in the areas of satellite communications, satellite-based information for management of natural resources and satellite meteorological applications. India has two operational space systems – (1) the Indian National Satellite (INSAT) System; providing services for telecommunication, telemedicine, TV and meteorology including disaster warning support, and (2) the Indian Remote Sensing Satellite (IRS) Systems for natural resources monitoring and management.

In addition, ISRO initiated development of "thematic satellites" for communication and remote sensing applications in education, health, map-making and ocean research. Several of such thematic satellites are as follows:

- EDUSAT for educational services (India's first thematic satellite),
- CARTOSAT for Earth observation in a sun-synchronous orbit,
- RESOURCESAT for the remote sensing data services,

- OCEANSAT to study surface winds and ocean surface temperature, observation of chlorophyll concentrations, monitoring of phytoplankton blooms, study of atmospheric aerosols and suspended sediments in the water and
- RISAT (Radar Imaging Satellite) for cloud penetration and day-night imaging capability.

The objectives of these thematic satellites were to use the space technology for a specific application and use this for socio-economic and cultural developments in India.

To harness the benefits of space technology, the United Nations General Assembly suggested the establishment of Centres for Space Science and Technology Education at the regional level in the developing countries and one such centre for training is at Ahmedabad, India. This has trained over 100 scientists from the south Asian region such as in the use of space technology.

SPACE BASED TECHNOLOGIES IN EDUCATION: AN INDIAN PERSPECTIVE

India has taken a pragmatic and grass-root approach to utilization of its space infrastructure for the societal needs. In 1960s' NASA field tested the impact of direct broadcast of television programmes via satellite, as a part of its programme on Applications of Satellite Technology. They shortlisted India, Brazil and the People's Republic of China as potential sites for the same. These countries were chosen on the basis of their size, proximity to the equator for a direct-broadcast. Of the three, political consideration ruled out China and inhomogeneous population distribution ruled out Brazil. As a consequence, India emerged as the only suitable candidate for such an experiment.

Simultaneously India, initiating its own national space programme, was interested in the role of satellites for the purpose of communication and asked UNESCO to undertake a feasibility study for this purpose. During November-December, 1967 UNESCO sent an expert mission to India which concluded that such a project would be feasible and that India could meet the technical requirements for the project that resulted in establishment of the National Satellite Communications Group (SATCOM) in 1968 to explore uses of a synchronous communications satellite for India.

This group consisted of representatives from various ministries, ISRO and All India Radio (AIR) and Doordarshan and recommended ATS-6, a second generation satellite from NASA for educational television applications.

In the Memorandum of Understanding signed between India and US, stated the general objectives as to:

- gain experience in the development, testing and management of a satellite-based instructional television system in rural areas and to determine optimal system parameters;
- demonstrate the potential value of satellite technology in the rapid development of effective mass communications in developing countries;

- demonstrate the potential value of satellite broadcast TV in the practical instruction of village inhabitants; and
- Stimulate national development in India, with important managerial, economic, technological and social implications.

Figure 1. Dr. Vikram A. Sarabhai, (left) Chairman of the ISRO and head of India's Department of Atomic Energy and Dr. Thomas O. Paine, NASA Administrator, sign an agreement to cooperate in an unprecedented experiment using a space satellite to bring instructional television programmes to some 5,000 Indian villages. The Applications Technology Satellite 6 (ATS 6) satellite launched May 30, 1974, became available to the ISRO for one year on August 1, 1975. ATS 6 acted as a broadcasting station in space, sending health and educational television programmes to inexpensive ground stations in remote villages of India. (Image Source: http://grin.hq.nasa.gov /ABSTRACTS/GPN-2002-000081.html).

The social objectives for India were to educate the populace about issues such as family planning, agricultural practices and national integration. Additional objectives were to impart general school and adult education, train teachers, and improve other occupational skills and general health and hygiene. Besides these, India also wanted to gain experience in the technology of broadcasting and reception facilities. In this Memorandum of Understanding the primary US objective was to test the design and functioning of an efficient, medium-power, wide band space-borne FM transmitter, operating in the 800–900 MHz band and gain experience on the utilisation of this space application. Arnold Frutkin, and Prof. Sarabhai signed an agreement regarding *Satellite Instructional Television Experiment* (SITE) in 1969 (Figure 1). The experiment was launched on 1 August 1975.

About SITE

The project enabled informational television programmes to be seen by rural India. The main objectives of the experiment were to educate the people of India on various societal issues. The ground hardware consisted of Direct Reception Systems (DRS) for community viewing of the TV programmes. They were installed in six States of the country in "clusters" of about 400 each (i.e. total 2400 DRSs). The instructional programmes (prepared by ISRO) were broadcast for 4 hours daily and covered programmes on science education, teachers training programme, programmes on good agricultural practices, and information on climate. These programmes trained over 50000 teachers in two 2-week sessions.

The experiment ran for one year from August 1975 to July 1976, covered over 2500 villages and was supported by United Nations Development Programme (UNDP), United Nations Educational, Scientific and Cultural Organization (UNESCO), United Nations Children's Fund (UNICEF) and International Telecommunication Union (ITU). The experiment was successful, as it played a major role in helping develop India's own satellite programme, INSAT.

Impact of SITE

The SITE programme ended in July, 1976 and NASA shifted its satellite away from India. The SITE programmes transformed Indian villages. Studies on the social impact of the experiment and on viewership trends indicated a viewerhip of 200 to 600 people per TV set reducing later to 60 to 80 people per TV set. This decline was due to faults in the television and hardware, interruptions in electricity supply, and pre-occupation with agricultural work. The most discernible impact of the SITE programme was in the fields of agriculture and family planning and nearly 52% of viewers applied the new knowledge gained via SITE.

Before SITE, the focus was on the use of terrestrial transmission of television signals. But SITE showed that India could use, Satellite Communication System for this purpose. The Indian National Satellite System (INSAT) was launched in 1982 and in September 2004, India launched its Tele Education satellite EDUSAT, the first satellite anywhere to serve the educational sector exclusively. EDUSAT is now used for an interactive satellite-based distance education. It is providing networks capable of transmission of live lectures, web based learning, interactive training, virtual laboratory, database access for reference material/library and has nearly transformed education in India.

Satellite Communication (Satcom) technology

Satcom technology offers the unique capability of simultaneously reaching out to very large numbers spread over large distances even in the most remote corners of the country. The hallmark of Indian Space Programme has been the application oriented efforts and the benefits that have accrued to the country. In the past two and a half decades Indian National Satellite (INSAT) system has revolutionized the country's telecommunications, TV broadcasting, DTH services, business communications, rural area connectivity, Tele-education, Tele-medicine, Village Resource Centres, Search and Rescue operations and Emergency Communications.

INSAT System

After SITE, the Indian national satellite (INSAT) has been the major catalyst in the rapid expansion of terrestrial television coverage in India. INSAT system is a joint venture of the Department of Space, Department of Telecommunications, India Meteorological Department, All India Radio and Doordarshan. Established in 1983, INSAT system is one of the largest domestic communication satellite systems in the Asia Pacific Region. INSAT is used for a variety of societal applications in the area of Tele-education, Tele-medicine and support to the Disaster Management System (DMS). Village Resource Centre (VRC), a single window agency providing the services offered by INSAT and IRS satellites to provide information on natural resources, land and water resources management, Tele-medicine, Tele-education, adult education, vocational training, health-care and family welfare programmes has been established.

Meteorological data from INSAT is used for weather forecasting and specially designed disaster warning receivers have been installed in vulnerable coastal areas for direct transmission of warnings against impending disaster like cyclones and tsunamis.

Educational TV Services

INSAT is being used to provide Educational TV (ETV) service for primary school children in Tamil, Marathi, Oriya, Telugu and Hindi. A general enrichment programme on higher education (college sector) is telecast on the national network. These programmes, provided by the University Grants Commission (UGC), are a part of its countrywide classroom programme. The Indira Gandhi National Open University (IGNOU) broadcasts half an hour curriculum based lectures daily via the national network for the students.

EDUSAT

To help meet this challenge of education and technology upgradation on a continuing basis, ISRO initiated 'Tele-Education' by launching EDUSAT that has a C-band national beam, a Ku-band national beam, and five Ku-band regional beams for education in regional languages.

EDUSAT has strengthened education efforts by augmenting curriculum based teaching, providing effective teachers' training, and community participation. Networks based on EDUSAT consist of either receive only (one way communication) terminals or interactive (two way communication) terminals in both the national and regional networks. Besides standard educational functions these networks enable, on line examination and admissions processes, distribution of administrative information. The Network is IP based and does not need expensive studio facility. Equipment required at the teaching end or hub comprises two cameras, two PCs and DVD player (if needed) in addition to the indoor and outdoor units of the hub hardware. The equipment needed at the interactive classroom end consists of webcam, PC, LCD projector, speakers, microphone and the satellite terminal. The classroom consisting of receive-only terminal requires a PC, projector and speakers.

Special Networks

ISRO has also set up networks for users' with special requirements like:

- Blind People's Association of Ahmedabad, Gujarat – for Visually challenged
- Rehabilitation Council of India
- Network of Central Institute of Mentally Retarded in Kerala
- A network of Centre for the Development of Advanced Computing (C-DAC) for mentally challenged in Kerala

One of the innovative networks on EDUSAT is the network for "Blind schools". Blind People's Association, Ahmedabad is a leading organisation promoting education, training, employment and rehabilitation for blind persons. Considering the specific needs of the blind people, an altogether different kind of broadcast network configuration delivering live audio and data which is read by blind persons through its printed impression (Braille) was set up.

GRAMSAT Programme

The Gramsat Programme (GP) is an initiative to provide communication networks at the state level connecting the state capital to districts and blocks. The networks provide Computer Connectivity, Data Broadcasting and TV Broadcasting facilities having applications like e-Governance, National Resource Information System (NRIS), Development Information, Tele-conferencing, Disaster Management, Tele-medicine and Distance Education.

Telemedicine

Telemedicine Programme is an innovative process of synergising benefits of Satellite communication technology and information technology with Biomedical Engineering and Medical Sciences to deliver the health care services to the remote, distant and underserved regions of the country. Providing healthcare to India's over one billion population of which about 75 percent live in villages, is a formidable task. About 75 percent of the doctors practice in urban areas and 23 percent in semi-urban areas. This leaves just 2 percent of the qualified doctors, who are attached to about 23,000 primary health and 3000 community health centers, to attend to 70 percent of the population living in villages.

ISRO's telemedicine pilot project was started in the year 2001 with the aim of introducing the telemedicine facility to the grass root level population as a part of proof of concept technology demonstration. The telemedicine facility connects the remote District Hospitals/Health Centres with Super Specialty Hospitals in cities, through the INSAT Satellites for providing expert consultation to the needy and underserved population.

INDIA'S INITIATIVES FOR SPACE EDUCATION AND CAREER OPPORTUNITIES IN SPACE TECHNOLOGY

Besides being an astronaut, a career in space also includes working as space scientist, technologist, biologist, aerospace medicine, robotics etc. Another promising option is a career in space laws, which like elsewhere is slowly developing in India.

India entered the space club with the launch of its first indigenously-built satellite *Aryabhatta* in 1975. Active research in space science is being pursued at several universities, research institutions and organisations such as the ISRO Satellite Centre, Bangalore; Physical Research Laboratory, Space Applications Centre, and UN center for Space Science and Technology Education in Asia and the Pacific, Ahmedabad; Space Physics Laboratory, Thiruvananthapuram; National Mesosphere-Stratosphere-Troposphere Radar Facility, Gadanki, Andhra Pradesh.

The present focus of the space programme is on promoting spinoffs in the discipline of human resource development, space science and technology education, industry–academia interface and international cooperation. Considering the need to provide an impetus to studies and research in the critical area of space science, space technology and also in atmospheric research and with a view to create appropriate man power base for future Indian programmes, the Indian Institute of Space Technology at the Thiruvananthapuram has been established with a view to bring in bright students at 12th standard and then impart them formal training in several aspects of engineering to sciences. This is the only institute of its kind in India, which imparts undergraduate and postgraduate education in Space Science and Technology, research programmes in Space Science and Technology and integrated Masters in Space Science, and subjects suitable to the requirements of Indian Space Programme.

ISRO's partnership with Indian industries is high with a majority of the production, fabrication, testing and software development activities; routine production jobs needing capital intensive infrastructure are outsourced. Only specialized jobs and jobs of strategic and security significance are done in house by ISROs production centres.

Several Indian Universities and other Institutions offer graduate / postgraduate / doctorate courses in the disciplines of relevance to space science / space technology and space applications.

Centre for Space Science and Technology Education in Asia and Pacific (CSSTEAP): UN programme for Asia Pacific

The Centre for Space Science and Technology Education for Asia and the Pacific (CSSTE-AP) was set up in India under the initiative of the UN Office for Outer Space Affairs (UN OOSA) and offers a ten month post graduate diploma course in Remote Sensing and Geographical Information Systems (every year starting in October), Satellite Communication (every alternate year starting in July), Satellite Meteorology and Global Climate (every alternate year starting in July) and Space and Atmospheric Studies (every alternate year starting in July). Following the course, candidates have the opportunity to carry out research in their own country for one year leading to the award of a Master's Degree. Several international agencies apart from the Government of India support the candidates to participate in the CSSTE-AP courses.

The Indian government provides facilities and faculty to the Centre through the Indian Institute of Remote Sensing (IIRS), Dehradun, Space Applications Centre (SAC), Ahmedabad and Physical Research Laboratory (PRL), Ahmedabad. A set of standard curricula developed by the United Nations is adapted for the educational programmes. These courses are open to young scientists of countries in Asia –Pacific Region. These include,

- Post Graduate programme in Remote Sensing and Geographical Information System (First phase of nine months at the centre followed by second phase of one year in the home country) offered by Indian Institute of Remote Sensing, Dehradun.
- Post Graduate programme in Satellite Communication (First phase of nine months at the centre followed by second phase of one year in the home country) offered by Space Applications Centre, Ahmedabad.
- Post Graduate programme in Satellite Meteorology and Global Climate (First phase of nine months at the centre followed by second phase of one year in the home country) offered by Space Applications Centre, Ahmedabad.
- Post Graduate programme in Space and Atmospheric Sciences (First phase of nine months at the centre followed by second phase of one year in the home country) offered by Physical Research Laboratory, Ahmedabad.

In all the above programmes, completion of the first phase leads to the award of a post-graduate diploma or certificate from the centre. The completion of both phases leads to the award of Master's degree from Andhra University, India.

Sponsored Research Programme (RESPOND)

The Indian Space Research Organisation (ISRO) supports research and development activities in Space-Science, Technology and Applications. This programme of Research Sponsored by ISRO is called RESPOND and works on support after peer reviews of proposals. The aim of RESPOND is to encourage quality research in areas of relevance to the Indian space programme. The primary objectives of the RESPOND programme are;

- to harness the advanced research areas of space science and technology for national development and to derive the maximum benefit for the people of India. The Indian Space Programme includes the following major elements:
- Demonstration of the feasibility of deriving benefits from space science and technology, applications in space communications, long distance education, earth resources survey, meteorology and geodesy.
- Development of indigenous capability for design and development of orbiting satellites for scientific research and space applications, sounding rockets and satellite launch vehicles.

The RESPOND Programme established ISROs links with academic Institutions to enhance quality Research and developmental projects of relevance to Space and derives useful outputs of such R and D to space based activities. RESPOND programme will enhance academic base, generate quality human resources and infrastructure at the academic Institutions to support the space programme.

The RESPOND Programme Supports:

Space Sciences

Physics of the ionosphere and magnetosphere, meteorology, dynamics of the atmosphere, geophysics, geology, astronomy, cosmology, astrophysics, planetary and interplanetary space physics and climatology.

Space Technology

Rocket and satellite technology, propulsion systems design and optimization, aerodynamics and heat transfer problems related to space vehicles, guidance and control systems for launch vehicles and spacecraft, polymer chemistry, propellant technology, ultra-light-weight structure, satellite energy systems, space electronics, Space communication systems, orbital mechanics, computer sciences and new material development.

Space Applications

Remote sensing of the earth's resources, space communication, satellite geodesy, image processing, satellite meteorology including weather forecasting, Space Education and Ecology.

Planetary Science and Exploration Programme (PLANEX)

Planetary research in India had a limited base except for a very strong group nucleated at the Tata institute of Fundamental Research, Mumbai which later on moved to PRL, Ahmedabad. ISRO initiated the Planetary Exploration Programme in 2001 to form viable groups of scientists in the country for research in planetary sciences. The possibility of embarking on a planetary exploration programme emerged with the successful performance of Polar Satellite Launch Vehicle (PSLV) in 1995 but it took about 5 years for ISRO to make this paradigm shift from Earth centric studies to a meaningful planetary programme. The success of Chandrayaan-I gave a boost to this initiative and currently exploration of Mars, satellites of Jupiter, comets and asteroids are under consideration. The Planetary Sciences and Exploration Programme, called PLANEX was conceived at Physical Research Laboratory, Ahmedabad and was supported by ISRO. It had many components including pooling of resources. Firstly, it was realized that no national academic programme can succeed without participation of universities. Therefore periodic workshops with different themes of planetary sciences were planned to enthuse and reorient the university students. To attract the university faculty to nucleate planetary research in their centres, a scheme of supporting scientific proposals financially, collaborating and interacting with them academically was also initiated.

Much interest was shown by many universities in this programme. Realizing that experimental work requires state of the art equipments which are expensive and difficult to maintain or operate by small groups, another vital element was added to this programme of building and operating a National facility where all possible equipments required for planetary studies are available which different universities and academic institutions can freely use. To facilitate frequent visits and joint work, a Visiting Professors programme was included. Thus the vision guiding the formation of PLANEX programme was to grow a viable planetary science community in the country which can take up the challenges of the ambitious planetary exploration programme of ISRO. Such a multi-pronged effort will strengthen the PLANEX Programme and make important scientific contributions in understanding the formation and evolution of planets and their interaction and influence on earth and life on earth, which are the main science objectives. No country has embarked on planetary exploration without a large number of scientists devoting their time and many institutions taking it up as their main activity. It is therefore imperative that PLANEX Programme quickly grows into a large centre as envisaged, if ISRO has to successfully conduct its Planetary Exploration Programme. With such a wide science base and combining it with high-end technology being developed by ISRO, the future of Planetary Exploration Programme looks bright.

CASE STUDIES

In India, there are a large number of people, who cannot undergo 'defined academic curriculum' for enhancing their education in space sector, due to various reasons. Also, one more obstacle is multiple languages. The highly technical language - which is being used at the institutional level, cannot be very effective at local / regional level, where there are more enthusiastic intellectuals. For enhancing capacity building, various projects are required to be placed at regional level in terms of seminars, workshops, short courses etc. Such objectives can be achieved by some small groups of highly motivated people. "Aryavarta Space Organization (ASO)", registered as Non Government Organization (NGO), is one such group with permitted associations to carry out research, development and innovation in aerospace technologies through effective transfer of knowledge and technology for the benefit of the people. ASO works in coordination with different domestic and international academic and industrial collaborations. The major operations carried out at ASO are research in Science and Technology, Management and Policy Issues, Aeromedicine and Life Science, along with project management, business support, consultancy in aerospace industry and organization challenges and competitions and promotion of the advancement and dissemination of knowledge related to Aerospace Field.

Very recently (in January 2011), ASO signed Memorandum of Understanding (MoU) with Central American Association of Space and Aeronautics (ACAE). It is the first private association in India in field of space technology and is expected to go a long way in capacity building in all areas of science, industry and technology, especially those related with space, space rights, astronomy and aeronautics. The MoU talks about mitigation, planning of natural disasters, emergency attendance and management of natural resources using space applications.

SRMSAT, JUGNU and STUDSAT: Institute-Academia Collaboration

A new saga in the Indian Space Programme was written with the launch of India's Polar Satellite Launch Vehicle (PSLV-C18). It put four satellites in orbit which included two nano satellites, first is SRMSat, built by students of SRM University, near Chennai and the second is Jugnu, put together by Indian Institute of Technology, Kanpur students.

The students and academicians of innumerable educational institutions of the country have partnered the research efforts in participating in many research projects involving extensive theoretical and experimental efforts. StudSat, first of its kind, Indian satellite of Pico-Category was developed by a group of 35 students from Andhra Pradesh and Karnataka, India, and was launched on 12 July 2010 by Polar Satellite Launch Vehicle. StudSat was the centre of attraction as the students had designed and built this pico satellite with an imaging camera and had employed several frontline technologies. They had also built a clean room to test the satellite and a ground station in Bangalore to receive its signals. The mission was experimental in nature and the major objective is for the students to have hands on experience on the design, fabrication and realisation of a space mission at a minimum cost. This is an exemplary incidence of the increasing awareness of space education in India.

Conclusion and Future Road Map for Space Technology Education in India

The vision for the Indian Space Programme, from the very inception has been to be 'second to none' in the development and applications of space technology for solution of the real societal problems. India is among the first few countries to realize the importance of space technology to solve the real problems of society and took initiatives to develop the space technology for the benefit of the nation. The emphasis on Self-reliance has been an important component of the vision, with which India undertook development of satellites, launch vehicles and associated ground segment indigenously in a progressive manner. Today, India's core competence in space is its ability to conceive, design, build and operate complex space systems and use them in various frontiers of national development. The space technology institutions/centres created in the country and the overall system/project management practices evolved in the course of implementation have been vital outcomes of the space programme.

Space industry is inherently technology intensive and is dependent on long term R and D for its future viability. Space has vast potential for advancement in technology and applications and hence the Human Resource development. This requires strengthening the Space based Industry-Academia interface and forging International collaborations. Since its inception, ISRO has been very careful in spotting and recruiting the highly skilled manpower and the best of the minds to take up its challenging tasks in hi-tech area. Their innate talent and technical competence together with the challenging complexities of space programme and organizational culture has been bringing out their best in demystifying SPACE.

The organization has achieved very high degree of success in the space programme during the last three decades and this could be attributed to the dedicated efforts of scientific and technical staff who enjoyed the excellent support from administrative and auxiliary staff.

At present, the ISRO is supported by nearly 16000 staff working in various projects in different parts of the country, which involves substantial research and developmental efforts. The organization is also well supported by Indian industries in realization of very critical space worthy hardware and software.

A Tripartite Model for Space Education

For the immediate future for India in the domain of Scientific and technological advancement, academia has a very major role to play in carrying out cutting age research. The time has come when the academia gets involved more strongly to the key technological sectors like Space, defence, health, energy, urban development, robotics and instrumentation. Also, the triad of Public-Academia-Industry needs further reinforcement for the benefit of the nation. In Figure 2, a triadic model for space education is suggested and discussed on the aspects as mentioned. Any education system and Indian education system in particular needs more efforts in implementing the suggestions made after observing the prevailing scenario. There are three major components of this model i.e. training, networking and mentoring, which are discussed separately.

Training

This component is comprised of the combination of formal and informal organisations, non-governmental organizations, information flow, financial incentives and infrastructures. It is within the educational and research institutions that the training of future experts will be done. In order for the training to have the desired impact, each of these elements must be properly structured. Under this component, seminars, workshops and conferences should be organized to teach the students and the small-scale farmers the basic concepts of space technology.

The institutions should invite international experts from international organizations such as National Aeronautics and Space Administration (NASA), European Space Agency (ESA), National Centre for Space Studies (CNES), Japan Aerospace Exploration Agency (JAXA), Canadian Space Agency (CSA) and other non-governmental organizations to provide training in the methods and tools for assessing adaptation and constructing scenarios of space education and climate and associated socioeconomic conditions as they relate to various applications ranging from communication to farming activities to defence and many more.

The best time to administer such training is at the second or third year for the undergraduate students, after the completion of coursework for postgraduate students and all the year round for the high school students. The success of this training should be assessed by the level of understanding of students in the basic concepts of space technology, the quality of potential projects that students are able to propose, including the scientific design of their proposed dissertations/projects. This training should also provide opportunities for students to learn from each other.

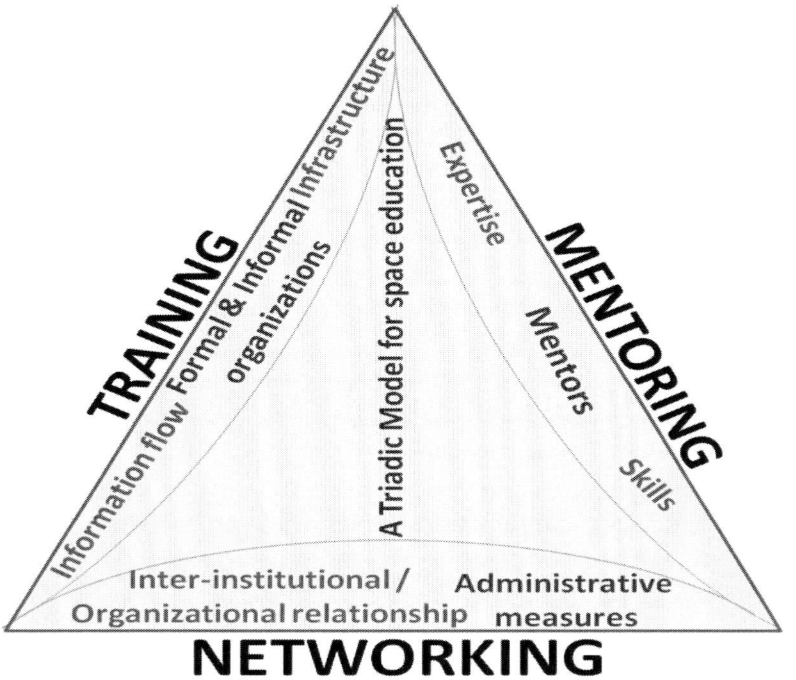

Figure 2. A triadic model of capacity building.

It is important to also include other important stakeholders (such as governments, agricultural input companies, credit societies, farmers' organisations, developmental organisations and other international partners) in this process. The stakeholders are very important because they provide both knowledge and hardware supports. The PLANEX and CSSTEAP programmes as discussed in earlier sections are the perfect initiatives in this direction.

Mentoring

Another approach in strengthening capacity building in space education is mentoring. This component involves the experts in this field who could identify problems and provide solutions. The point of focus for these experts is to stimulate interest of students in the study of space technology. This can be achieved by inviting or linking students with reputable scientists or policy makers whose core responsibilities are in the area of space communication and space technology. These mentors could also help the students plan and initiate their dissertations/theses. The universities, research institutes, international donors, developmental agencies and non-governmental organisations, should also make available a database of notable scientists in the field of space science and technology to the students, which they can choose from.

Mentors should also teach the students about publishing, especially in peer reviewed journals. Mentors should be able to inform students of both local and international conferences and workshops which can enhance students' capacities.

Networking

Capacity development is time dependent. Some societies may possess skills to solve a particular problem, while those skills may be inadequate to overcome emerging and future challenges. As a result of this, there is the need to create a conducive policy environment where students may interact with the experts. This environment provides conditions to enable economic, political, socio-cultural, general infrastructure, inter-institutional/organizational configuration, adequate policies, laws and administrative systems.

In order for students to have interactions with the experts, it is important to organize regional workshops. These regional workshops could also bring together the project mentors, researchers, university teachers and stakeholders from developing and developed countries to discuss current issues, techniques and methodology on space science and technology. University teachers and students may present experiences/papers from their research activities, while mentors advise students and teachers to collaborate across study teams to solve common problems and plan for joint publications with the students.

For sustainability of networking, there should be dedicated online discussion groups where further interaction can take place. Students and teachers could also link up with the experts or mentors via communication gadgets provided, and presently EDUSAT is providing the same with the active support from National Informatics Centre. The dedicated online discussion group would bring together and strengthen regional networks of scientists and students who can collaborate and develop effective teams in investigations futuristic requirements from space technology. At the same time, there should always be a focal point that links up with the students even after they had completed their projects for the purpose of follow up.

The students should also be allowed to make presentations about their personal experiences in international fora. In the present assessment of the activities vis-à-vis the existing stock of human resources, it is essential to strengthen the development teams with addition of nearly 700-800 scientists/engineers according to the survey published in the 11th plan report of Planning Commission of India. Concerted efforts will be required to direct towards transforming the organization towards more effective one with best composition of stabilized staff, well in tune with the long term vision, immediate mission requirements and also in synchronization with the stated policies of the government of India. Thrust on continuous improvement in human resources is the prime focus of the India's space programme.

Continuous training for middle and senior level management personnel on management, leadership and HR skills apart from professional updates along with re-education programme through sponsorship in reputed management and technical institutes should be the main thrust area. Specialised training on new vistas of technologies and science should be provided to the young engineers/scientists so as to keep them updated in their areas for a better performance in their assignments. Similar training programmes need to be planned for administrative and auxiliary services for providing necessary impetus for making administrative and auxiliary services more dynamic and effective in the era of e-governance.

Future of Indian Space Programme

Indian space programme now finds itself on the crossroads. The Indian space programme designs, builds, and launches its own satellites onboard home-grown rockets. Rather than the national prestige, the Indian Space Programme focuses on entirely pragmatic applications like, communications satellites which provides services to far-flung regions of a vast country, meteorology packages (often carried on the same geosynchronous satellites that perform communications missions), and remote-sensing satellites to map India's natural resources.

In time to come ISRO is moving beyond with space missions that focuses on immediate practical space applications. The space programme is all set to move on to the next phase, the natural next step is the planetary exploration. India's future space ambitions include robotic missions to Moon and Mars, human spaceflight, and an atmospheric re-entry capsule that could eventually be used to ferry people into space and perhaps even for lunar sample return missions.

Space Vision India 2025

- Satellite based communication and navigation systems for rural connectivity, security needs and mobile services
- Enhanced imaging capability for natural resource management, weather and climate change studies
- Space science missions for better understanding of solar system and universe
- Planetary Exploration
- Development of Heavy lift launcher
- Reusable Launch Vehicles - Technology demonstrator missions leading to Two Stage To Orbit Human Space Flight

"There is plenty of room in the space...sky is the limit...Exploration through human resource development is the carrier to achieve it...."!!

Acknowledgment

The author is extremely grateful to Prof. N. Bhandari for his insightful comments and suggestions during the preparation of this chapter. I am indebted to Prof. A. K. Singhvi for his astute editing of this chapter.

References

Dr. A.P.J Abdul Kalam, Address and interaction with Students at Judge Business School, University of Cambridge, UK, London, Jun 9 2009, http://www.abdulkalam.com/kalam/jsp/print_content1.jsp

Satellite Instructional Television Experiment. http://en.wikipedia.org/wiki/Satellite_Instructional_Television_Experiment#cite_note-0

Aryavarta Space Organization. http://aryavartaspaceorg.blogspot.in/

PRIMER: Indian Space Programme, compiled by Susmita Mohanty. http://www.earth2orbit.com/publications/publications.html

Indian Space Research Organization, India. http://www.isro.gov.in

Eleventh Five Year Plan (2007–2012) Inclusive Growth Volume I, Published by Planning Commission, Government of India, Oxford University Press (2008).

Paterson, Chris. "Satellite". Museum of Broadcast Communications. Retrieved 2006-09-13.

PLANEX Newsletter, Volume -1, Issue-1.

In: The Space Education Phenomenon at NASA …
Editor: Norma Teresinha Oliveira Reis

ISBN: 978-1-62618-188-5

Chapter 6

NASA GSFC OPPORTUNITIES FOR STEM PROFESSIONALS USING THE VANTAGE POINT OF SPACE

Blanche W. Meeson[1*] and Geraldine B. Robbins[2]
[1]NASA Goddard Space Flight Center, Greenbelt, Maryland, US
[2]Triangle Coalition for Science and Technology Education, NASA Goddard Space Flight Center, Greenbelt, Maryland, US

ABSTRACT

NASA has a variety of learning opportunities for STEM professionals. Three opportunities at GSFC are examined in this chapter: 1) standard summer research and development internship for undergraduates, 2) senior internship for undergraduate and graduate students and 3) a workshop series for informal learning center professionals. We describe these programs, examine their evolution with respect to most effective education practices and their assessment and evaluation, and identify the similarities and differences between them. The internship programs highlight authentic project-based research and development experiences with the senior internship providing a richer, deeper, and more demanding experience that has greater professional value. The workshops for informal learning center professionals on-the-other hand, focus on building knowledge of GSFC's science and engineering strengths among these professionals, and on building enduring partnerships between individuals (participants and GSFC scientists, engineers and educators) and between organizations (GSFC and the informal learning centers). Finally, we examine the characteristics of these programs from a design and management perspective. Through this examination we identify a general structure that provides insight into more effective design and management of similar education programs.

* Blanche.w.meeson@nasa.gov; gbrobbins@hotmail.com

INTRODUCTION

Learning opportunities have been a key part of NASA's educational activities since the beginning of the agency. When NASA was established in 1958, its charter directed the agency to *provide for the widest practicable and appropriate dissemination of information concerning its activities and the result thereof* [1]. Today, this charge includes a full range of education and public engagement activities for adults and children.

Initially much of NASA education focused on faculty and graduate student participation in NASA's research and engineering endeavors via fellowships and internships. That focus continues today with expanded participation of undergraduates in authentic engineering and scientific research and development (RandD) projects. Additionally, in recent years NASA has recognized the importance of informal education both to capture the interests and imagination of children and adults when linked to formal education and vice versa [2]. With that recognition have come strategies to expand the use of NASA's knowledge products in informal settings and to provide coordination and coherency between the formal and informal offerings for all audiences, from elementary to college students, parents and the public.

As the science and technology (SandT) workforce for the entire country approaches retirement age, and the demand for SandT professionals grows, the issue of developing individuals to meet current and pending workforce needs increases in importance [3, 4, 5].

When coupled with demographic shifts (the U.S. is fast becoming majority minority) and the low participation of minority populations in SandT careers, these concerns heighten [6, 7, 8]. The ability of the U.S. to retain its competitive edge in this vital economic arena is dependent upon its ability to produce a highly talented and innovative SandT workforce in sufficient quantity to fulfill the demand [9, 10].

To do this, we must apply all of our resources in ways that are known to be effective for

1) Inspiring and engaging all children in and about science and technology, and
2) Motivating engaged children, young adults and college students from diverse backgrounds to excel in STEM courses of study and develop the habits-of-mind required of computer scientists, engineers and scientists.

Capitalizing on the linkage between formal and informal education is one way that NASA can inspire, engage and motivate students, and help students develop the habits-of-mind they will need to successfully prepare for careers in science and technology [11, 12].

The scholarly work in education has matured since NASA's formation. With that maturity has come a deeper understanding of highly effective education practices in informal and formal settings and the recognition of the interconnection between them [13, 14].

As one might expect, a student's learning path meanders between formal and informal education environments and adult students engaged in NASA Goddard Space Flight Center's (GSFC's) internships are neither engaged in traditional formal nor in traditional informal education (unstructured "free choice" learning environments) [15, 16]. They are in fact engaged in a learning experience somewhere between these two extremes.

In their NASA GSFC internship, students are applying the knowledge they have acquired in their formal curriculum-structured academic setting to their research project in an informal, research-structured setting. This transference of knowledge from a curriculum to a research-structured environment tests the students' depth of understanding and their ability to extend their understanding in new and different ways. This transference and extension begins and deepens the development of the habits-of-mind required of successful scientists and technologists [17, 18, 19].

Linking this learning to the student's curriculum and continuing the development of these habits-of-mind is a critical element in shaping future scientists and technologists. This linkage is especially important for underrepresented and underserved populations since they often come from environments where there is little exposure to individuals with careers in science or technology, or those who possess these habits-of-mind [15, 20].

With this context in mind, we will focus on two education endeavors at GSFC for STEM learners; 1) authentic research and development internships for college students and 2) authentic STEM partnerships with educators at science and technology centers and museums who develop and deploy NASA educational programs or create exhibits for the public.

Research and Development Internship Opportunities for College Students

The NASA Goddard Space Flight Center has for many years operated summer internship opportunities for undergraduate and graduate students. These opportunities have provided authentic research or development experiences designed to benefit both the intern and the mentor. These experiences have also provided a set of extensions that vary with the internship opportunity or the research or development experience. Internship extensions have included field experiences, coaching on preparation of a poster or oral presentation, career networking events, and team projects.

Historically, these opportunities focused on one-time internships for individuals without explicit consideration of workforce supply or on-going development of the individual. They typically provided an informal research experience with little attention to the on-going career development of the intern once they return to the formal curriculum-structured education setting. The general rational was passive. It was thought that exposure to research or development efforts at GSFC was sufficient to motivate students to excel in their academic studies [21, 22]. Even if NASA or GSFC never saw the intern again, it was hoped that he or she would become a high performing contributor to the SandT workforce pool.

That perspective was refocused in NASA's 2011 Strategic Plan [23] that recognized *hands-on experiences are unparalleled in their ability to develop a student's advanced STEM skills and prepare them for a career.* Consequently, NASA seeks to support formal higher education degree programs by *providing undergraduate and graduate students with hands-on opportunities to contribute to our current missions* (Goal 5.1), and to "*work hand-in-hand with colleges and universities to provide student research and engineering experiences that contribute to our missions*" (Goal 6.1).

In all areas of vital interest to GSFC, this means encouraging and mentoring promising undergraduates and graduate students, especially those from fast growing underrepresented populations, to excel in their degree programs. In the sciences, this means encouraging promising students to matriculate to graduate school and eventually qualify for the NASA hiring pool. In engineering, this means the focus is on the development of professionals at the baccalaureate level.

Intern placement records and focus group discussions reflect the existence of multiple-year placements with no coordinated broadening of perspective or more formal development of those individuals. Mentors, in some cases, recognized the value of a continuing research experience when integrated with an intern's academic program as a way to promote the matriculation of and academic development of a particularly promising intern. Clearly, students who return for a second or third year, especially to the same mentor, have impressed that mentor with their performance. Such individuals would likely benefit from a deeper more demanding research or development experience, especially when linked to their academic program. Likewise, mentors would benefit from that intern's work. Ultimately, one would expect these interns to rise to the top of the workforce pool from which GSFC and its contractors draw.

Standard College Internship Opportunity

In 2008 and 2009 GSFC began experimenting with the structure and organization of its student internship programs by incorporating improvements designed to strengthen the linkage between an intern's curriculum-structured academic background and their interests, and a mentor's research and development project needs. GSFC began an improvement path grounded in educational research and on feedback and input from the GSFC mentor and intern community [24]. Previously, most interns were accepted into an internship program and then placement was sought with a mentor (a scientist, engineer, or other staff). Of the 26 internship programs at GSFC-Greenbelt, only half have participated in the improvement efforts described here. With time, it is anticipated that most internship programs will migrate to a model that improves the linkage between the intern's curriculum-structured academic background and the mentor's project.

Initially two fundamental changes were made —

1) Instead of GSFC Office of Education placing interns, mentors posted authentic research or development project descriptions and students applied to them by writing a short essay describing their interest in the posted project and their qualifications for that project, and
2) Selection of students for placement with a mentor was made by the mentor; not by the education program. Mentors selected multiple interns who applied to their project and the education office using advice from a placement panel placed one of the selected interns with the selecting mentor.

These changes addressed a concern voiced both by mentors and by interns. Mentors were given a critical role in the selection of interns and students were allowed to apply to projects of interest to them.

As the selection and placement processes changed opportunities for coherency among the internship programs emerged. Beginning in summer 2010, all interns at GSFC-Greenbelt, irrespective of the education program in which they were placed, were offered the opportunity to participate in a set of coherent extensions beyond their research or development project (Table 1).

Beginning in 2011, a single coordinated opening ceremony on the first day of the internship and a single closing ceremony at the end of the summer experience were added to the activities. All 340 interns from 26 programs were strongly encouraged to participate in these events and in the professionally organized poster session for interns at the end of the session.

To facilitate their participation in the poster session, interns were offered the opportunity to attend a special training session on preparation of oral presentations and poster sessions for execution at scientific or technical conferences. About half of the interns at GSFC took advantage of the opportunity to prepare a poster and to interact with the science and engineering community at GSFC and with their peers in this forum. They participated in the culture of science and engineering and furthered the development of habits-of-mind and the behavioral and communication skills required of scientists and engineers [15, 19, 25].

Table 1. Extensions Offered to All Interns at GSFC

Orientation: A half-day ceremony in which interns meet each other and their Center Advisor; receive an introduction to NASA GSFC from the Center Director and others; an introduction to their summer experience from their Center Advisor; meet their mentor; and visit the laboratory where they will carry-out their research or development project.
Career Networking Event: A day-long event in which interns are introduced to the range of careers at NASA GSFC via a panel session, the employment opportunities and processes, and a structured networking event where interns can meet and speak with employees and management from various organizations within GSFC.
Professional Seminar: A renowned NASA scientist or engineer gives an engaging seminar on their research or development work. The seminar is designed for a science/engineering attentive audience and assumes the listener is interested in the topic but is not an expert. In 2010 and 11, Dr. John Mather, NASA's only Nobel Laureate was the speaker.
How to Prepare a Professional Seminar or Poster: An interactive presentation providing very practical guidance on preparing and presenting a professional oral presentation or a poster. Authentic examples of best practices, and issues to consider when preparing your poster or presentation are provided. The presenter is a GSFC science and engineering professional that gives presentations and posters as part of their professional responsibilities.
Professional Poster Session: Preparation and participation in a professional poster session using standards of a professional meeting and attended by the science and engineering community at GSFC.
Closing Ceremony: A half-day ceremony in which interns receive their certificate of participation and say farewell to their peers, mentors, and new colleagues.

As these changes were occurring a multi-year formative evaluation was conducted to assess the demographic make-up of the intern participants and the effectiveness of these changes for both mentors and interns. In addition to demographics, student questions focused on program logistics, the student's perception of their experience and their working relationship with their mentor. The response of interns to these questions has guided program improvements since 2009.

Answers to questions about intern performance and the interns' perception of their internship experience provide insight into the overall value of the internship opportunities to both mentors and interns. All mentors were offered the opportunity to provide input on the performance of their intern. The percent of mentors who responded increased over the 3-year period (2009-2011) from 40% to 75%. Of these mentors over 60% thought highly of their intern, considering the intern to be a *top performer*. Moreover, more than 90% of mentors who responded were satisfied with the intern's performance and would like the intern to return. Likewise interns generally had very favorable experiences, with about 60% interested in returning for a second internship or beginning a career with NASA GSFC. In all years, about 11% of the interns were uninterested or ambivalent about returning or beginning a career with NASA.

The demographic data collected during the formative evaluation indicates that the GSFC intern programs are quite effective at attracting high quality female and minority students. This is of special interest to NASA since GSFC's disciplines of interest are those that have the lowest representation of women and minorities in national statistics; physical sciences (e.g., physics, chemistry, and astronomy), engineering (aerospace, chemical, electrical, material, mechanical), computer science, and Earth sciences (ocean, land, and atmosphere) [6, 7]. Nationally, in 2011 25% of the U.S. baccalaureate degrees in these fields were women, while the participation rate of females in GSFC summer internship programs (dominated by these disciplines) was 36%, 11% higher than the national graduation rate/participation rate.

Table 2. Diversity by race/ethnicity in GSFC internship programs compared to the US population and to science and technology baccalaureate degrees in GSFC disciplines*

Race/Ethnicity	2008§ % of U.S. Population	2008§ Degrees Awarded (% of total in category)	2008 GSFC-Greenbelt Internship Participation (%)	2011 GSFC-Greenbelt Internship Participation (%)
White	66	64	49	61
Asian American	4.3	10	10	8
African American	12.2	6	17.1	20
Hispanic	15.5	7	15.5	7.4
Native American/Native Peoples	2.4	0.5	2.3	0.6

* GSFC related science and technology disciplines (computer science, engineering, physical sciences (physics, chemistry, astronomy, other), Earth sciences (land, atmosphere, ocean), mathematics and statistics)

§ Data from National Science Foundation, 2011a

The ethnic and racial diversity of the interns was also quite broad and reflects the multi-year effort GSFC has undertaken to identify and recruit diverse, high-quality internship applicants within these science and engineering disciplines that are critical to GSFC's work. For example in 2008, GSFC's internship programs were highly successful at recruiting African American and Hispanic students (two groups traditionally underrepresented in the sciences and engineering) [6] into its internship programs when compared to the proportion of degrees awarded (Table 2).

In addition to the 2009 survey, in 2011, a more detailed exit survey of interns was administered, one that provided insight into qualitative aspects of an intern's summer experience at GSFC. The survey results highlight several strengths and identified areas for improvement. The research and development project focus of internships at NASA GSFC is their primary strength. Over 70% of the interns were *extremely satisfied* with their internship experience, while suggesting several areas where improvements were needed. Most interns (95%) indicated that the extension activities offered to all interns were beneficial to them. Comments from individual interns indicated that the *career-networking event* and the *presentation and poster preparation session* were most beneficial.

Interns identified three major areas for improvement. The first and most requested area spoke to logistical concerns covering housing, transportation, receipt of announcements, pre-arrival information, and security including access to GSFC facilities needed to carry-out their project. The second area of concern was tied to adequate preparation for the intern's arrival or mentoring for the intern when the mentor was unavailable. The third area was a request for more exposure to GSFC—its facilities, research and development activities—beyond their project, and more forums for social engagement among the interns.

In addition to the responses discussed above, high interest was expressed by both mentors and interns for an opportunity focused on returning interns. Accordingly, the NASA-GSFC Office of Education undertook a needs assessment of GSFC stakeholders[1] in 2011 and a test of extensions previously recommended for returning interns. The assessment sought to identify the interest and specific needs of an internship opportunity that targets returning interns. The test sought to identify effective extensions from previous recommendations.

Table 3. Summer 2011 Test of Extension Experience for Returning Interns

Science and Technology Research and Development Extensions
Small group discussions with senior NASA leadership • Agency Leadership for Education, Technology, Science, and Scientific International Partnerships • GSFC Leadership for Science and Technology, Science and Exploration, a Nobel Laureate Behind the scenes facility tours of GSFC guided by the facility leadership • Scientific Visualization Facility • Integrated Design Center • Rocket design, build and launch facilities • Aircraft facilities • Balloon design and test facilities

[1] GSFC stakeholders in this context are employees of NASA who lead organizations, mentor undergraduate or graduate interns or fellows, or highlight issues related to diversity and equity in the workforce.

The extensions (Table 3) exposed interns to a range of unique facilities and expertise at NASA GSFC and HQ, not offered via the standard internship.

The interns who participated in this test were most inspired by the visit with NASA's Chief Technology Officer and NASA's Chief Scientist, and the behind-the-scenes, guided tour of GSFC's Wallops Flight Facility. The tour of Goddard facilities included rare access to NASA facilities and discussions with NASA scientists, engineers and managers not available to interns or most NASA employees.

Senior Internship Opportunity

The findings from the needs assessment and the extension experiment led to the formation of a *Senior Internship Opportunity*. An experimental pilot is planned for summer 2012. The unique and innovative aspects of this internship are 1) creation of a professionally meaningful product by the intern, 2) coupling of a summer research experience with an academic year segment that allows completion of the research begun during the summer, and 3) joint mentoring of the intern during the academic year by the GSFC mentor and a local faculty advisor. Research has shown that linking students' academic programs to their summer internship, both before and after the internship, can improve their academic outcomes. Students gain scientific and cultural understanding as they participate in the culture of science with a mentor that helps move them toward full membership in the science community as they gain knowledge and self-confidence to deal with the inherent challenges of STEM research [25, 22, 18, 15, 17].

To qualify for this opportunity (Table 4), applicants must have participated in a previous GSFC internship while a college student and must provide the following as part of their application:

1) A strong letter of support from their previous NASA GSFC mentor
2) The name of an academic faculty advisor to oversee the academic portion of the research or development project
3) A well thought-out outline for their proposed research or development project and a detailed outline for a final *professionally-meaningful product.*

As designed, the Senior Internship opportunity requires a greater commitment from the mentor and the intern than other GSFC internships. As well, there is an added commitment of a faculty advisor at the intern's home institution. The mentor will be engaged with the intern during the entire lifecycle of the internship (summer segment and 6-9 month academic year segment). The intern works independently while seeking advice and guidance from the GSFC mentor during the summer segment. The mentor, in-turn, guides and challenges the intern's critical thinking and analysis skills, and encourages reading or examination of the literature.

The mentor will also interact with and establish a collaborative working relationship with the intern's faculty advisor at the intern's home academic institution. With the intern, they author an agreement at the beginning of the internship describing the deliverable the intern is to produce. The faculty advisor and the GSCF mentor work hand-in-hand to provide advice, guidance and assistance to the intern as the project is completed and the final deliverable is prepared. An additional benefit of these student-focused, collaborative relationships are new, mutually beneficial research partnerships between GSFC mentors and academic advisors of interns. In the long-term, it is envisioned that these colleague-to-colleague relationships will lead to sustainable institutional partnerships, especially with small or minority serving institutions.

Table 4. Senior Internship Opportunity

Opportunity Components
Internship Requirements 1. Participated in a previous college internship at GSFC. 2. Provide letter of recommendation from a previous NASA GSFC mentor. 3. Write a well thought out outline for the research or development project they wish to complete during their second (continuing) internship and a detailed outline for the final professionally meaningful product. 4. Provide name of a faculty advisor at home institution for the project to which they are applying. 5. Produce a professionally meaningful product at end of academic segment; mentor, advisor and student formally agree on product at beginning of summer segment. 6. Participate in a ten-week summer segment and 6-9 month academic year segment. 7. GSFC mentor supports and provides guidance to intern during entire project lifecycle; mentor, advisor and intern work together to provide coordinated advice, guidance and local support to intern during academic segment of internship.
Summer Segment Extensions 1. Overview of a NASA field center (NASA GSFC) from senior management's perspective. 2. Exposure to interpersonal networking and team building skill development. 3. Informal seminar series among participants and their GSFC mentors. 4. Insiders tour of unique facilities at GSFC (e.g., scientific visualization; instrumentation integration facilities; aircraft, balloon and rocket facilities).

SCIENCE AND ENGINEERING WORKSHOPS FOR INFORMAL LEARNING CENTER PROFESSIONALS

Separate from and in addition to the internship opportunities described thus far, GSCF informal education is involved in a three year pilot study with select small- to medium-sized museums in the Goddard Region (Virginia to Maine). The purpose of the study is to infuse current NASA content into and to develop long lasting partnerships with the museums.

Small- to medium-sized museums and science and technology informal learning centers (ILCs) were targeted because of their significance in the community. Many are important sources of engagement and inspiration, as well as being places where underserved populations can experience and build understanding of the innovations in science and technology [26, 14, 12].

To grow this vision, these centers must be able to routinely infuse their facility with new and engaging programs and materials, which is a significant challenge for most. Many lack funds to hire experts to assist in the design or creation of new exhibits and programs or to update or refurbish existing services [27].

Recognizing these centers serve an audience that NASA may not reach and their fiscal constraints can be addressed by NASA, GSFC saw an opportunity for a fruitful partnership. NASA GSFC has science and technology innovations, and the communication and education resources that are of interest to these centers, while these informal learning centers serve an audience that GSFC would like to reach. Consequently, in 2010 GSFC piloted the first in a series of science and engineering workshops for the professional staff at these learning centers.

The purpose of the workshop was to support these centers by providing timely and engaging content for their exhibits and programs, and to build enduring partnerships with them. The model included a two-day workshop for their exhibit and education program professionals at GSFC and periodic updates and communications. The workshop focused on information and resources related to GSFC science and technology. Information and resources that could invigorate exhibit content and was packaged and ready for use.

As with NASA's internship programs, informal learning center professionals applied to the workshop and met a set of application requirements. This process was used to ensure the best match of potential partners—informal learning center size, objectives, interests, and needs—with GSFC offerings. Like with the internships, applicants were required to provide a recommendation reflecting their capability and interest. Unlike the internships, an endorsement was required by both the applicant and the supervisor. The endorsement agreed to commit the organization to using NASA content in their facility. It further agreed to evaluate the new or improved program or product. As with GSFC internships, successful applicants were provided a stipend. In this case, the stipend was used to implement within their center a new or refurbished program or product. The intent was to encourage use by reducing the cost to refresh a program or create a new exhibit or program using the NASA content.

Two pilot workshops were held in 2010 and 2011 and a third is planned for 2012. Ten to 15 informal learning centers participated each year with four of the centers from 2010 returning in 2011 (Table 5). The centers represent the spectrum of small- to medium-sized informal learning centers located in the northeastern U.S. (GSFC's service area).

They range from science centers, to museums, to observatories. They serve different primary audiences (e.g., school groups, the public, educators), from urban, rural or both. All service diverse populations.

Each workshop was built upon the assessed effectiveness of the previous one. In 2010 and 2011 the workshops were divided into subject matter tracks so participants could focus on areas most relevant to the mission and current initiatives of their facility. Moreover, the tracks were rotated from year-to-year so the workshop is of interest to a broad range of informal learning centers in the northeastern United States and to returning participants. About one third of the 2010 participants returned for the 2011 pilot.

Table 5. Attributes and interests of participating informal learning centers

Year	Attendees (Num)	Institution	Location	Facility Visitors	Area of Interest	Track Placement	URL
2010	2	Carnegie Science Center	PA, Pittsburg	~700,000 visitors/yr;	Earth Science Heliophysics	Climate change Life cycle of a star	http://www.carnegie *sciencecenter.org/*
2010	2	DaVinci Science Center	PA, Allentown	More than 80,000 participants/yr	Earth Science Planetary Science	Climate change Lunar Science	http://www.davinci sciencecenter.org/
2010	2	Delaware Aerospace Education Foundation (DASEF)	DE, Symrna	19,000 participants from 101 schools; 450 programs for teachers and 4500 for the public	Planetary Science	Lunar Science	*http://www.dasef.org/*
2010	2	Discovery Museum and Planetarium	CT, Bridgeport	~55,000 children participants/yr	Planetary Science Astrophysics	Lunar Science Space Weather	http://www.discovery *museum.org/*
2010	2	Kopernik Observatory and Science Center	NY, Vestal	Not Available	Earth Science Heliophysics	Climate Change Life cycle of a star	http://www.kopernik.org/
2010	2	Maryland Science Center	MD, Baltimore	~500,000 visitors/yr	Earth Science Astrophysics	Climate Change Space Weather	*http://www.mdsci.org/*
2010	2	McAuliffe Shepard Discovery Center	Not Available	Not Available	Earth Science Heliophysics	Climate Change Life cycle of a star	*http://www.starhop.com/*

Table 5. (Continued)

Year	Attendees (Num)	Institution	Location	Facility Visitors	Area of Interest	Track Placement	URL
2010	2	Museum of Natural History and Planetarium RWP	RI, Providence	Primarily school groups	Heliophysics Astrophysics	Life cycle of a star Space Weather	http://providenceri.com/*museum/*
2010	1	New England Air Museum	CT, Hartford	Not Available	Earth Science	Climate change	*http://www.neam.org*
2010	1	Rochester Museum and Science Center	NY, Rochester	Not Available	Heliophysics	Life cycle of a star	*http://www.rmsc.org*
2010	2	Wings of Eagles Discovery Center	NY, Horseheads	Not Available	Earth Science Planetary Science	Climate change Lunar Science	*http://www.wingsofeagles.com*
2011	1	Clay Center for the Arts and Sciences of West Virginia	WV, Charleston	Not Available	None Specified	NASA Mission Design	*http://theclaycenter.org/*
2011	1	DaVinci Science Center	PA, Allentown	More than 80,000 participants/yr	Engineering Planetary	NASA Mission Design MESSENGER/ Mercury	http://www.davincisciencecenter.org/
2011	1	Discovery Museum and Planetarium	CT, Bridgeport	~55,000 children participants/yr	Astrophysics Engineering	NASA Mission Design Our Universe	*http://www.discoverymuseum.org/*
2011	2	Fairbanks Museum and Planetarium	VT, St. Johnsbury	Not Available	Heliophysics/ Earth Science Planetary/Astro-physics	Climate Change MESSENGER/ Mercury	*http://www.fairbanksmuseum.org/*

Year	Attendees (Num)	Institution	Location	Facility Visitors	Area of Interest	Track Placement	URL
2011	2	Kopernik Observatory and Science Center	NY, Vestal	Not Available	Planetary/ Engineering Heliophysics/ Astrophysics	NASA Mission Design Our Universe	http://www.kopernik .org/
2011	2	Maryland Science Center	MD, Baltimore	~500,000 visitors/yr	Heliophysics Astrophysics	MESSENGER/ Mercury Our Universe	*http://www.mdsci.org/*
2011	1	Montshire Museum of Science	VT, Norwich	150,000 visitors/yr; about 72,000 are school children; rural	Earth Science	Climate Change	*http://montshire.org/*
2011	4	Morehead Planetarium and Science Center	NC, Chapel Hill	160,000 visitors/yr; 85,000 school children	Earth Science None specified None specified Astrophysics	Climate Change NASA Mission Design Our Universe	http://www.morehead *planetarium.org*
2011	2	Museum of Science and Technology	NY, Syracuse	180,000 visitors/yr; about 30,000 school children; urban, inner-city	None specified None specified	Climate Change Climate Change	*http://www.most.org/*
2011	2	The Robert J Novins Planetarium	NJ, Toms River	Located at a community college	None specified None specified	Our Universe Our Universe	*http://www.ocean.edu/ planet.htm*

Table 5. (Continued)

Year	Attendees (Num)	Institution	Location	Facility Visitors	Area of Interest	Track Placement	URL
2011	1	Sciencenter	NY, Ithaca	97,000 visitors/yr; about 4,700 are school children; underserved community	None specified	MESSENGER / Mercury	http://www.sciencenter.org/
2011	2	Science Museum of Virginia	VA, Richmond	NO INFO on website	Earth Science Heliophysics	Climate Change NASA Mission Design	*http://www.smv.org/*
2011	1	Tellus Science Museum	GA, Cartersville	125,000 visitors/yr; about 45,000 school children	Planetary Science	MESSENGER / Mercury	*http://www.tellusmuseum.org/*
2011	1	William Miller Sperry Observatory	NJ, Cranford	Located at a community college	Astrophysics	Our Universe	http://www.asterism.org/

Placement of participants in tracks was based on their interest and participant-limit for each track. Each morning, unifying topics and community building were tackled in a plenary session. In that session, GSFC scientists and engineers engaged participants in over-arching and cross-cutting topics relevant to the tracks such as available multimedia resources, (http://*svs.gsfc.nasa.gov/*) and the digital learning network (http://dln.nasa.gov/). Participants engaged in community building efforts such as peer-to-peer networking among participants, reflection of morning speakers and the previous days' events, and a share-a-thon. In the afternoon they delved into the subject matter of their track, such as climate change, the universe, engineering, the sun, the moon, Mercury/MESSENGER.

The focus of the 2011 and the upcoming 2012 workshops were identified in part by formative assessments from the 2010 and 2011 workshops, respectively. The intent of this approach is to engage participants and is well supported by the research [28]. Based on participant input, an engineering track, designed to integrate GSFC scientific study areas, was added in 2011. The other three 2011 tracks focused on recent scientific discoveries and emerging scientific concepts related to our universe, planet Mercury, and climate change on Earth.

For each track, GSFC scientists and engineers led the forum. They provided engaging activities, behind the scenes tours of relevant facilities, timely science highlights, engaging visual resources to illustrate these highlights, and illustrations to unpack complex scientific concepts. As an example, the overarching aspects of the engineering track were introduced in the plenary session—highlighting the diversity and breadth of engineering at GSFC. The track then provided an in-depth view into the engineering lifecycle, mission operations, instrument development at GSFC, and the integrated mission and instrument design center. Track participants engaged in interactive discussions with GSFC engineers while addressing each of these topics. These interactions included "up-close-and-personal" guided visits to facilities where workshop participants could see and experience first-hand the engineering principles presented in the discussions. In all cases, participants were provided with presentation materials and on-line resources they could utilize in their programming when they returned to their informal learning centers.

A critical aspect of the 2011 workshop, and for the one planned for 2012, is the community-building component. The intent is two fold: 1) to build deep relationships with organizations that will endure as individuals at the informal learning centers and GSFC change, and 2) to foster development of a community of practice among professionals at small- to medium-sized informal learning centers including the professionals at the GSFC Visitors Center.

To frame the community-building efforts, the workshop organization, and the application and selection processes were designed to foster the development of these deep relationships between GSFC and the small- to medium-sized informal learning centers. Specifically, requiring applicants to provide endorsement of participation and commitment from their management to utilize workshop resources in their programming was designed to begin building this relationship.

The acceptance and follow-through by participating centers to integrate workshop content into a program or exhibit within the following 12 months and to undertake an assessment and evaluation of that effort, once deployed, demonstrated commitment on their part to a relationship with GSFC. Likewise, GSFC offered to provide follow-through with each of the informal learning centers as they integrated content into a program or exhibit. To that end,

GSFC provided a primary point of coordination and facilitated the development of multiple linkages between these organizations and the GSFC through the now-familiar subject matter experts that had presented and guided workshop activities.

Connecting the informal learning centers and their education professionals with the GSFC subject matter experts and with GSFC educators fostered the development of numerous professional connections between GSFC and these centers. This relationship model uses these linkages, between GSFC professional staff and the informal learning centers, as a means for the relationship 1) to endure the vagaries of staffing changes and 2) to provide expertise to the participants as they utilize the resources provided during the workshop to create or refurbish programs and exhibits.

Another mechanism to deepen the GSFC/ILC relationship was the reservation of workshop seats, each year, for those informal learning centers that participated in a previous workshop. In this way, ILCs could stay engaged with GSFC and receive continuous update for programs and exhibits. Between workshops, the GSFC coordinator provided regular updates to participants and kept in touch via a number of formal and informal mechanisms. The coordinator tracked activities at the ILCs resulting from the workshop experience, supplied learning resources and materials for special events (such as sun-viewing glasses for the once-in-a-100-years Transit of Venus event), and provided a virtual professional development opportunity for participants and their local colleagues.

The share-a-thon and the peer-to-peer networking events were added in 2011 in response to 2010 participant input. These too were intended to build and support the newly established community of practice among the informal learning center education participants.

Table 6. Nature of NASA materials utilized as a result of the workshop experience

Center+	Technology	Web	Learning Resources	Printed Materials	Subject Matter Experts	Education Techniques
2010						
1	✓	✓	✓			
2	✓	✓	✓			
3			✓		✓	
4	✓	✓	✓		✓	
5			✓			
6	✓	✓	✓		✓	
7			✓		✓	
2011						
1	✓	✓	✓		✓	
2			✓	✓		
3		✓				
4	✓	✓	✓	✓	✓	
5	✓		✓	✓	✓	✓
6		✓	✓		✓	
7			✓		✓	
8		✓	✓	✓	✓	✓

+ numbers indicate a different informal learning center.

Table 7. Informal learning center programming that benefited from 2010 workshop participation*

Center	Activity Changed/Created	Frequency
1	New Planetarium show	Long-term exhibit
1	Science-on-a-Sphere show	Two-days
1	One day event	Repeat during school year
2	Middle school activities on the life of stars	Regular repeat cycle
3	Planetary Walk	Long-term exhibit
3	Summer camps (6) on various topics	Repeat in summer
4	Summer programs for children: 1) Space exploration 2) Space and Engineering Rocketry	Repeat during school year
4	Summer camp for children: The Milky Way	Repeat in summer
4	Public program on moon rocks	Two-days
5	Space Day	Repeat once
6	Summer camps (3) for children on different topics	Repeat in summer
7	Space and Aviation Day	Repeat once
8	Space weather programs	Repeat in summer

* Data for 2011 not available at time of publication.

These events provided expanded professional opportunities, for these educators from similar professional environments, thereby fostering professional relationships and future partnerships. With seats reserved for participants returning from a previous workshop, the first time attendees were linked to previous workshop organizations and content through the peer-to-peer networking with these returning participants. Moreover, GSFC provided both a workshop website where workshop materials could be shared and a list serve for participants to communicate with each other.

As a result of these partnership-building workshops, participants in 2010 and 2011 reported using a wide range of NASA resources in their programming. Those resources varied in nature from technology resources to education techniques (Table 6) with the learning resources and the subject matter expert resources being used the most. Of those who created or refurbished a program or an exhibit incorporating the NASA content, some made major additions to a permanent exhibit (existing or newly created) while others made additions to educational programs for children or for educators (Table 7). Several infused the new content into all categories of their center's programming: exhibits, education programs, and special events. All facilities targeted middle school aged (7 to 13 year olds) children, their teachers or their families. Some also targeted elementary and high school students.

The NASA content delivered though these small- to medium-sized informal learning centers touched the lives of many hundreds of thousands of individuals through permanent exhibits where visitor engagement lasted anywhere from a few seconds to a few minutes. In 2010, sustained engagement also occurred with about 260 children (elementary through high school) who participated in summer camps (three to 5 days) and in all day science programs at these centers.

The centers that participated in the workshops in 2010 and 2011, generally, were able to offer these programs and events only because of their participation in these NASA Goddard workshops. Clearly, being able to access timely resources and subject matter experts at NASA GSFC to update programs and exhibits is highly valued by these centers. As one workshop participant stated, *we (small informal learning centers) make a lot happen with a little.*

As we move forward, the long-term value of these workshops and the resulting partnerships between NASA GSFC and small- to medium-sized informal learning centers will continue to be assessed. To date, our research indicates that small- to medium-sized informal learning centers make effective partners for GSFC and vice versa and that the ILCs appreciate and include NASA content as they are able in their exhibits and programs.

CONCLUSION

NASA Goddard has a long history developing STEM professionals. As Goddard's educational activities have diversified and matured over the years so has the understanding of the relationship between those activities and the ability to move individuals along the learning pathway from exposure and inspiration to deep exploration and learning in an area of science and engineering. Today GSFC recognizes the critical role of STEM professionals in that pathway—from professional research scientists and engineers to professional science educators in the classroom and in informal learning venues.

Here we have presented three of our STEM development programs, two for individuals preparing to become scientific researchers or engineers and one for informal education professionals. When these learning experiences are compared (Table 8) it is clear that they are characterized by a common set of criteria.

These criteria provide a common structure to the learning experiences—structure that is derived from the research on most effective education practices [14, 13, 15]. When criteria are compared across the learning experiences unifying concepts emerge even through the specifics for each learning experience may be different. As an example consider, *the Nature of Engagement* criteria. In all learning experiences, *interaction* between the participant and those guiding the learning is a key element of the experience. The differences arise in the specific nature of those interactions; in the case of the informal learning center professionals they are interacting with GSFC scientists and engineers, with GSFC educators, and with their peers. The participants in the internship programs on-the-other-hand are engaged in an intensive research activity with a mentor, the mentor's research or develop team, and in the advanced internship, the intern's academic advisor. It is important to recognize these similarities and differences so that maximum value can be obtained from both.

There is an increasing recognition of the important role human networks and community building play in the STEM profession and, therefore, their importance in the development of STEM professionals [15, 25, 19]. A key element of a successful learning experience that

Table 8. Comparison of learning experiences for STEM professionals

Criteria	Learning Experiences			Unifying Concepts
	Informal Learning Center—Professional Development	**Internship-Standard**	**Internship-Senior**	
Environment	✓ Small group workshop setting ✓ Continuing interaction during following year and beyond	✓ Individual research or development laboratory setting	✓ Individual research or development laboratory setting at academic institution and at SandT research facility	Small groups
Nature of Engagement	✓ Hands-on engagement with guest speakers	✓ Hands-on summer research or development experience	✓ Hands-on summer and academic year research or development experience.	Interactive
Product	✓ Institution commitment to use within one year in ✧ museum education programs ✧ public displays ✧ interactive exhibits ✓ Assessment and evaluation plan, survey instruments and collected data	✓ Some portion of a product	✓ Professional product appropriate to discipline	Produce a product
Anticipated Outcome	✓ Expand specific knowledge of NASA content ✓ Acquired education resources ✓ Information on effectiveness of deployed assets ✓ Know what GSFC does, and facilities required to accomplish that work	✓ Engage in authentic scientific research or technology development ✓ Know what NASA does and the facilities to accomplish its work ✓ Expanded scientific or technical knowledge	✓ Participation in and completion of an authentic scientific research or technology development project ✓ The culture of science and engineering at NASA ✓ Expanded scientific or technical knowledge	Expand knowledge

Table 8. (Continued)

Criteria	Learning Experiences			Unifying Concepts
	Informal Learning Center—Professional Development	**Internship-Standard**	**Internship-Senior**	
Assessing and Evaluating Achievement of Objectives	✓ Post experience survey of participants ✓ Report of programs created, revised, or extended ✓ Report of participants and the extent of their engagement with new or enhanced programs at learning center	✓ End-of-internship on-line survey (SOLAR system) ✓ End of internship exit interviews	✓ Pre- and post-survey of mentors and students ✓ Exit interviews of mentors and students ✓ Report of effective processes, areas for improvement, and achievement of objectives	Measure effectiveness
Network and Community Building	✓ Relationships with NASA engineers, scientists and educators ✓ Peer-to-peer relationships between museum professionals in northeast ✓ Ad hoc interactions via email, phone, and events calendar ✓ Formal community gathering via web and NASA's digital learning network	✓ Mentor-Student ✓ Research or development team ✓ Peer-to-peer relationships	✓ Mentor-Student-Academic Advisor ✓ Peer-to-peer relationships	Culture and community

provides long-term career value for informal educators and young scientists or engineers is one that provides mechanisms to develop the socialization and behavioral skills necessary to be successful in their chosen field or to bridge the culture differences between fields. The modification of the informal learning center professionals' workshop in 2011 to include more opportunities for networking and community building was a direct response to this increased awareness. Likewise, the standard internship provides a career networking opportunity so all interns have an opportunity to practice professional networking skills. The Senior Internship opportunity builds upon this event and goes a step further. It provides formal training in networking and in team building, recognizing that the academic programs of SandT students often do not provide opportunities to develop these skills.

Furthermore, immersing interns or informal learning center professionals in research and development settings where partnerships are key to a project's completion helps them understand the key role partnerships play in today's science and technology. And it exposes them to the networking and team building skills required for partnerships to be successful.

Taken together these programs demonstrate how GSFC is continuing to do what NASA does best: offer opportunities for cutting edge science and engineer experiences in out-of-school and lifelong learning environments, and the application of our science and engineering habits-of-mind to the practice of education. All opportunities immerse participants in the NASA culture on a NASA center where participants work side-by-side or interact frequently with NASA scientists and engineers. These opportunities employ the latest knowledge of most effective practices for development of STEM professionals. They seek to advance the understanding of those practices by carrying out assessment and evaluation that identifies which aspects of the science and engineering experiences are effective, which are not, and why. And finally, they provide insights into more effective design and management of education programs by the identification of common structural elements and unifying concepts among successful programs.

ACKNOWLEDGMENTS

The authors would like to acknowledge the contribution of Marcianne Delaney and Carmel Conaty who were instrumental to the formulation, implementation, and evaluation of the 2010 workshop for informal learning center education professionals. The work presented here was made possible by their efforts. The authors would also like to acknowledge the contribution of Phyllis Barton who was instrumental in the evaluation of the internship programs at NASA GSFC. Finally, the authors would like to acknowledge the generous support of GSFC's Visitor Center by NASA's informal science education program that sponsored the workshops for informal education professionals and GSFC's Office of Education that supported the evaluation of the internship program.

REFERENCES

[1] National Aeronautics and Space Act, Public Law 85-568, (1958).

[2] M. Syvertson, J. Koke, S. Voirol, M. Kirkpatrick, J. Schoemer, T. Nolan, and N. Normandy, *NASA is Listening—Collaborations between NASA Earth Science Enterprise and the Informal Education Community,* National Aeronautics and Space Administration internal report (2000).

[3] Business Roundtable, *Tapping America's Potential: The Education for Innovation Initiative*, Business Roundtable, Washington (2005). http://businessroundtable.org

[4] National Science Board, *America's Pressing Challenge—Building a Stronger Foundation: A Companion to Science and Engineering Indicators–2006,* (2006), NSB-06-02.

[5] C. Coble and M. Allen, *Keeping America Competitive: Five Strategies To Improve Mathematics and Science Education*, Education Commission of the States, Denver (2005). http://ecs.org

[6] National Science Foundation, *Women, Minorities, and Persons with Disabilities in Science and Engineering: 2011.* Division of Science Resources Statistics, National Center for Science and Engineering Statistics, Arlington (2011), Special Report NSF 11-309.

[7] National Science Foundation, *Characteristics of Scientists and Engineers in the United States: 2006.* Division of Science Resources Statistics, National Center for Science and Engineering Statistics, Arlington (2011), Special Report NSF 11-318.

[8] F. Hobbs and N. Stoops, *Demographic Trends in the 20th Century.* U.S. Census Bureau, Census 2000 Special Reports, Series CENSR-4. U.S. Government Printing Office, Washington (2002).

[9] Task Force on the Future of American Innovation. *The Knowledge Economy: Is the United States Losing Its Competitive Edge?* (2005). http://www.futureofinnovation.org

[10] N. Augustin, *Is America Falling Off the Flat Earth*, The National Academies Press, Washington (2007).

[11] J. H. Falk and L. D. Dierking, American Scientist 98, 486 (2010).

[12] M. Fenichel and H. A. Schweingruber, *Surrounded by Science: Learning Science in Informal Environments*, National Research Council, Washington (2010).

[13] J. Bransford, A. L. Brown, R. R. Cocking, M.S. Donovan, and J.W. Pellegrino, *How People Learn: Brain, Mind, Experience, and School Expanded Edition*, The National Academies Press, Washington (2000).

[14] P. Bell, B. Lewenstein, A. W. Shouse, and M. A. Feder, *Learning Science in Informal Environments: People, Places, and Pursuits,* Committee on Learning Science in Informal Environments, National Research Council, Washington (2009).

[15] M. Pender, D. E. Marcotte, M. R. Sto. Domingo and K. I. Maton, *Education Policy Analysis Archives* 18, 1 (2010).

[16] J. Bransford, N. Vye, R. Steverns, P. Kuhl, D. Schwartz, P. Bell, A. Meltozoff, B. Barron, R. Pea, B. Reeves, J. Roschelle, and N. Sabelli, *Learning Theories and Education: Toward a Decade of Synergy,* In: P. Alexander, P. Winne (Eds.) *Handbook of Educational Psychology* (2nd Edition), Lawrence Erlbaum Associates Publishers, Mahwah (2006).

[17] S. J. Carey, *Peer Review* 12, 1 (2010).

[18] J. Brownell and L. E. Swaner, *Five High-Impact Practices—Research on Learning Outcomes, Completion, and Quality*, Association of American Colleges and Universities (2010).

[19] E. Seymour, A. Hunter, S. L. Laursen, and T. Deantoni, *Science Education* 88, 493 (2004).

[20] R. L. Morley, J. J. Havick, and G. S. May, *Journal of Engineering Education* 87, 321 (1998).

[21] K. Kendricks and A.Arment, *Journal of College Science Teaching* 41, 22 (2011).

[22] L. Sharp, B. Kleiner, and J. Frechtling, A *Description and Analysis of Best Practice Findings of Programs Promoting Participation of Underrepresented Undergraduate Students in Science, Mathematics, Engineering and Technology Fields*, (2000), NSF 01-31.

[23] National Aeronautics and Space Administration, *2011 NASA Strategic Plan*, National Aeronautics and Space Administration, Washington (2011), NP-2011-01-699-HQ.

[24] B. Meeson, NASA GSFC internal publication (2008).

[25] A. Hunter, L.S. Laursen, and E. Seymour, *Science Education* 91, 36 (2007).

[26] L. D. Dierking, *Linking After-School Programs and STEM Learning: A View from Another Window*. Coalition for After-School-Science (CASS), New York (2007). http://www.afterschoolscience.org/pdf/ member-publications/Linking After-School Programs and STEM Learning.pdf

[27] J. R. Hill, M. J. Hannafin, and D. P. Domizi, *Resource-based Learning and Informal Learning Environments: Prospects and Challenges*, In: L.W. Hin Tan and R. Subramaniam (Eds.) *E-Learning and Virtual Science Centers,* Information Science Publishing, Hershey (2005).

[28] P. H. Rossi, M. W. Lipsey, and H. E. Freeman, *Evaluation: A Systematic Approach, 7th ed.* Sage Publication, Thousands Oaks (2004).

In: The Space Education Phenomenon at NASA …
Editor: Norma Teresinha Oliveira Reis
ISBN: 978-1-62618-188-5

CONTRIBUTORS

Blanche Meeson (Blanche.W.Meeson@nasa.gov)

Dr. Blanche Meeson is a physical scientist at NASA Goddard Space Flight Center (GSFC) where she has been applying her background in ocean science and information science to STEM education. She has been actively engaged in science education for the last 15 years working in a range of capacities including national planning for the Integrated Ocean Observing System, creation of on-line STEM learning resources for high school and college students, information systems to support application to and information about education programs, and design and management of STEM educational programs for NASA. Today she is helping others learn and apply most effective practices derived from the education research and using those practices in the design and implementation of educational programs. She was instrumental in the planning, design and implementation of the Senior Internship described in this chapter.

Bhushit G. Vaishnav (bhushit@prl.res.in)

Dr. Bhushit G. Vaishnav is serving as Scientist and Head, Academic Services at Physical Research Laboratory (PRL), Ahmedabad, India. Dr. Vaishnav's area of specialization is the theoretical Atomic – Molecular Physics and more specifically electron scattering process with atomic and molecular targets of applied interest. Prior to joining PRL he served (during 2000-2008) as Physics faculty at science, engineering and diploma colleges. He has written one book entitled "Engineering Physics" for the students of 1st year engineering affiliated to the Gujarat Technological University. Besides this he has published 07 papers in the international peer reviewed journals in the areas of his research interest. He is also interested in science outreach activities and encouraging young students for taking interest in science education.

Charles A. Wood, Ph.D.

Dr. Charles A. Wood, director of the Center for Educational Technologies at Wheeling Jesuit University, and PI of NASA-sponsored Classroom of the Future, which funded the development of the BLiSS Sim iPad app. In addition to serving as PI and lead scientist on NASA, NIH, NSF, and Benedum Foundation education development projects at the center, he is also an investigator in the Cassini mission to Saturn.

Geraldine Robbins (gbrobbins@hotmail.com)

Geraldine Robbins is a Senior Albert Einstein Distinguished Educator Fellow in the Office of Education at NASA Goddard Space Flight Center, Greenbelt, MD, where she serves in the capacity of Deputy Lead for Informal Education and Lead Evaluator for Goddard Informal Education Programs. In addition to her experiences in informal education, Robbins has over ten (10) years experience as a mathematics and science educator and almost fifteen (15) years experience as a chemist and environmental contractor. Robbins is formally trained in inquiry-based teaching and learning and in the constructivist theory of cognitive learning. Her interest in informal education and evaluation evolved as she researched practices in education and solidified when she heard Gandhi's call, *"You must be the change you want to see in the world"*.

Jim Thieman (james.r.thieman@nasa.gov)

Dr. Thieman is an astrophysicist (radio astronomer) at NASA/Goddard Space Flight Center as well as a Data Systems Manager for the National Space Science Data Center. He has spent more than 20 years making NASA data findable and usable by researchers and the general public. Jim spends a part of his time as the government coordinator for the Radio Jove education and outreach program. This program is designed to provide students and other interested groups and individuals with a hands-on opportunity to learn about radio astronomy. The participants build a radio telescope from an inexpensive kit and this introduces them to the basics of radio astronomy, radio receivers, antennas, etc. It is a stepping-stone to the understanding that astronomy is more than just what can be seen through an optical telescope. The non-profit Radio Jove program has distributed more than 1700 kits around the world in 12 years.

João Batista Garcia Canalle, Ph.D (joaocanalle@gmail.com)

Dr. João Batista Garcia Canalle is an astrophysicist at State University of Rio de Janeiro, where he uses all his experience to popularize the Astronomy and Astronautics Sciences. Canalle is best known for his work as the Brazilian coordinator of the Brazilian Olympiad of Astronomy and Astronautics. It is the first Astronomical Olympiad in all South America and it has been organizing since 1998. On 2011 over 800.000 students took part in its 14th edition. On the other hand he is the head of an ambitious program of training teachers of elementary and high school for all over Brazil. Dr. Canalle has dozens of papers published in the Brazilian Journals dedicated to Physics and Astronomy Education. Canalle's interest about education and popularization on Astronomy in Brazil begun when observed that there were many didactics books of elementary school with many conceptual mistakes on Astronomy.

Laurie Ruberg, Ph.D

Laurie Ruberg, Ph.D., serves as the associate director of the Center for Educational Technologies at Wheeling Jesuit University, project manager and featured blogger for the

NASATalk web-based collaborative. She received her Ph.D. in Curriculum and Instruction from Virginia Tech. Her research interests include problem-based learning instructional design in the area of climate science issues and program evaluation specializing in STEM education and health and safety research projects. Ruberg has received several awards for her design and evaluation work.

Kendyl Morris Kimberly

Kendyl Morris Kimberly, graduated from Wheeling Jesuit University in 2011 and earned a B.A. in Professional Communications. As a Project Associate for the Center for Educational Technologies at Wheeling Jesuit University, she primarily focused on the NASATalk website's social media plan.

Marcos Cesar Pontes (astronauta@marcospontes.com.br)

Marcos Cesar Pontes (astronauta@marcospontes.com.br) born in Bauru, SP, Brazil, in 1964, Marcos Pontes is currently an Astronaut at Brazil's disposal, waiting for governmental assignment for his second spaceflight, he is Motivational Lecturer, Specialist in Performance and Professional Development Coach, Master of Systems Engineering, Aeronautical Engineer by Brazil's Technological Institute of Aeronautics (ITA), Professor and invited Researcher of the Advanced Studies Institute of Sao Paulo University – Sao Carlos (USP-SC), Technical Director of the National Institute for Space and Aeronautical Development, Businessman, Technical Consultant, United Nations Industrial Development Ambassador, President of the Marcos Pontes Foundation's President and Author of two books: "Mission Accomplished. The complete history of the first Brazilian space mission" and "It is Possible! How to transform your dreams into reality," published by Chris McHilliard publisher, Brazil.

Norma Teresinha Oliveira Reis (norma.reis@gmail.com)

Norma Teresinha Oliveira Reis is a pedagogue and teacher who works since 2004 in the Brazilian federal government, specifically in the country's education ministry. In 2008, she pursued a graduate degree at the International Space University (ISU), and during that academic year she visited NASA Goddard Space Flight Center as a summer intern, where she carried out research on Eclipses Through the Centuries and NASA space science education. Norma is an independent researcher of the 'space education phenomenon,' and contributes with lectures and publications to disseminate this area among teachers and students in the scope of her academic and professional activities. At the Brazilian education ministry, Norma has worked in different areas such as advisory, didactic book evaluation, educational technologies, project management, educational research, and digital content production on subjects such as Astronomy and Astronautics.

INDEX

Q

R

S

T

U

V

W

Y

Z